A TEXT BOOK ON

COORDINATE GEOMETRY

A TEXT BOOK ON

COORDINATE GEOMETRY

[For B.A. & B.Sc. Classes of Indian Universities]

by

BRIJ MOHAN M.A., Ph.D. (Liv)

COX MEDALLIST

KRISHNA KUMARI DEVI GOLD MEDALLIST

Former Professor and Head of the Department of Mathematics

ARUNODAYANPRAKASHAN

ISBN : 81-88473-77-4

ARUNODAYANPRAKASHAN
21-A, Daryaganj New Delhi-110002
Ph. : 0091#11#23273167, 23275710
Fax : 0091#11#23275710
Second (Arunodaya Prakashan) Edition : 2009
Second Edition : 2010

Cover Design : Vani Prakashan
Printed by : R. Teck. Printers, Delhi-95

Rs. : 450.00

A TEXT BOOK ON
COORDINATE GEOMETRY
by : Brij Mohan

अरुणोदय प्रकाशन, वाणी प्रकाशन की सहयोगी संस्था

PREFACE

This book has been written to meet the requirements of the B. A. and B. Sc. Students of Indian Universities. The subject matter has been graded properly and is presented in a manner so as to be understood by the average student. Numerous examples have been solved to illustrate the technique of solving questions. Alternate solutions to examples have been given wherever possible. The examples have been selected from standard books on the subject and from Examination Papers of various Indian Universities.

Suggesstions for improvement of the book will be thankfully received.

—AUTHORS

CONTENTS

CHAPTER I

COORDINATES

1·1. Cartesian Coordinates.

Let XOX' and YOY' be two fixed straight lines in the plane of the paper intersecting at right angles at the point O. Let P be any point in this plane and let PM and PN be two straight lines drawn through P parallel to XOX' and YOY' respectively. If the lengths ON and NP are known, the position of the point P is uniquely fixed in the plane. The lengths ON and NP usually denoted by x, y respectively are called the cartesian coordinates or briefly the coordinates of the point P. The lines XOX', YOY' of indefinite length are called the coordinate axes. The point of intersection O of the axes is called the **origin**. The length ON is called **abscissa** or the x-coordinate of the point P and the length NP is called the ordinate or the y-coordinate of the point. The coordinates of the point P are written as (x, y).

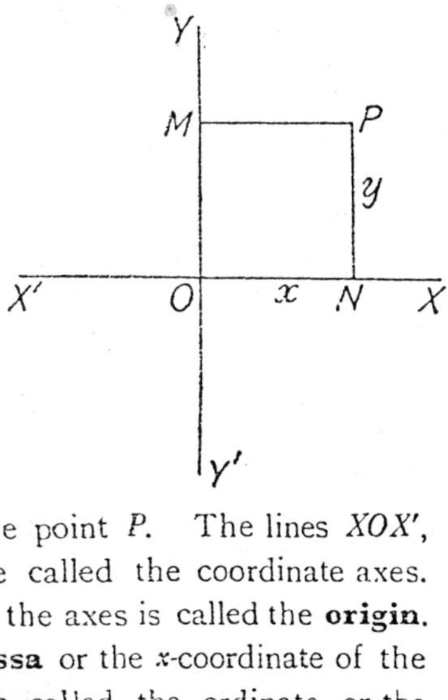

If the angle between the axes is a right angle, the axes are said to be rectangular; otherwise they are called oblique, and are inclined at any angle ω to each other.

Distances measured from O along OX and OY are considered positive and those measured along OX' and OY' negative. Thus the coordinates of P in the first quadrant are both positive, in the second quadrant x is negative, y positive ; in the third quadrant x and y are both negative and in the fourth quadrant x is positive, y negative.

Henceforth the axes will be regarded rectangular, unless stated otherwise.

1·2. Distance between two points.

Let the axes XOX' and YOY' be inclined to each other at an angle ω.

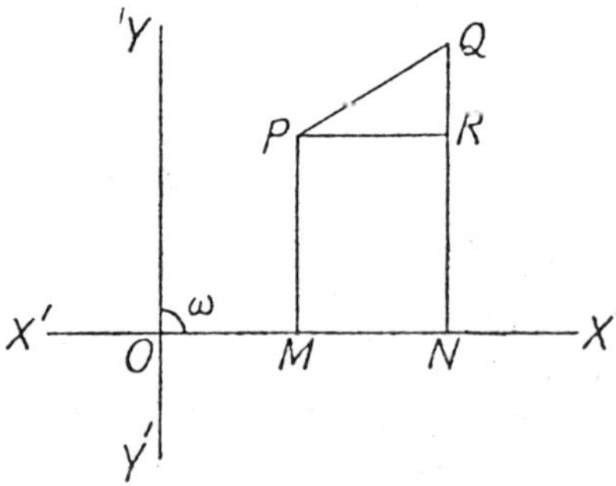

Let P and Q be two points whose coordinates are (x_1, y_1) and (x_2, y_2) respectively. Draw PM and QN parallel to OY to meet OX in M and N respectively and PR parallel to OX to meet QN in R.

Now, $OM=x_1$, $PM=y_1$, $ON=x_2$ and $QN=y_2$.

$$\therefore \quad PR=x_2-x_1 \text{ and } RQ=y_2-y_1.$$

The angle QRP is $\pi-\omega$.

Hence, from the triangle PQR,

$$PQ^2=PR^2+RQ^2-2PR.RQ.\cos(\pi-\omega)$$
$$=(x_2-x_1)^2+(y_2-y_1)^2+2(x_2-x_1)(y_2-y_1)\cos\omega.$$

$$\therefore \quad PQ=\sqrt{\{(x_2-x_1)^2+(y_2-y_1)^2+2(x_2-x_1)(y_2-y_1)\cos\omega\}}.$$

If the axes are rectangular, $\omega=90°$ and in that case $PQ=\sqrt{[(x_2-x_1)^2+(y_2-y_1)^2]}$.

Corollary. If the axes are rectangular, the distance of the point (x_1, y_1) from the origin $(0, 0)$ is $\sqrt{(x_1^2+y_1^2)}$.

Note. The above formula has been derived for points lying in the first quadrant, but it will be true for all points in the plane if due regard is paid to the signs of the coordinates.

Ex. 1. *Show that the three points (1, 1), (−1, −1) and $(-\sqrt{3}, \sqrt{3})$ are the angular points of an equilateral triangle, the axes being rectangular.*

Distance between (1, 1) and (−1, −1)

$$=\sqrt{[(1+1)^2+(1+1)^2]}$$
$$=\sqrt{8}=2\sqrt{2}.$$

Distance between (−1, −1) and $(-\sqrt{3}, \sqrt{3})$

$$=\sqrt{[(-1+\sqrt{3})^2+(-1-\sqrt{3})^2]}$$
$$=[4-2\sqrt{3}+4+2\sqrt{3}]$$
$$=\sqrt{8}=2\sqrt{2}.$$

Distance between $(-\sqrt{3}, \sqrt{3})$ and (1, 1)

$$=\sqrt{[(-\sqrt{3}-1)^2+(\sqrt{3}-1)^2}$$
$$=\sqrt{8}=2\sqrt{2}.$$

Hence the coordinates are the vertices of an equilateral triangle of side $2\sqrt{2}$.

1.3. To find the coordinates of the point dividing the line joining two given points in a given ratio.

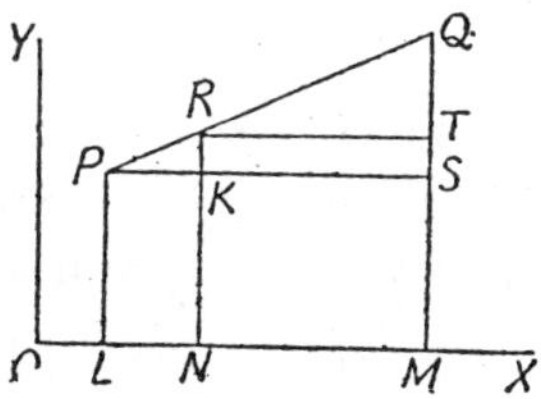

Let the coordinates of P and Q be (x_1, y_1) and (x_2, y_2) respectively and let R be the point which divides PQ in the ratio $m_1 : m_2$.

Draw PL, RN and QM parallel to OY to meet OX in L, N and M respectively and draw PS parallel to OX to intersect RN and QM in K and S respectively. Draw RT parallel to OX to meet QM in T.

Let (x, y) be the coordinates of R.

As triangles PKR and RTQ are similar, their sides are proportional. Hence

$$PK : RT = PR : RQ$$

or
$$\frac{x - x_1}{x_2 - x} = \frac{m_1}{m_2}$$

or
$$m_2 x - m_2 x_1 = m_1 x_2 - m_1 x$$

or
$$(m_1 + m_2)\, x = m_1 x_2 + m_2 x_1.$$

$$\therefore \quad x = \frac{m_1 x_2 + m_2 x_1}{m_1 + m_2}.$$

Similarly as $RK : QT = PR : RQ$,

$$y = \frac{m_1 y_2 + m_2 y_1}{m_1 + m_2}.$$

If the point R divides PQ in the ratio $m_1 : m_2$ externally, the ratio $PR : RQ$ will become $m_1 : -m_2$ as the direction of RQ now becomes opposite to PR and the coordinates (x, y) of R will be obtained from the above formula by substituting $-m_2$ for m_2, so that

$$x = \frac{m_1 x_2 - m_2 x_1}{m_1 - m_2}$$

and
$$y = \frac{m_1 y_2 - m_2 y_1}{m_1 - m_2}.$$

Corollary. If PQ is bisected by R, it divides the line in the ratio 1 : 1. Therefore, the coordinates of the middle point of PQ are given by

$$\frac{x_1 + x_2}{2} \text{ and } \frac{y_1 + y_2}{2}.$$

Ex. *Find the points which divide the line joining the points (5, 2) and (2, −3) in the ratio 2 : 3 internally and externally.*

The point of internal division is given by

$$x=\frac{2\times 2+3\times 5}{5}=\frac{19}{5},$$

$$y=\frac{2\times(-3)+3\times 2}{5}=0.$$

The point of external division is given by

$$x=\frac{2\times 2-3\times 5}{2-3}=11,$$

$$y=\frac{2\times(-3)-3\times 2}{2-3}=12.$$

1·4. Area of a triangle in terms of the coordinates of its vertices.

Let (x_1, y_1), (x_2, y_2), (x_3, y_3) be the coordinates of the vertices A, B, C of the triangle ABC respectively. Draw AL, BM, CN perpendicular to OX through A, B and C.

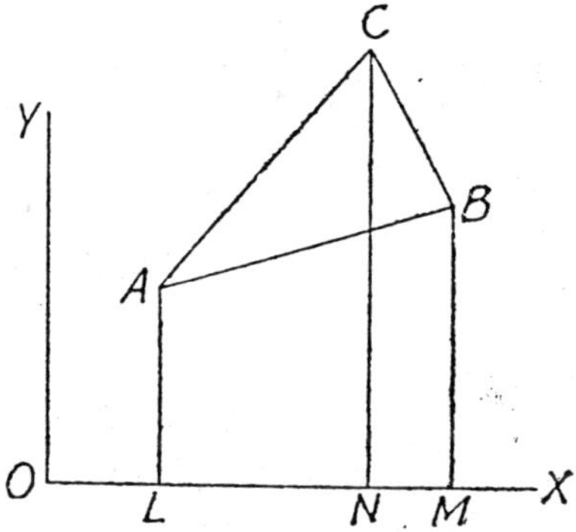

Then the area of $\triangle ABC$

$$=\text{area of trapezium } AN+\text{area of trapezium } BN-\text{area of trapezium } AM.$$

$$=\tfrac{1}{2}LN\,(AL+CN)+\tfrac{1}{2}NM\,(CN+BM)-\tfrac{1}{2}LM\,(AL+BM)$$

$$=\tfrac{1}{2}\,(x_3-x_1)\,(y_1+y_3)+\tfrac{1}{2}\,(x_2-x_3)\,(y_3+y_2)-\tfrac{1}{2}\,(x_2-x_1)\,(y_1+y_2)$$

$$=\tfrac{1}{2}x_1\,(y_1+y_2-y_1-y_3)+\tfrac{1}{2}x_2\,(y_3+y_2-y_1-y_2)+\tfrac{1}{2}x_3\,(y_1+y_3-y_3-y_2)$$

$$=\tfrac{1}{2}x_1\,(y_2-y_3)+\tfrac{1}{2}x_2\,(y_3-y_1)+\tfrac{1}{2}x_3\,(y_1-y_2)$$

$$=\tfrac{1}{2}\,[(x_1y_2-x_2y_1)+(x_2y_3-x_3y_2)+(x_3y_1-x_1y_3)].$$

The above result can also be written in the determinant form as,

$$\frac{1}{2}\begin{vmatrix} x_1 & y_1 & 1 \\ x_2 & y_2 & 1 \\ x_3 & y_3 & 1 \end{vmatrix}.$$

Note :—The above result for the area of the triangle will always be positive if in travelling from A to B and then to C the area lies on the left side of the person travelling ; otherwise it will be negative.

However, in practice the area should always be taken to be positive.

Corollary. The three points (x_1, y_1), (x_2, y_2) and (x_3, y_3) will lie in a straight line if the area enclosed by the triangle formed by them is zero, that is

$$\begin{vmatrix} x_1 & y_1 & 1 \\ x_2 & y_2 & 1 \\ x_3 & y_3 & 1 \end{vmatrix}=0.$$

Hence the above equation is the condition for three points (x_1, y_1), (x_2, y_2) and (x_3, y_3) to lie in a straight line.

1·5. Area of a quadrilateral in terms of the coordinates of its vertices.

Let the coordinates of A, B, C and D be (x_1, y_1), (x_2, y_2), (x_3, y_3) and (x_4, y_4) respectively. Join AC.

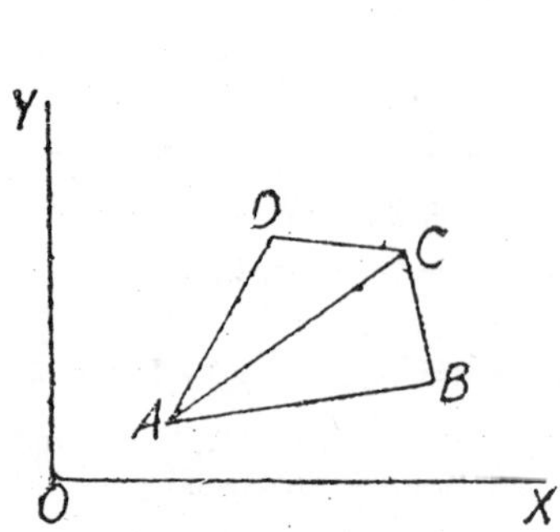

Now, the quadrilateral is divided into two triangles.

Hence the area of the quad. $ABCD$

$=$area of the $\triangle ABC+$area of the $\triangle ACD$

$$=\tfrac{1}{2}x_1(y_2-y_3)+\tfrac{1}{2}x_2(y_3-y_1)+\tfrac{1}{2}x_3(y_1-y_2)$$
$$+\tfrac{1}{2}x_1(y_3-y_4)+\tfrac{1}{2}x_3(y_4-y_1)+\tfrac{1}{2}x_4(y_1-y_3)$$
$$=\tfrac{1}{2}x_1(y_2-y_4)+\tfrac{1}{2}x_2(y_3-y_1)+\tfrac{1}{2}x_3(y_4-y_2)+\tfrac{1}{2}x_4(y_1-y_3)$$
$$=\tfrac{1}{2}[(x_1y_2-x_2y_1)+(x_2y_3-x_3y_2)+(x_3y_4-x_4y_3)$$
$$+(x_4y_1-x_1y_4)].$$

In general, if the figure is a polygon of n sides, the area will be given by

$$\tfrac{1}{2}[(x_1y_2-x_2y_1)+(x_2y_3-x_3y_2)+\ldots+(x_ny_1-x_1y_n)].$$

Ex. 1. *The coordinates of A, B, C, D are (6, 3), (−3, 5), (4, −2) and (x, 3x) respectively and* $\frac{\triangle DBC}{\triangle ABC}=\frac{1}{2}$, *find x.*

(C. University 1949)

Area of $\triangle ABC$

$$=\tfrac{1}{2}[\{(6\times 5)-3(-3)\}+\{(-3)\times(-2)-4\times 5\}$$
$$+\{4\times 3-6(-2)\}]$$
$$=\tfrac{1}{2}[30+9+6-20+12+12]$$
$$=\tfrac{49}{2}.$$

Area of $\triangle DBC$

$$=\tfrac{1}{2}[\{-3\times(-2)-4\times 5\}+\{4\times 3x-x(-2)\}$$
$$+\{x\times 5-(-3)\times 3x\}]$$
$$=\tfrac{1}{2}[6-20+12x+2x+5x+9x]$$
$$=\tfrac{1}{2}[28x-14]$$
$$=14x-7.$$

But $$\frac{\triangle DBC}{\triangle ABC}=\tfrac{1}{2}.$$

$$\therefore\quad \frac{2(14x-7)}{49}=\tfrac{1}{2}$$

or $$4(2x-1)=7$$

or $$8x-4=7.$$

$$\therefore\quad x=\tfrac{11}{8}.$$

Ex. 2. *Prove that the mid-point of the hypotenuse of a right-angled triangle is equidistant from the vertices.*

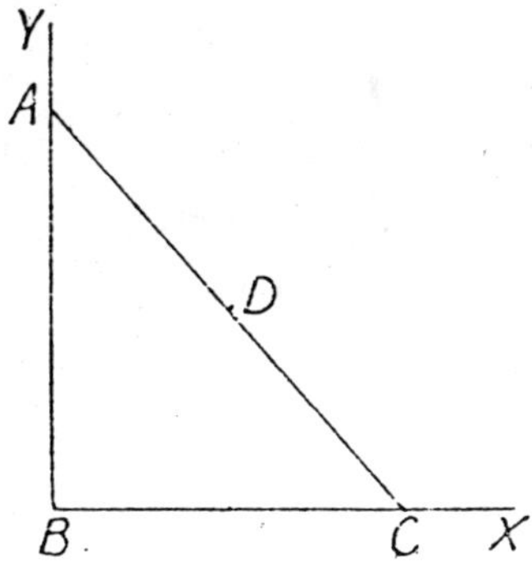

Let ABC be a triangle, rt. angled at B. Take BC and BA as the axes of x and y respectively. Let the coordinates of C be $(x_1, 0)$ and those of A $(0, y_1)$. The coordinates of D, the mid-point of AC, are

$$\left(\frac{x_1}{2}, \frac{y_1}{2}\right).$$

Now, $$BD=\sqrt{\left(\frac{x_1^2}{4}+\frac{y_1^2}{4}\right)}=\tfrac{1}{2}\sqrt{(x_1^2+y_1^2)},$$

$$DA=\sqrt{\left(\frac{y_1^2}{4}+\frac{x_1^2}{4}\right)}=\tfrac{1}{2}\sqrt{(x_1^2+y_1^2)}.$$

Also $$DC=\sqrt{\left(\frac{x_1^2}{4}+\frac{y_1^2}{4}\right)}=\tfrac{1}{2}\sqrt{(x_1^2+y_1^2)}.$$

$$\therefore \quad BD=DA=DC.$$

Ex. 3. *Find the coordinates of the incentre of a triangle.*

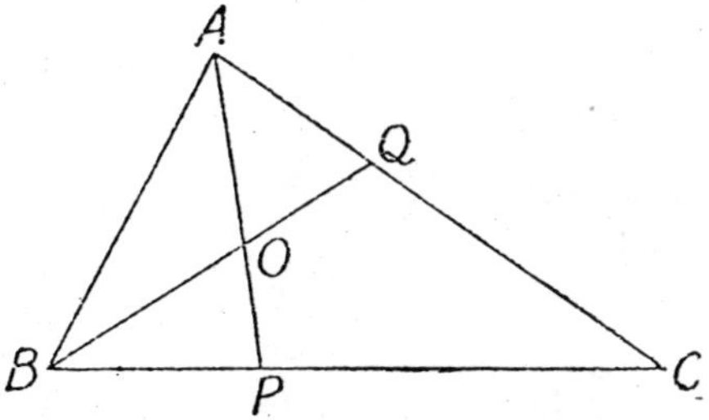

Let ABC be a triangle, the coordinates of whose vertices A, B, C are (x_1, y_1), (x_2, y_2) and (x_3, y_3) respectively and the lengths of whose sides are a, b and c.

As AP bisects the angle BAC, $\dfrac{BP}{PC}=\dfrac{BA}{AC}=\dfrac{c}{b}$.

Hence the point P divides BC in the ratio $c : b$.

Therefore, the coordinates of P are

$$\frac{cx_3+bx_2}{c+b} \text{ and } \frac{cy_3+by_2}{c+b}.$$

The length of BC is a; hence the length of $BP=\dfrac{ac}{b+c}$.

Now, as BQ bisects the angle ABC,

$$\frac{PO}{OA}=\frac{BP}{BA}=\frac{\frac{ac}{b+c}}{c}=\frac{a}{b+c},$$

so that O divides PA in the ratio $PO : OA$, that is, $a : b+c$.

Therefore, the coordinates of O are given by

$$\frac{ax_1+(b+c)\dfrac{(cx_3+bx_2)}{c+b}}{a+b+c} \text{ and } \frac{ay_1+(b+c)\dfrac{(cy_3+by_2)}{c+b}}{a+b+c}$$

i.e.,
$$\frac{ax_1+bx_2+cx_3}{a+b+c} \text{ and } \frac{ay_1+by_2+cy_3}{a+b+c}.$$

1·6. Polar Coordinates.

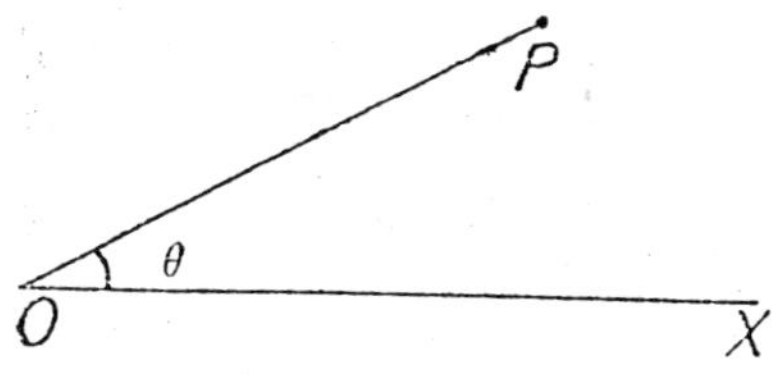

Let O be a fixed point and OX a fixed line. Let $\angle XOP$ be θ and let the distance OP be r. Then the polar coordinates of P are (r, θ). The angle θ is positive when measured counter-clockwise from the line OX. The distance r is positive when measured from O along the line bounding the angle XOP. The fixed point O is called the **pole** and the fixed line OX the **initial line.** The angle θ is called the vectorial Angle of the point P. The coordinates of the point P in the above position can also be denoted by $(-r, \theta+\pi)$, $(r, 2\pi+\theta)$ etc. according to the convention of signs given above.

1·7. Relation between rectangular and polar coordinates.

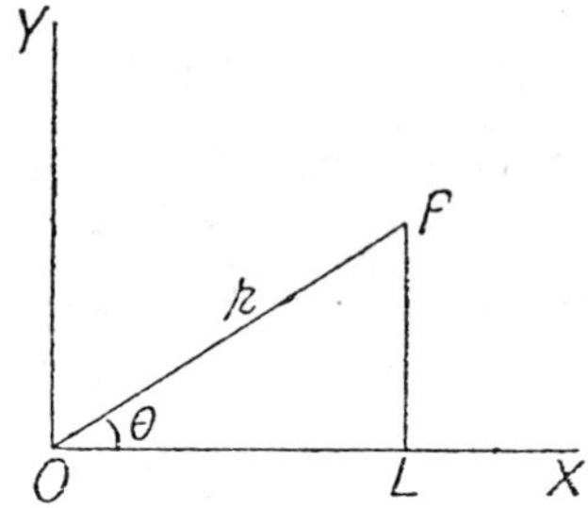

Let O, the origin, be the pole and OX the initial line. Then the rectangular and polar coordinates of the point P are (x, y) and (r, θ) respectively.

Now, $OL = x = r \cos \theta$ ..(1)

and $PL = y = r \sin \theta$ ..(2)

Squaring and adding (1) and (2), we get

$$x^2 + y^2 = r^2. \qquad \ldots \quad \ldots \quad \ldots (3)$$

Dividing (2) by (1), we get

$$\tan \theta = \frac{y}{x}$$

or

$$\theta = \tan^{-1} \frac{y}{x}. \qquad \ldots \quad \ldots \quad \ldots (4)$$

With the above relations, we can change any expression in x and y into another one in r and θ and vice versa.

1·8. Distance between two points in polar coordinates.

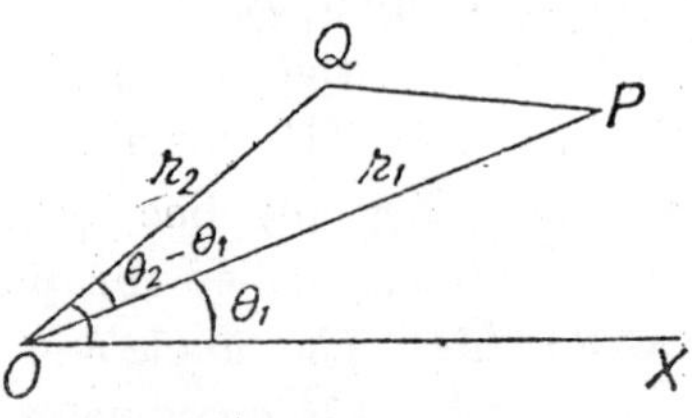

Let the coordinates of P and Q be (r_1, θ_1) and (r_2, θ_2) respectively.

Then the angle POQ is $\theta_2 - \theta_1$.

Now, from $\triangle OPQ$,

$$PQ^2 = OP^2 + OQ^2 - 2OP.OQ \cos POQ$$

$$= r_1^2 + r_2^2 - 2r_1 r_2 \cos (\theta_2 - \theta_1).$$

or
$$2r_1r_2 \sin\left(\frac{\theta_2-\theta_1}{2}\right)\cos\left(\frac{\theta_2-\theta_1}{2}\right)$$
$$=r \sin\frac{\theta_2-\theta_1}{2}(r_1+r_2);$$

$$\therefore\quad r=\frac{2r_1r_2\cos\left(\frac{\theta_2-\theta_1}{2}\right)}{r_1+r_2}=\frac{2r_1r_2\cos\left(\frac{\theta_1-\theta_2}{2}\right)}{r_1+r_2}.$$

1·10. Locus and its equation.

The **Locus** of a moving point is the path traced out by it when it moves according to some given condition. A point moves so that its distance from the y-axis always remains 'a'. The locus in this case is a straight line parallel to y-axis at a distance a from it. Its equation is $x=a$.

Equation to a locus is the algebraic relation which exists among the coordinates of any point on the locus. The points not lying on the locus do not satisfy the aforesaid algebraic relation. In the case given above the relation $x=a$ is satisfied by all points lying on the locus whose equation is $x=a$ and is not satisfied by any point not lying on the locus.

To find the equation to a locus the following steps should be carried out.

1. Select the origin in such a manner that the relation between the coordinates of the moving point is the simplest.

2. Take the coordinates of the moving point as (x, y).

3. Find the relation between x and y on the basis of the given condition.

1·9. Area of a triangle in polar coordinates.

Let the points P, Q, R be (r_1, θ_1), (r_2, θ_2) and (r_3, θ_3) respectively.

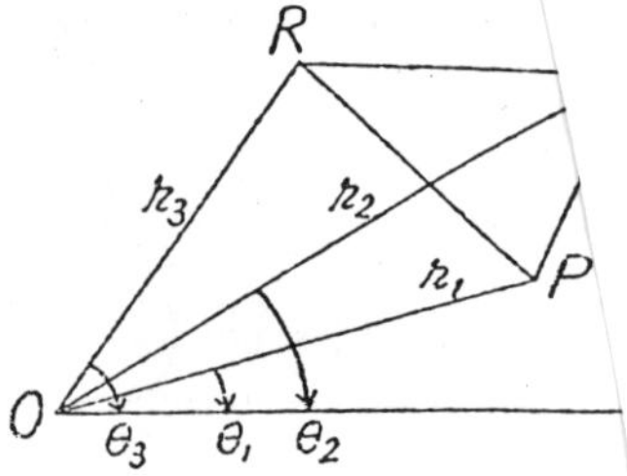

Area of $\triangle$ PQR

$=$area of $\triangle$ OPQ
$+$area of $\triangle$ OQR
$-$area of $\triangle$ OPR

$=\frac{1}{2}OP.OQ \sin POQ+\frac{1}{2}OQ.OR \sin QOR$
$-\frac{1}{2}OP.OR \sin PC$

$=\frac{1}{2}r_1r_2 \sin(\theta_2-\theta_1)+\frac{1}{2}r_2r_3 \sin(\theta_3-\theta_2)-\frac{1}{2}r_1r_3 \sin(\theta_3-\theta_1)$

$=\frac{1}{2}\{r_2r_3 \sin(\theta_3-\theta_2)+r_3r_1 \sin(\theta_1-\theta_3)+r_1r_2 \sin(\theta_2-\theta_1)\}.$

Ex. *The vectorial angle of the point P on the join of the points P_1 (r_1, θ_1) and P_2 (r_2, θ_2) is $\frac{1}{2}(\theta_1+\theta_2)$. Show that the length of the radius vector of the point is*

$$\frac{2r_1r_2 \cos\frac{(\theta_1-\theta_2)}{2}}{r_1+r_2}.$$

(C. University 1948)

As the vectorial angle of P is $\frac{\theta_1+\theta_2}{2}$, the angle FOP_1 is $\frac{\theta_2-\theta_1}{2}$. Also the angle POP_2 is $\frac{\theta_2-\theta_1}{2}$. Let OP be r.

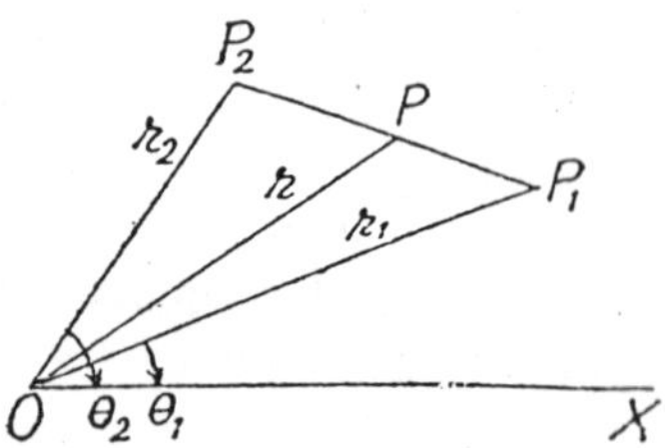

Now, the area of $\triangle$ P_1OP_2

$=$area of $\triangle$ P_1OP+area of $\triangle$ POP_2

or $\quad \frac{1}{2}r_1r_2 \sin(\theta_2-\theta_1)=\frac{1}{2}rr_1 \sin\left(\frac{\theta_2-\theta_1}{2}\right)$

$+\frac{1}{2}rr_2 \sin\frac{\theta_2-\theta_1}{2}$

Ex. 1. *Find the locus of points where a given straight line of length 2a subtends a right angle.*

Let AB be the given straight line of length $2a$ and let P be the point such that $\angle APB$ is a right angle.

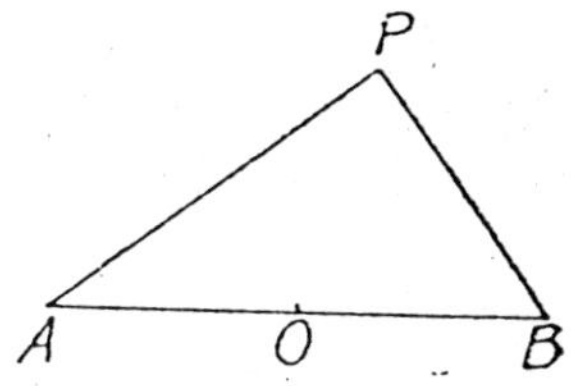

It is required to find the locus of P. Select O, the middle point of AB, as the origin. Take AB as the x-axis. Then the coordinates of B and A are $(a, 0)$ and $(-a, 0)$ respectively. Let (x, y) be the coordinates of the moving point P. According to the given condition,

$$AP^2+PB^2=AB^2$$

or $$(x+a)^2+y^2+(x-a)^2+y^2=4a^2$$

or $$2x^2+2a^2+2y^2=4a^2$$

or $$x^2+y^2=a^2.$$

Ex. 2. *A rod of length l rests against the floor and a wall of a room. If the rod begins to slide on the floor, find the locus of its middle point.*

Let AB be the rod, OA the floor and OB the wall. Let O be the origin and OA and OB the axes of x and y respectively.

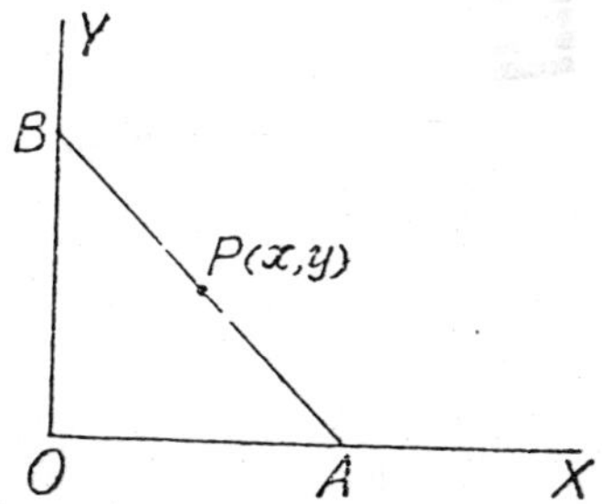

Let the coordinates of P, the middle point of AB, be (x, y).

With O as origin let the coordinates of A and B be $(a, 0)$ and $(0, b)$ respectively.

Then, as P is the middle point of AB,

$$x=\frac{a}{2} \qquad \ldots \quad \ldots (1)$$

$$y=\frac{b}{2} \qquad \ldots \quad \ldots (2)$$

Also, as $\angle AOB$ is a right angle,

$$AB^2=OA^2+OB^2$$

or
$$l^2=a^2+b^2 \qquad \ldots \quad \ldots(3)$$

Eliminating a and b from (1), (2) and (3), we get

$$l^2=4x^2+4y^2$$

or
$$x^2+y^2=\frac{l^2}{4}$$

which is a relation between x and y, the coordinates of the middle point of AB, and hence is the equation of its locus.

Exercise I.

1. Proove that the points $(2a, 4a)$, $(2a, 6a)$ and $(2a+\sqrt{3}a, 5a)$ are the vertices of an equilateral triangle whose side is $2a$. *(C. U. 1952)*

2. The vertices of a triangle are (x_1, y_1), (x_2, y_2) and (x_3, y_3). Show that the coordinates of its centroid are

$$\left(\frac{x_1+x_2+x_3}{3}\right), \left(\frac{y_1+y_2+y_3}{3}\right).$$

3. Find the coordinates of points which divide, internally and externally, the line joining the points $(l+m)$, $(l-m)$ and $(l-m, l+m)$ in the ratio $l : m$.

4. If G is the centroid of a triangle ABC and P any other point, prove that

$$3\,(GA^2+GB^2+GC^2)=BC^2+GA^2+AB^2$$

and $PA^2+PB^2+PC^2=GA^2+GB^2+GC^2+3GO^2$.

5. The cartesian coordinates (x, y) of a point on a curve are given by

$$x : y : 1=t^3 : t^2-3 : t-1,$$

where t is a parameter ; show that the points given by $t=a, b, c$ are collinear if $abc-(bc+ca+ab)+3\,(a+b+c)=0$. *(M. T. 1946)*

6. Prove that the line joining the middle points of any two sides of a triangle is half of the third side.

7. Prove that the lines joining the middle points of the opposite sides of a quadrilateral and the line joining the middle points of its diagonals meet in a point and bisect one another.

8. If (1, 1), (7, −3), (12, 2) and (7, 21) are the coordinates of the vertices of a quadrilateral, prove that its area is 132 square units. *(Roorkee 1945)*

9. If the area of the quadrilateral whose angular points A, B, C, D taken in order, are (1, 2), (−5, 6), (7, −4) and $(k, -2)$ be zero, find the value of k. *(C. U. 1945)*

10. Show that the area of a triangle formed by joining the middle points of the sides of a given triangle is one-fourth of the area of the latter.

11. Find the length of the straight line joining the points whose polar coordinates are $\left(l, \frac{\pi}{2}\right)$ and $\left(3l, \frac{\pi}{6}\right)$.

12. Find the area of the triangle whose vertices in polar coordinates are

$$\left(-c, \frac{\pi}{6}\right), \left(c, \frac{\pi}{2}\right) \text{ and } \left(-2c, -\frac{2\pi}{3}\right).$$

13. Transform the following equations to polar coordinates :

(i) $x^2+y^2=a^2$ (ii) $y=px$ (iii) $x^2+y^2=2mx$
(iv) $(x^2+y^2)^2=a^2(x^2-y^2)$.

14. Transform the following equations to cartesian coordinates :—

(i) $r=a$ (ii) $\theta=\tan^{-1} p$ (iii) $r=a \sin 2\theta$
(iv) $r^{1/2}=a^{1/2} \cos \theta/2$ (v) $r^2 \cos 2\theta=a^2$.

15. The sum of the squares of the distances of a moving point from two fixed points $(c, 0)$ and $(-c, 0)$ is equal to a constant quantity $2a^2$. Find the equation to its locus.

16. A and B are the fixed points $(c, 0)$ and $(-c, 0)$ respec-

tively ; find the equation of the locus of P, when

(i) $PA+PB=k$, a constant quantity.

(ii) $PA^2+PB^2=$a constant (k).

(iii) $PA=k.PB$, k being a constant.

17. Find the locus of a point whose distance from $(3, \pi/3)$ is k.

18. The distance of a moving point from a fixed point is equal to its distance from a fixed straight line. Obtain the equation of the locus of the moving point.

CHAPTER II

THE STRAIGHT LINE

2·1. Equation to straight lines parallel to coordinate axes.

Let AB be a straight line parallel to the y-axis drawn at a distance h from the origin

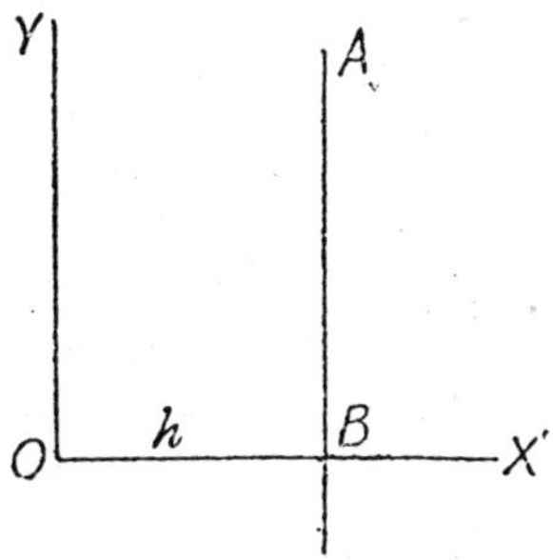

When a point moves on the locus AB, its abscissa always remains equal to h. Hence, if (x, y) are the coordinates of any point on AB, the condition which is satisfied by its coordinates is

$$x=h \qquad \ldots \quad \ldots(1)$$

Equation (1) does not involve any variable. Therefore the above relation (1) is the equation to the straight line AB.

Similarly, the equation to a straight line parallel to the x-axis at a distance k from the origin is $y=k$ ... ...(2)

2·2. Equation to a straight line which cuts off an intercept c on the y-axis and is inclined at an angle θ to the x-axis.

Let the coordinates of any point P on AB be (x, y). Draw PN perpendicular to OX. Now, $ON=x$ and $PN=y$. Also, from $\triangle OAB$, $AO=c \cot \theta$.

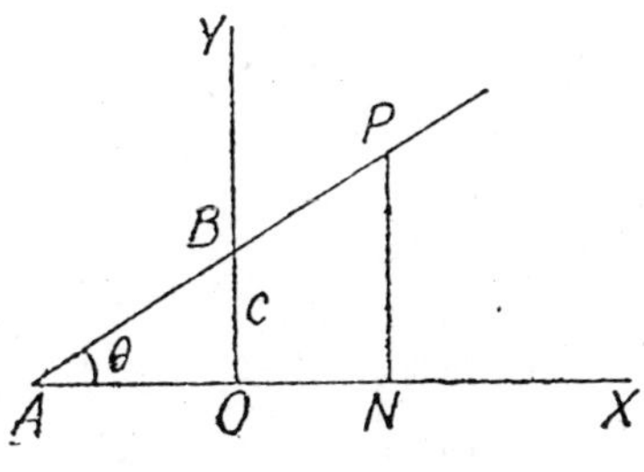

$$\therefore \quad AN=AO+ON$$
$$=c \cot \theta+x.$$

Now from $\triangle PAN$,

$$PN = AN \tan\theta,$$

or $$y = (c \cot\theta + x)\tan\theta$$

$$= c + x\tan\theta. \qquad \ldots \quad \ldots (1)$$

The above is clearly a relation between x and y and does not involve any variable quantity. Hence

$$y = x\tan\theta + c$$

is the equation to the straight line AB. $\tan\theta$ is the tangent of the angle θ which gives the direction of the straight line AB with reference to the axes OX and OY. Hence it is called the direction-tangent of the straight line AB. The direction-tangent is also denoted by m, so that

$$\tan\theta = m \quad \text{or} \quad \theta = \tan^{-1} m$$

and the equation to the straight line AB reduces to $y = mx + c$.

Note :—The angle θ is taken positive if measured from OX in the anticlockwise direction and the intercept c is considered positive if taken along the positive direction of the y-axis.

Corollary. If the straight line passes through the origin, $c=0$ and the equation becomes $y = mx$.

2·3. Equation to a straight line which cuts off intercepts a and b from the axes of X and Y respectively.

Let P be a point (x, y) on the straight line AB.

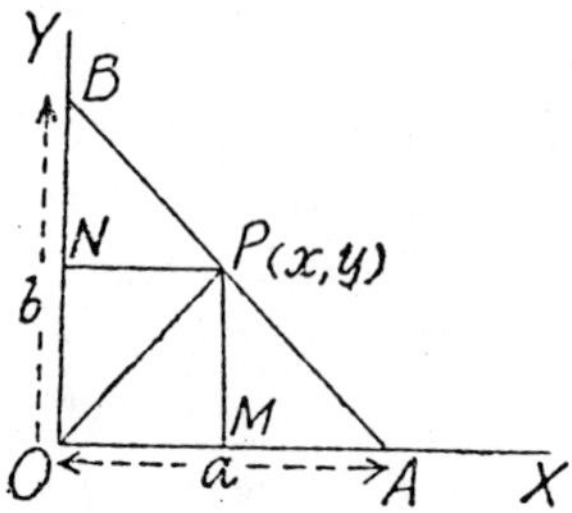

Now,

$$\triangle BOA = \triangle POA + \triangle POB,$$

or $$\tfrac{1}{2}ab = \tfrac{1}{2}ay + \tfrac{1}{2}bx$$

or $$1 = y/b + x/a,$$

i.e. $$x/a + y/b = 1.$$

As this is a relation between

x and y not involving variables, it represents the equation to the straight line AB.

2·4. Equation to a straight line, given the length of the perpendicular on it from the origin and the inclination of this perpendicular to the x-axis.

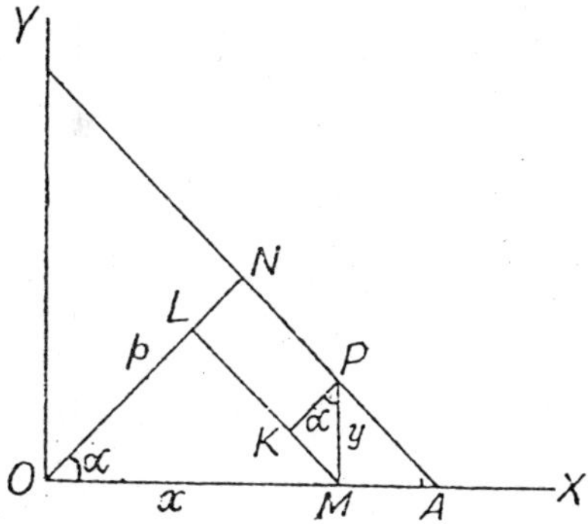

Let the length ON of the perpendicular from the origin O on the straight line AB be p and let α be the angle which it makes with the x-axis in the anticlockwise direction from the axis. Let $P(x, y)$ be a point on AB.

$$OM=x,\ PM=y,\ \angle PMK=\alpha.$$

From $\triangle OLM$, $OL=OM\cos\alpha=x\cos\alpha$.

From $\triangle PMK$, $PK=PM\sin\alpha=y\sin\alpha$.

$$\therefore\quad p=ON=OL+LN$$
$$=OL+PK$$
$$=x\cos\alpha+y\sin\alpha.$$

Hence the equation to the straight line AB is

$$x\cos\alpha+y\sin\alpha=p.$$

2·5. Equation of a straight line passing through two given points.

Let the straight line AB pass through two given points A and B whose coordinates are (x_1, y_1) and (x_2, y_2) respectively. Let P be a point on AB whose coordinates are (x, y). From A, B and P draw perpendiculars AK, BL and PM on OX and draw AS parallel to OX cutting BL and PM in R and S respectively.

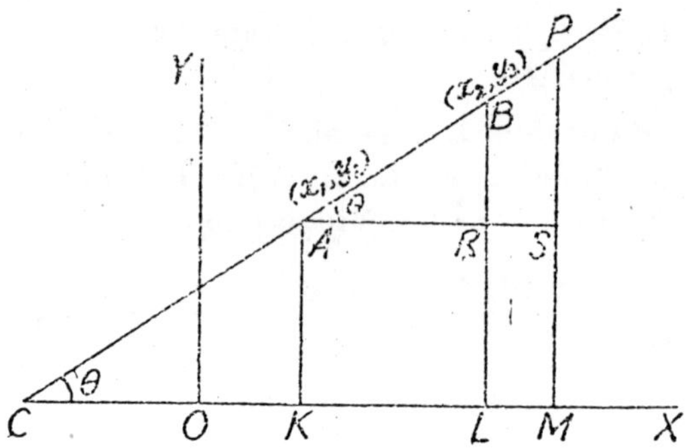

Now, $AR = x_2 - x_1$, $AS = x - x_1$,

$BR = y_2 - y_1$, $PS = y - y_1$.

From similar triangles ARB and ASP,

$$\frac{AR}{AS} = \frac{BR}{PS},$$

or
$$\frac{x_2 - x_1}{x - x_1} = \frac{y_2 - y_1}{y - y_1},$$

or
$$y - y_1 = \frac{y_2 - y_1}{x_2 - x_1}(x - x_1),$$

which is the equation of the straight line AB.

The above equation can also be written as

$$y = \frac{y_2 - y_1}{x_2 - x_1}x + y_1 - \frac{y_2 - y_1}{x_2 - x_1}.x_1.$$

Comparing the above equation with $y = mx + c$, we find that the direction tangent m of this line is

$$\frac{y_2 - y_1}{x_2 - x_1}.$$

It is also clear from the figure that if angle $BAS = \theta$, from the triangle BAR, $\tan\theta = \dfrac{BR}{AR} = \dfrac{y_2 - y_1}{x_2 - x_1}$.

Hence the direction tangent of the straight line joining the points (x_1, y_1) and (x_2, y_2) is given by $\dfrac{y_2 - y_1}{x_2 - x_1}$.

Corollary. The equation to the straight line joining (x_1, y_1) to the origin is $y=\frac{y_1}{x_1}.x$.

2·6. We have found the equation to a straight line in the following forms :—

$$y=mx+c \qquad \ldots \quad \ldots(1)$$

$$\frac{x}{a}+\frac{y}{b}=1 \qquad \ldots \quad \ldots(2)$$

$$x \cos \alpha+y \sin \alpha=p \qquad \ldots \quad \ldots(3)$$

$$y-y_1=\frac{y_2-y_1}{x_2-x_1}(x-x_1). \qquad \ldots \quad \ldots(4)$$

It appears from the above results that the equation to a straight line is of the first degree in x and y. We shall now show that the equation to a straight line is always of the first degree and conversely, that every equation of the first degree always represents a straight line.

(The equation to a straight line is of the first degree in x and y).

A straight line is completely determined if we know two points on it. Let (x_1, y_1) and (x_2, y_2) be the two points on it. Let (x, y) be any other point on the line. As these three points lie on a straight line, the area enclosed by the straight lines joining them should be zero. Hence

$$\begin{vmatrix} x & y & 1 \\ x_1 & y_1 & 1 \\ x_2 & y_2 & 1 \end{vmatrix}=0$$

or $\qquad x(y_1-y_2)+y(x_2-x_1)+x_1y_2-x_2y_1=0$

which can be written in the form

$$ax+by+c=0 \qquad \ldots \quad \ldots(1)$$

where a, b and c are constants. (1) is an equation of the first degree. Therefore, the equation to a straight line is of the first degree in x and y.

(An equation of the first degree in x and y always represents a straight line.)

Let $ax+by+c=0$ be the equation to a straight line. Let (x_1, y_1), (x_2, y_2) and (x_3, y_3) be three points on it.

Hence these points must satisfy the above equation,

so that $$ax_1+by_1+c=0,$$
$$ax_2+by_2+c=0$$
and $$ax_3+by_3+c=0.$$

Eliminating a, b and c, we get

$$\begin{vmatrix} x_1 & y_1 & 1 \\ x_2 & y_2 & 1 \\ x_3 & y_3 & 1 \end{vmatrix}=0,$$

which means that the area of the triangle obtained by joining the points (x_1, y_1), (x_2, y_2) and (x_3, y_3) is zero, that is to say, the points lie on a straight line.

Therefore every equation of the first degree represents a straight line.

2·7. Reduction of the general equation of the first degree to standard forms.

Gradient Form—

$$ax+by+c=0$$

or $$y=-\frac{a}{b}x-\frac{c}{b}.$$

Hence the gradient m is $-\frac{a}{b}$, so that the straight line makes an angle θ with the positive direction of the x-axis, such that $\tan\theta = -\frac{a}{b}$ and it cuts off an intercept $\frac{c}{b}$ from the negative side of the y-axis.

Intercept Form—

$$ax+by+c=0$$

or

$$\frac{x}{-\frac{c}{a}}+\frac{y}{-\frac{c}{b}}=1.$$

The straight line cuts off intercepts equals to $\frac{c}{a}$ and $\frac{c}{b}$ from the negative sides of the x and y-axis respectively.

Perpendicular Form—

$$ax+by+c=0.$$

Comparing it with $x\cos\alpha+y\sin\alpha=p$, we get

$$\frac{\cos\alpha}{a}=\frac{\sin\alpha}{b}=\frac{p}{-c}=\pm\frac{\sqrt{(\cos^2\alpha+\sin^2\alpha)}}{\sqrt{(a^2+b^2)}}$$

$$=\pm\frac{1}{\sqrt{(a^2+b^2)}} \qquad \ldots \quad \ldots(1)$$

The sign of $\frac{1}{\sqrt{(a^2+b^2)}}$ may be positive or negative. It should be such that the length of the perpendicular p is positive. From (1) above, we find that if c is negative, the sign of $\frac{1}{\sqrt{(a^2+b^2)}}$ should be positive because then $p=\frac{c}{\sqrt{(a^2+b^2)}}$, a positive quantity; but if c is positive the sign of $\frac{1}{\sqrt{(a^2+b^2)}}$ should be negative, because only then p will be positive.

Therefore, $\cos\alpha=\pm\frac{a}{\sqrt{(a^2+b^2)}}$,

$$\sin\alpha = \pm\frac{b}{\sqrt{(a^2+b^2)}},$$

and $$p = \frac{c}{\sqrt{(a^2+b^2)}}.$$

Ex. 1. *If p is the length of the perpendicular from the origin on the line $\frac{x}{a}+\frac{y}{b}=1$, show that $\frac{1}{p^2}=\frac{1}{a^2}+\frac{1}{b^2}$.*

Comparing the equation $\frac{x}{a}+\frac{y}{b}=1$ with $x\cos\alpha+y\sin\alpha=p$, we get

$$\frac{\cos\alpha}{\frac{1}{a}}=\frac{\sin\alpha}{\frac{1}{b}}=\frac{p}{1}=\pm\frac{1}{\sqrt{\left(\frac{1}{a^2}+\frac{1}{b^2}\right)}},$$

that is $$p^2=\frac{1}{\frac{1}{a^2}+\frac{1}{b^2}}$$

or $$\frac{1}{a^2}+\frac{1}{b^2}=\frac{1}{p^2}.$$

Ex. 2. *Find the equations to the straight lines which pass through the point $(2, -3)$ and cut off equal intercepts from the axes.*

Here two cases arise, one when the intercepts are of the same sign and the other when they are of opposite signs.

In the first case let the equation to the straight line be $\frac{x}{a}+\frac{y}{a}=1.$

As it passes through $(2, -3)$,

$$2-3=a,$$

or $$a=-1.$$

So, the equation becomes

$$x+y+1=0.$$

In the second case let the equation to the straight line be $\frac{x}{a}+\frac{y}{-a}=1.$

As it passes through $(2, -3)$,

$$2+3=a$$

or

$$a=5,$$

so that the equation becomes

$$x-y=5.$$

2·8. Equation to a straight line passing through a given point and inclined to the x-axis at a given angle.

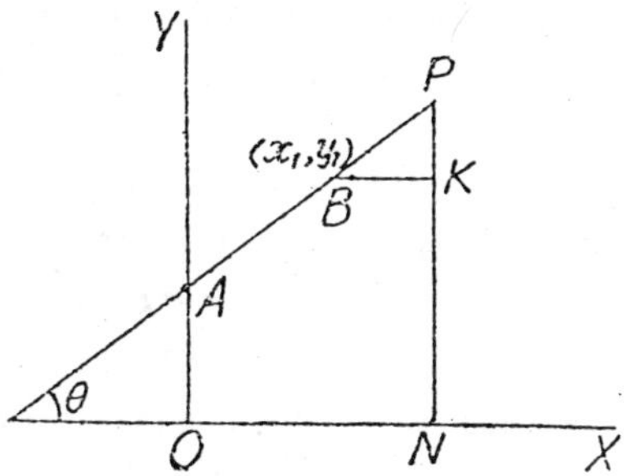

Let AB be a straight line passing through a point $B(x_1, y_1)$ and let it be inclined to the x-axis at an angle θ.

Let P be any other point on the straight line whose coordinates are (x, y).

From P draw PN perpendicular to OX.

Draw BK parallel to OX cutting PN at K.

$$\angle PBK=\theta,\ BK=x-x_1 \text{ and } PK=y-y_1.$$

From $\triangle PBK$, $\tan\theta=\frac{PK}{BK}=\frac{y-y_1}{x-x_1}$,

or

$$\frac{x-x_1}{\cos\theta}=\frac{y-y_1}{\sin\theta},$$

which is the equation to the straight line AB.

If r is the distance of any point $P(x, y)$ on the straight line from B, we have

$$x-x_1=r\cos\theta$$

and

$$y-y_1=r\sin\theta,$$

so that $$\frac{x-x_1}{\cos\theta}=\frac{y-y_1}{\sin\theta}=r.$$

The coordinates of any point (x, y) on the line whose distance from (x_1, y_1) is r, are therefore given by

$$x=x_1+r\cos\theta,$$
$$y=y_1+r\sin\theta.$$

Ex. *Find the direction in which a straight line must be drawn through (1, 2) so that its point of intersection with the line $x+y=4$ may be at a distance $\frac{1}{3}\sqrt{6}$ from this point.*

Let the direction of the straight line with the x-axis be θ. The equation to the straight line passing through (1, 2) and drawn in the direction θ with the x-axis is given by

$$\frac{x-1}{\cos\theta}=\frac{y-2}{\sin\theta}.$$

The coordinates of the point (x, y) whose distance from (1, 2) on this line is $\frac{1}{3}\sqrt{6}$ are therefore given by

$$x=1+\tfrac{1}{3}\sqrt{6}\cos\theta,$$
$$y=2+\tfrac{1}{3}\sqrt{6}\sin\theta.$$

But this point lies on the line

$$x+y=4.$$

Hence $1+\frac{1}{3}\sqrt{6}\cos\theta+2+\frac{1}{3}\sqrt{6}\sin\theta=4$

or $$\cos\theta+\sin\theta=\frac{3}{\sqrt{6}}$$

or $$(\cos\theta+\sin\theta)^2=\tfrac{9}{6}=\tfrac{3}{2},$$

or $$2\sin\theta\cos\theta=\tfrac{1}{2},$$

or $$\sin 2\theta=\tfrac{1}{2},$$

or $$2\theta=30^\circ \text{ or } 150^\circ.$$

$$\therefore\ \theta=15^\circ \text{ or } 75^\circ.$$

Exercise II

1. Find the equation to the straight line —

(a) making an angle θ with the x-axis and cutting off an intercept c from the negative side of the y-axis;

(b) through (a, b) parallel to the axes,

(c) through (1, 2) and making an angle of 45° with the x-axis.

2. Find the length of the perpendicular from the origin and the angle which it makes with x-axis in the case of the following straight lines :—

(a) $x\sqrt{3}+y=10$.

(b) $x-\sqrt{3}y+1=0$.

3. Find the equation to the line—

(a) which cuts off intercepts 5 and -6 from the x and y-axes respectively;

(b) which passes through the point $(2, -1)$ and cuts off equal intercepts from the axes.

4. The line $x-y\sqrt{3}+6=0$ meets the axes in A and B. Determine (i) $\angle XAB$, (ii) area of $\triangle AOB$, and (iii) perpendicular distance of AB from O.

5. Find the equation to the line joining the points :—

(i) $(a \cos \phi, b \sin \phi)$ and $(a \cos \theta, b \sin \theta)$;

(ii) $(at_1^2, 2at_1)$ and $(at_2^2, 2at_2)$.

6. Find the direction in which a straight line must be drawn through the point $(\sqrt{3}, 2)$ so that its point of intersection with the line $\sqrt{3}x+y=9$ may be at a distance 2 from this point.

7. A straight line is drawn through the point (2, 1) and is inclined to the x-axis at an angle of 45°. Find the length intercepted on the line between the given point and the line $y-2x+6=0$.

2·9. Angle between two straight lines.

(a) Let the equations to the straight lines AB and CD be $y=m_1x+c_1$ and $y=m_2x+c_2$ respectively. Let α be the

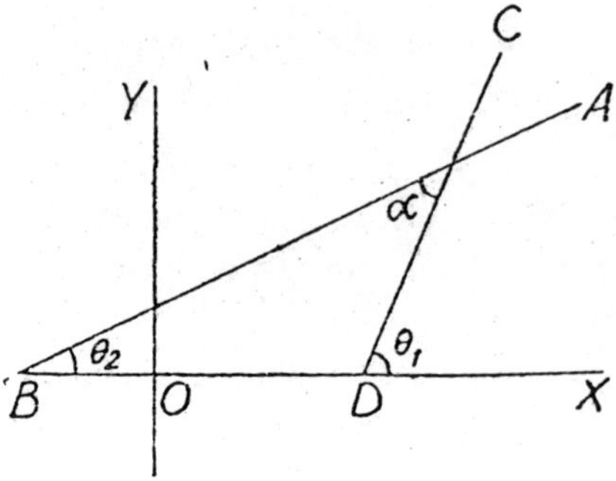

angle between them.

As θ_1 is the angle which CD makes with the x-axis,

$$\tan \theta_1 = m_1.$$

Similarly, $$\tan \theta_2 = m_2.$$

Also $$\alpha = \theta_1 - \theta_2,$$

or
$$\tan \alpha = \tan (\theta_1 - \theta_2)$$
$$= \frac{\tan \theta_1 - \tan \theta_2}{1 + \tan \theta_1 \tan \theta_2}$$
$$= \frac{m_1 - m_2}{1 + m_1 m_2}.$$
$$\therefore \quad \alpha = \tan^{-1} \frac{m_1 - m_2}{1 + m_1 m_2}.$$

Note. If the value of $\tan \alpha$ is negative, it means that it is the value for the obtuse angle of intersection between the lines.

Cor. 1. The lines are parallel if the angle between them is zero, that is

$$m_1 - m_2 = 0 \quad \text{or} \quad m_1 = m_2.$$

Cor. 2. The lines are perp. to each other if the angle between them is 90° that is

$$1 + m_1 m_2 = 0 \quad \text{or} \quad m_1 m_2 = -1.$$

(b) If the equations to the straight lines are

$$a_1 x + b_1 y + c_1 = 0$$

and $$a_2x+b_2y+c_2=0,$$

their direction tangents are $\left(-\frac{a_1}{b_1}\right)$ and $\left(-\frac{a_2}{b_2}\right)$ respectively and the angle between the lines is given by

$$\tan\alpha=\frac{-\frac{a_1}{b_1}+\frac{a_2}{b_2}}{1+\frac{a_1a_2}{b_1b_2}}$$

$$=\frac{a_2b_1-a_1b_2}{a_1a_2+b_1b_2}.$$

The condition that the lines be parallel is that

$$a_2b_1-a_1b_2=0.$$

The condition for the lines to be at right angles is

$$a_1a_2+b_1b_2=0.$$

(c) If the equations to the lines are

$$x\cos\alpha+y\sin\alpha=p$$

and $$x\cos\beta+y\sin\beta=k,$$

the angle between the lines is the same as the angle between the perpendiculars drawn to them from the origin, which is equal to $\alpha\sim\beta$.

2·10. Straight lines parallel and perpendicular to the line $ax+by+c=0$.

(i) If m is the slope of the line parallel to the above line,

$$m=-a/b.$$

Hence the equation to the line is

$$y=-\frac{a}{b}x+k$$

or $$ax+by+bk=0$$

or $$ax+by+\lambda=0 \qquad \text{where } \lambda=bk.$$

Hence, to find the equation to any line parallel to $ax+by+c=0$ we simply change the constant c and the equation is $ax+by+\lambda=0$.

(ii) If m is the slope of the line perpendicular to

$$ax+by+c=0,$$

$$m\times\left(-\frac{a}{b}\right)+1=0 \quad \text{or} \quad m=\frac{b}{a}.$$

Hence the equation of the line is

$$y=\frac{b}{a}x+k$$

or $$bx-ay+ak=0$$

or $$bx-ay+\lambda=0, \qquad \text{where } \lambda=ak.$$

Hence, to find the equation to any line perpendicular to $ax+by+c=0$ we interchange the coefficients of x and y, change the sign of one of these coefficients and also change the constant term in the given equation.

2·11. Point of intersection of two straight lines.

Let the two straight lines be

$$a_1x+b_1y+c_1=0$$

and $$a_2x+b_2y+c_2=0.$$

Let (x_1, y_1) be the point of intersection. As the point of intersection lies on both the straight lines, it satisfies both the equations and therefore

$$a_1x_1+b_1y_1+c_1=0 \qquad \ldots \quad \ldots(1)$$

$$a_2x_1+b_2y_1+c_2=0 \qquad \ldots \quad \ldots(2)$$

Solving the above simultaneous equations for x_1 and y_1, we get

$$\frac{x_1}{b_1c_2-b_2c_1}=\frac{y_1}{a_2c_1-a_1c_2}=\frac{1}{a_1b_2-a_2b_1},$$

which give $$x_1=\frac{b_1c_2-b_2c_1}{a_1b_2-a_2b_1}$$

and $$y_1=\frac{a_2c_1-a_1c_2}{a_1b_2-a_2b_1}.$$

Note :—If $a_1b_2-a_2b_1=0$, the point of intersection lies at infinity. But this is also the condition for the lines to be parallel. Hence parallel lines intersect at the point at infinity.

2·12. Straight lines passing through the intersection of two given straight lines.

Let the given straight lines be

$$a_1x+b_1y+c_1=0$$

and

$$a_2x+b_2y+c_2=0.$$

The point of intersection (x_1, y_1) of the above straight lines is given by

$$x_1=\frac{b_1c_2-b_2c_1}{a_1b_2-a_2b_1},$$

$$y_1=\frac{a_2c_1-a_1c_2}{a_1b_2-a_2b_1}.$$

Let the equation to the straight lines passing through (x_1, y_1) be $y=mx+c$.

As (x_1, y_1) lies on it, we have

$$y_1=mx_1+c \quad \text{or} \quad c=y_1-mx_1.$$

Hence the equation to the straight line passing through (x_1, y_1) is

$$y=mx+y_1-mx_1 \quad \text{or} \quad y-y_1=m\,(x-x_1).$$

Aliter :—Let (x_1, y_1) be the point of intersection of the straight lines

$$a_1x+b_1y+c_1=0 \qquad \ldots \quad \ldots(1)$$

$$a_2x+b_2y+c_2=0 \qquad \ldots \quad \ldots(2)$$

Then, we have

$$a_1x_1+b_1y_1+c_1=0 \qquad \ldots \quad \ldots(3)$$

$$a_2x_1+b_2y_1+c_2=0 \qquad \ldots \quad \ldots(4)$$

Let us consider the equation

$$a_1x+b_1y+c_1+\lambda\,(a_2x+b_2y+c_2)=0 \quad \ldots \quad \ldots(5)$$

Equation (5) is of the first degree in x and y; hence it represents a straight line. Also, if a point (x_1, y_1) satisfies equations (1) and (2), it automatically satisfies equation (5) for all values of λ, by virtue of the relations (3) and (4). Therefore, equation (5) is the equation to all straight lines which pass through the intersection of (1) and (2).

Cor. 1. If the straight lines $a_1x+b_1y+c_1=0$ and $a_2x+b_2y+c_2=0$ pass through a common point, the straight line $a_1x+b_1y+c_1+\lambda(a_2x+b_2y+c_2)=0$ also passes through that point for all values of λ. In other words, the straight lines

$$a_1x+b_1y+c_1=0,$$
$$a_2x+b_2y+c_2=0$$

and
$$a_1x+b_1y+c_1+\lambda(a_2x+b_2y+c_2)=0$$

meet at a point, for all values of λ.

Cor. 2. If a system of straight lines is represented by $a_1x+b_1y+c_1+\lambda(a_2x+b_2y+c_2)=0$, where λ is an arbitrary constant, then all the lines of the system pass through a fixed point, which is the point of intersection of the straight lines $a_1x+b_1y+c_1=0$ and $a_2x+b_2y+c_2=0$.

2·13. Condition for concurrence of three straight lines :—

Let the equations to the straight lines be

$$a_1x+b_1y+c_1=0, \qquad \ldots \quad \ldots \quad \ldots(1)$$
$$a_2x+b_2y+c_2=0, \qquad \ldots \quad \ldots \quad \ldots(2)$$
$$a_3x+b_3y+c_3=0. \qquad \ldots \quad \ldots \quad \ldots(3)$$

Let the point of intersection of (1) and (2) be (x_1, y_1).

Then
$$a_1x_1+b_1y_1+c_1=0, \qquad \ldots \quad \ldots \quad \ldots(4)$$
$$a_2x_1+b_2y_1+c_2=0. \qquad \ldots \quad \ldots \quad \ldots(5)$$

If the point (x_1, y_1) also lies on (3), the three straight lines meet at a point, that is

$$a_3x_1+b_3y_1+c_3=0. \qquad \ldots \quad \ldots \quad \ldots(6)$$

Eliminating (x_1, y_1) from (4), (5) and (6), we get

$$\begin{vmatrix} a_1 & b_1 & c_1 \\ a_2 & b_2 & c_2 \\ a_3 & b_3 & c_3 \end{vmatrix} = 0$$

or $$a_1 (b_2c_3 - b_3c_2) + b_1 (c_2a_3 - c_3a_2) + c_1 (a_2b_3 - a_3b_2) = 0$$

which is the condition of concurrence.

2·14. Another criterion for the concurrence of three straight lines.

If three non-zero constants λ, μ, ν can be found such that $\lambda (a_1x+b_1y+c_1)+\mu (a_2x+b_2y+c_2)+\nu (a_3x+b_3y+c_3)$ vanishes identically, then the lines represented by $a_1x+b_1y+c_1=0$, $a_2x+b_2y+c_2=0$ and $a_3x+b_3y+c_3=0$ are concurrent.

Proof. The coordinates of the point of intersection (x_1, y_1) of $a_2x+b_2y+c_2=0$ and $a_3x+b_3y+c_3=0$ will make both of them zero,

i. e. $$a_2x_1+b_2y_1+c_2y_1=0$$

and $$a_3x_1+b_3y_1+c_3y_1=0 \quad \ldots \quad \ldots \quad \ldots \text{(A)}$$

Now,

$$\lambda (a_1x+b_1y+c_1)+\mu (a_2x+b_2y+c_2)+\nu (a_3x+b_3y+c_3)=0$$

identically which means the expression vanishes for all values of x and y.

Substituting (x_1, y_1) in the above equation for (x, y), we get

$$\lambda (a_1x_1+b_1y_1+c_1)+\mu (a_2x_1+b_2y_1+c_2)+\nu (a_3x_1+b_3y_1+c_3)=0$$

or $$\lambda (a_1x_1+b_1y_1+c_1)=0 \qquad \text{[by (A)]}.$$

But $$\lambda \neq 0.$$

$$\therefore \quad a_1x_1+b_1y_1+c_1=0,$$

which means that the straight line $a_1x+b_1y+c_1=0$ passes through (x_1, y_1) also, which is the point of intersection of the other two. Hence the three lines meet in a point.

Ex. *Prove that the perpendiculars drawn from the vertices of a triangle to the opposite sides are concurrent.*

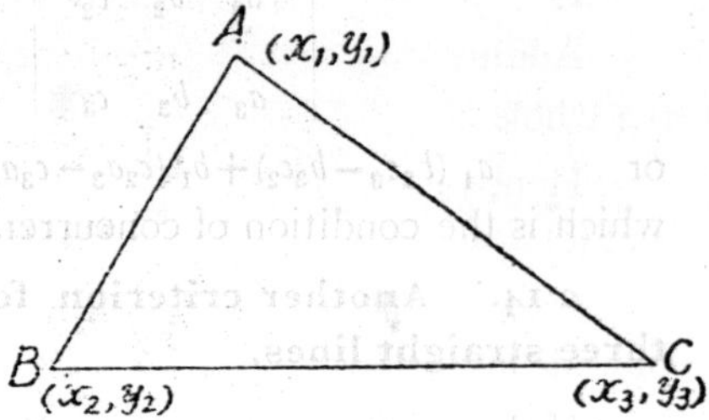

Let ABC be a triangle whose vertices A, B and C are respectively (x_1, y_1), (x_2, y_2) and (x_3, y_3).

The equation to the straight line BC, as it passes through (x_2, y_2) and (x_3, y_3), is

$$y-y_2=\frac{y_3-y_2}{x_3-x_2}(x-x_2).$$

The equation to the straight line perpendicular to this will have its direction tangent m such that

$$m\times\frac{y_3-y_2}{x_3-x_2}=-1;$$

$$\therefore \quad m=-\frac{x_3-x_2}{y_3-y_2}.$$

The equation to a straight line whose slope is m and which passes through (x_1, y_1) is

$$y-y_1=m(x-x_1)$$

$$=-\frac{x_3-x_2}{y_3-y_2}(x-x_1)$$

or $\quad x(x_3-x_2)+y(y_3-y_2)-x_1(x_3-x_2)-y_1(y_3-y_2)=0 \quad \ldots(1)$

which is the equation to the straight line passing through A and perpendicular to BC. Similarly, the equation to the straight line through B perpendicular to CA will be

$$x(x_1-x_3)+y(y_1-y_3)-x_2(x_1-x_3)-y_2(y_1-y_3)=0 \quad \ldots(2)$$

and the equation to the straight line through C perpendicular to AB will be

$$x(x_2-x_1)+y(y_2-y_1)-x_3(x_2-x_1)-y_3(y_2-y_1)=0\ldots \quad \ldots(3)$$

Adding (1), (2) and (3) together and adding, we find that the whole expression vanishes.

Hence the three straight lines represented by (1), (2) and (3) are concurrent.

2·15. Condition that the point (x_1, y_1) may lie on one side or the other of the straight line $ax+by+c=0$.

Let AB be a st. line $ax+by+c=0$ and let P be the point (x_1, y_1).

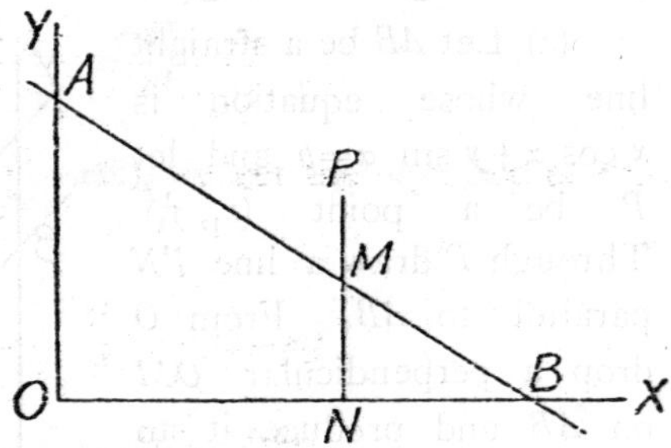

Draw PN perpendicular from P on the x-axis cutting the line AB at M. The abscissa of the point M is the same as that of P; hence the ordinate of M, y', is given by (as the point M lies on AB),

$$ax_1+by'+c=0,$$

or

$$y'=-\frac{ax_1+c}{b}.$$

Now the point P lies on the upper side of AB

if $$PN-PM>0, \text{ i. e.}$$

if $$y_1-y'>0,$$

i. e. if $$y_1+\frac{ax_1+c}{b}>0,$$

i. e. if $$ax_1+by_1+c>0,$$

and it lies on the lower side of AB

fi $$PN-PM<0,$$

or $$ax_1+by_1+c < 0.$$

Cor. 1. The point (x_1, y_1) and the origin are on the same side of the line $ax+by+c=0$ if ax_1+by_1+c and c have the same sign.

Cor. 2. Two points (x_1, y_1), (x_2, y_2) lie on the same side or opposite sides of the line $ax+by+c=0$, according as the expressions ax_1+by_1+c and ax_2+by_2+c have the same sign or opposite signs respectively.

2·16. Length of the perpendicular from a fixed point on a given straight line.

(a) Let AB be a straight line whose equation is $x \cos \alpha+y \sin \alpha=p$ and let P be a point (x_1, y_1). Through P draw a line PN parallel to AB. From O drop a perpendicular OM on AB and produce it to meet the line PN in N. Now, $OM=p$.

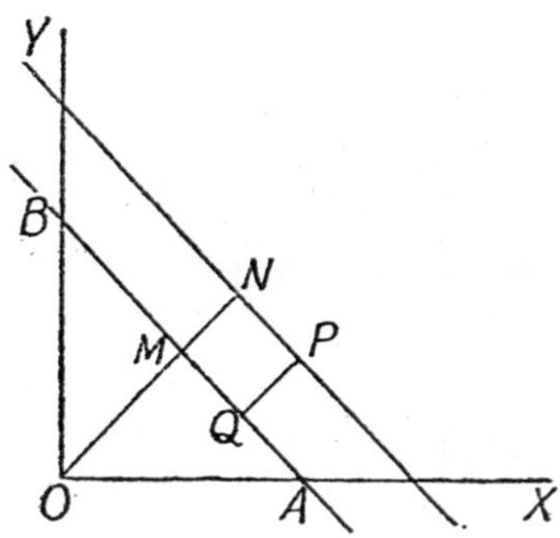

Any line parallel to $x \cos \alpha+y \sin \alpha=p$ is given by
$$x \cos \alpha+y \sin \alpha=k.$$

If it passes through (x_1, y_1),
$$x_1 \cos \alpha+y_1 \sin \alpha=k.$$

Therefore, the equation to line PN is
$$x \cos \alpha+y \sin \alpha=x_1 \cos \alpha+y_1 \sin \alpha.$$

Hence the length of the perpendicular ON on PN is
$$x_1 \cos \alpha+y_1 \sin \alpha.$$

But $$OM=p.$$

$$\therefore \quad PQ=MN=ON-OM$$
$$=x_1 \cos \alpha+y_1 \sin \alpha-p,$$

which is the length of the perpendicular from P on the line AB.

If the point $P(x_1, y_1)$ be on the origin side of the line, the length of the perpendicular

$$PQ = MN = OM - ON$$
$$= p - x_1 \cos\alpha - y_1 \sin\alpha$$
$$= -(x_1 \cos\alpha + y_1 \sin\alpha - p).$$

Hence, if the point (x_1, y_1) is on the same side of the line as the origin, the length of the perpendicular from (x_1, y_1) on the line is negative and if it is on the non-origin side, the length of the perpendicular is positive, its value being $x_1 \cos\alpha + y_1 \sin\alpha - p$.

Rule. The length of the perpendicular from (x_1, y_1) on the line $x \cos\alpha + y \sin\alpha = p$ is $\pm(x_1 \cos\alpha + y_1 \sin\alpha - p)$ which is obtained by substituting x_1 and y_1 for x and y respectively in the equation to the line.

(b) Let the equation to the given line be

$$ax + by + c = 0.$$

We have seen that it can be reduced to the form $x \cos\alpha + y \sin\alpha = p$ by putting

$$\cos\alpha = -\frac{a}{\sqrt{(a^2+b^2)}}, \cos\beta = -\frac{b}{\sqrt{(a^2+b^2)}}$$

and
$$p = \frac{c}{\sqrt{(a^2+b^2)}},$$

if c is positive.

The equation now becomes

$$\frac{-a}{\sqrt{(a^2+b^2)}}x - \frac{b}{\sqrt{(a^2+b^2)}}y = \frac{c}{\sqrt{(a^2+b^2)}}.$$

Hence, by (a), the length of the perpendicular from (x_1, y_1) on the above line is

$$\pm\left\{\frac{-ax_1 - by_1 - c}{\sqrt{(a^2+b^2)}}\right\} \text{ if } c \text{ is positive}$$

or $$\mp\left\{\frac{ax_1+by_1+c}{\sqrt{(a^2+b^2)}}\right\} \text{ if } c \text{ is positive.}$$

Negative sign for the point (x_1, y_1) to lie on the side of the line away from origin,

Positive sign for the point (x_1, y_1) to lie on the same side of the line as origin.

Rule :—The length of the perpendicular from (x_1, y_1) on the line $ax+by+c=0$ is obtained by substituting x_1 and y_1 for x and y respectively in the equation to the line and then dividing the expression by $+\sqrt{(a^2+b^2)}$ if (x_1, y_1) is on the same side of the line as the origin and by $-\sqrt{(a^2+b^2)}$ if $(x_1 y_1)$ in on the side of the line opposite to the origin, c being positive in both the cases.

2·17. Straight lines bisecting the angles between two given straight lines.

Let the equations to the two given straight lines AB and CD, intersecting at K, be

$$a_1x+b_1y+c_1=0 \qquad \ldots \quad \ldots(1)$$

and $$a_2x+b_2y+c_2=0. \qquad \ldots \quad \ldots(2)$$

Let PQ be a bisector of the angle AKC. Let the co-ordinates of any point P on the bisector PQ be (x, y).

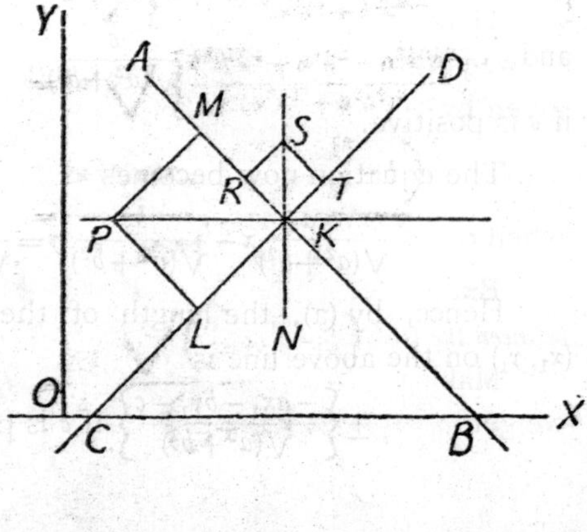

To find the equation to PQ we have to find a relation between x and y on the basis of the given condition. As P is on the bisector PQ, the lengths of the perpendiculars PL and PM are equal. The perpendicular from (x, y) on (1) is given by $+\frac{a_1x+b_1y+c_1}{\sqrt{(a_1^2+b_1^2)}}$, the sign being plus as it is on the same side of the line AB as the origin.

Similarly, the perpendicular from (x, y) on (2) is

$$+\frac{a_2x+b_2y+c_2}{\sqrt{(a_2^2+b_2^2)}},$$

assuming that c_1 and c_2 in the equations are both positive.

Hence $$\frac{a_1x+b_1y+c_1}{\sqrt{(a_1^2+b_1^2)}}=\frac{a_2x+b_2y+c_2}{\sqrt{(a_2^2+b_2^2)}},$$

which is the equation to the bisector PQ.

Taking the coordinates of S, any point on the other bisector SN, as (x, y),

$$\text{Perpendicular } SR=-\frac{a_1x+b_1y+c_1}{\sqrt{(a_1^2+b_1^2)}},$$

$$\text{Perpendicular } ST=\frac{a_2x+b_2y+c_2}{\sqrt{(a_2^2+b_2^2)}}.$$

Hence the equation to the bisector SN is given by

$$\frac{a_1x+b_1y+c_1}{\sqrt{(a_1^2+b_1^2)}}=-\frac{a_2x+b_2y+c_2}{\sqrt{(a_2^2+b_2^2)}},$$

assuming that c_1 and c_2 are both positive. If c_1 and c_2 are both negative, then also the above results will follow.

Hence the equations to the bisectors of the angles between the lines $a_1x+b_1y+c_1=0$ and $a_2x+b_2y+c_2=0$ are given by

$$\frac{a_1x+b_1y+c_1}{\sqrt{(a_1^2+b_1^2)}}=\pm\frac{a_2x+b_2y+c_2}{\sqrt{(a_2^2+b_2^2)}}$$

when c_1 and c_2 are either both positive or both negative.

Ex. 1. *Find the equations to the bisectors of the angles between the lines $3x+4y-5=0$ and $12x+5y+2=0$.*

Making the constants positive in the first equation, we get

$$-3x-4y+5=0.$$

Hence the equations to the bisectors are

$$\frac{-3x-4y+5}{\sqrt{(9+16)}}=\pm\frac{12x+5y+2}{\sqrt{(144+25)}}$$

or $$\frac{-3x-4y+5}{5}=\pm\frac{12x+5y+2}{13}$$

or $$-39x-52y+65=\pm(60x+25y+10)$$

or $$9x+7y-5=0 \text{ and } 7x-9y+25=0.$$

Ex. 2. *Find the area of the triangle formed by the lines*

$y=mx, y=nx$ *and* $ax+by+c=0.$ *(Utkal 1946)*

The point of intersection of

$$y=mx \text{ and } ax+by+c=0$$

is given by $$ax+bmx+c=0$$

or $$x=\frac{-c}{a+bm}.$$

Hence the point is $\left(\frac{-c}{a+bm}, \frac{-cm}{a+bm}\right)$.

Similarly, the point of intersection of $y=nx$ with $ax+by+c=0$ is

$$\left(\frac{-c}{a+bn}, \frac{-cn}{a+bn}\right).$$

The point of intersection of $y=mx$ and $y=nx$ is clearly (0, 0).

The area of the triangle whose vertices are the above points of intersection is given by

$$\begin{vmatrix} 0 & 0 & 1 \\ \frac{-c}{a+bm} & \frac{-cm}{a+bm} & 1 \\ \frac{-c}{a+bn} & \frac{-cn}{a+bn} & 1 \end{vmatrix}$$

or
$$\begin{vmatrix} \dfrac{-c}{a+bm} & \dfrac{-cm}{a+bm} \\ \dfrac{-c}{a+bn} & \dfrac{-cn}{a+bn} \end{vmatrix}$$

or
$$\frac{c^2n-c^2m}{(a+bm)(a+bn)}$$

or
$$\frac{c^2(n-m)}{(a+bm)(a+bn)} \text{ if } n>m.$$

Ex. 3. *Show that the equations of the lines which pass through* (x_1, y_1) *and are inclined at an angle* α *to the line* $ax+by+c=0$ *are*
$$\begin{vmatrix} x & y & 1 \\ x_1 & y_1 & 1 \\ a\sin\alpha\pm b\cos\alpha & b\sin\alpha\pm a\cos\alpha & 1 \end{vmatrix}=0.$$

Let the equation to the required straight line be
$$px+qy+r=0 \qquad \ldots \quad \ldots(1)$$
As it passes through the point (x_1, y_1),
$$px_1+qy_1+r=0 \qquad \ldots \quad \ldots(2)$$
Then, by the data,
$$\tan\alpha=\pm\frac{-\dfrac{p}{q}+\dfrac{a}{b}}{1+\dfrac{pa}{qb}}$$

or
$$\frac{\sin\alpha}{\cos\alpha}=\pm\frac{aq-bp}{qb+pa}$$

or
$$qb\sin\alpha+pa\sin\alpha=\pm(aq-bp)\cos\alpha$$

or
$$p(a\sin\alpha\pm b\cos\alpha)+q(b\sin\alpha\pm a\cos\alpha)=0 \qquad \ldots(3)$$

Eliminating p, q and r from (1), (2) and (3), we get
$$\begin{vmatrix} x & y & 1 \\ x_1 & y_1 & 1 \\ a\sin\alpha\pm b\cos\alpha & b\sin\alpha\pm a\cos\alpha & 0 \end{vmatrix}=0,$$
which is the equation required.

Ex. 4. *The sides BC, CA and AB of a triangle are*

$$u_1 \equiv a_1 x + b_1 y + c_1 = 0,$$

$$u_2 \equiv a_2 x + b_2 y + c_2 = 0,$$

$$u_3 \equiv a_3 x + b_3 y + c_3 = 0.$$

Show that the equations to the lines through A parallel and perpendicular to BC are

$$u_2 (a_3 b_1 - a_1 b_3) + u_3 (a_1 b_2 - a_2 b_1) = 0$$

and

$$u_2 (a_1 a_3 + b_1 b_3) - u_3 (a_1 a_2 + b_1 b_2) = 0$$

respectively.

A is the point of intersection of $u_2=0$ and $u_3=0$. Hence any line passing through A is

$$u_2 + \lambda u_3 = 0 \qquad \ldots \quad \ldots (1)$$

or

$$(a_2 + \lambda a_3) x + (b_2 + \lambda b_3) y + c_2 + \lambda c_3 = 0.$$

This line must be parallel to

$$u_1 \equiv a_1 x + b_1 y + c_1 = 0.$$

Hence

$$-\frac{a_2 + \lambda a_3}{b_2 + \lambda b_3} = -\frac{a_1}{b_1}$$

or

$$a_2 b_1 - a_1 b_2 = \lambda (a_1 b_3 - a_3 b_1)$$

or

$$\lambda = \frac{a_2 b_1 - a_1 b_2}{a_1 b_3 - a_3 b_1}.$$

The equation to the line is therefore obtained by substituting for λ in (1) which gives

$$(a_1 b_3 - a_3 b_1) u_2 + (a_2 b_1 - a_1 b_2) u_3 = 0$$

or

$$u_2 (a_3 b_1 - a_1 b_3) + u_3 (a_1 b_2 - a_2 b_1) = 0.$$

If the line (1) is perpendicular to $u_1=0$,

$$-\frac{a_2 + \lambda a_3}{b_2 + \lambda b_3} \times \left(-\frac{a_1}{b_1}\right) = -1$$

or

$$a_1 a_2 + b_1 b_2 = -\lambda (a_1 a_3 + b_1 b_3)$$

or

$$\lambda = -\frac{a_1 a_2 + b_1 b_2}{a_1 a_3 + b_1 b_3}.$$

Hence the equation to the line through A and perpendicular to BC is given by

$$(a_1a_3+b_1b_3)\, u_2-(a_1a_2+b_1b_2)\, u_3=0.$$

Ex. 5. *The line $lx+my+n=0$ bisects the angle between a pair of lines of which $px+qy+r=0$ is one. Show that the other line is given by the equation*

$$(px+qy+r)\,(l^2+m^2)=2\,(lp+mq)\,(lx+my+n).$$

(Andhra 1947)

The other line passes through the intersection of $px+qy+r=0$ and the bisector $lx+my+n=0$. Hence its equation will be given by

$$(px+qy+r)+\lambda\,(lx+my+n)=0 \qquad \dots \quad \dots(1)$$

where λ is yet to be determined.

Now the equation to the bisector of the angle between (1) and $px+qy+r=0$ is $lx+my+n=0$.

The equation to the bisector is given by

$$\frac{(p+\lambda l)\,x+(q+\lambda m)\,y+r+\lambda n}{\sqrt{[(p+\lambda l)^2+(q+\lambda m)^2]}}=\pm\frac{px+qy+r}{\sqrt{(p^2+q^2)}}$$

or
$$[\sqrt{(p^2+q^2)}\mp\sqrt{\{(p+\lambda l)^2+(q+\lambda m)^2\}}]\,(px+qy+r)$$
$$+\lambda\sqrt{(p^2+q^2)}\,(lx+my+n)=0.$$

If one of the above equations is the same as $lx+my+n=0$, these equations should not contain any terms of the type px, qy or r which is possible only if

$$\sqrt{(p^2+q^2)}=\pm\sqrt{[(p+\lambda l)^2+(q+\lambda m)^2]},$$

or
$$p^2+q^2=\lambda^2\,(l^2+m^2)+2\lambda\,(lp+mq)+p^2+q^2,$$

or
$$\lambda=-\frac{2\,(lp+mq)}{l^2+m^2}.$$

Substituting this value of λ in (1), we get the required equation as

$$(px+qy+r)\,(l^2+m^2)=2\,(lp+mq)\,(lx+my+n).$$

Exercise III

1. (a) If the join of the points (2, 3) and (−1, 5) subtends

a right angle at the point P, find the locus of P.

(b) Find the coordinates of the foot of the perpendicular from the point (2, 3) on the line $x+y-11=0$.

(C. U. 1948)

2. (a) If the sum of the perpendiculars dropped from a variable point P on two lines $x+y-5=0$ and $3x-2y+7=0$ be always equal to 10, prove that P must move on a right line.

(b) For all values of the parameter λ prove that the line $5x+y+11-\lambda(2x-7y+3)=0$ goes through a certain fixed point, and determine its coordinates.

(C. U. 1950)

3. (a) Find the equations to the diagonals of the rectangle the equations of whose sides are $x=a$, $x=a'$, $y=b$, $y=b'$.

(b) What are the points on the axis of x whose perpendicular distance from the straight line $\frac{x}{a}+\frac{y}{b}=1$ is a?

(C. U. 1951)

4. Show that the perpendiculars let fall from any point of the straight line $7x-9y+10=0$ upon the two straight lines $3x+4y=5$ and $12x+5y=7$ are equal to each other. *(C. U. 1952)*

5. (a) Find the equations of the two straight lines drawn through the point $(0, a)$ on which the perpendicnlars let fall from the point $(2a, 2a)$ are each of length a.

(b) Prove that the diagonals of the parallelogram formed by the four straight lines $\sqrt{3}x+y=0$, $\sqrt{3}y+x=0$, $\sqrt{3}x+y=1$ and $\sqrt{3}y+x=1$ are at right angles to one another. *(C. U. 1953)*

6. Show that the distance of the point (x_0, y_0) from the

lines $ax+by+c=0$ measured parallel to a line making an angle θ with the x-axis is $-\dfrac{ax_0+by_0+c}{a\cos\theta+b\sin\theta}$.

(C. U. 1954)

7. Find the internal bisectors of the angles of the triangle whose sides are $x=0$, $y=0$ and $3x+4y-12=0$.

(C. U. 1949)

8. Show that the product of the perpendiculars drawn from the two points $\{\pm\sqrt{(a^2-b^2)}, 0\}$ upon the straight line $\dfrac{x}{a}\cos\theta+\dfrac{y}{b}\sin\theta=1$ is b^2.

9. If p and p_1 be the perpendiculars from the origin upon the lines

$$x\sec\alpha+y\operatorname{cosec}\alpha=a$$

and $$x\cos\alpha-y\sin\alpha=a\cos 2\alpha,$$

show that $$4p^2+p_1^2=a^2.$$

10. Find the equation of the bisector of that angle between the lines

$$4x-3y+1=0 \text{ and } 12x+5y+13=0$$

in which the origin lies. *(Dacca 1946)*

11. Find the centre of the inscribed circle of the triangle the equations of whose sides are $4y+3x=0$, $12y-5x=0$ and $y-15=0$. *(Agra 1942)*

12. Obtain the equations of the lines passing through the intersection of $4x-3y-1=0$ and $2x-5y+3=0$ and equally inclined to the axes. *(Patna 1947)*

13. Show that the equation of the straight line joining the point (x_1, y_1) to the point of intersection of the lines $ax+by+c=0$ and $lx+my+n=0$ is

$$\frac{ax+by+c}{ax_1+by_1+c}=\frac{lx+my+n}{lx_1+my_1+n}.$$

14. Show that the perpendicular from the origin upon the straight line joining the points $(a\cos\alpha, a\sin\alpha)$ and $(a\cos\beta, a\sin\beta)$ bisects the distance between them.

15. Show that the area of the triangle formed by the lines whose equations are $y=m_1x+\frac{a}{m_1}$, $y=m_2x+\frac{a}{m_2}$ and $y=m_3x+\frac{a}{m_3}$ is $\frac{1}{2}a^2(m_2-m_3)(m_1-m_2)/m_1^2m_2^2m_3^2$.

16. The sides of a triangle are

$$u_r \equiv x\cos\alpha_r+y\sin\alpha_r-p_r=0\ (r=1, 2, 3).$$

Show that its orthocentre is given by

$$u_1\cos(\alpha_2-\alpha_3)=u_2\cos(\alpha_3-\alpha_1)=u_3\cos(\alpha_1-\alpha_2).$$

(Travancore 1947)

17. Show that the area of the triangle formed by the lines $y=m_1x+c_1$, $y=m_2x+c_2$ and $y=m_3x+c_3$ is

$$\frac{1}{2}\frac{(c_2-c_3)^2}{m_2-m_3}+\frac{1}{2}\frac{(c_3-c_1)^2}{m_3-m_1}+\frac{1}{2}\frac{(c_1-c_2)^2}{m_1-m_2}.$$

Hint. Find the areas of the triangles formed by any two lines and $x=0$ and add the areas so obtained.

Suppose we consider $y=m_1x+c_1$, $y=m_2x+c_2$ and $x=0$. The points of intersection are $(0, c_1)$, $(0, c_2)$ and $\left(\frac{c_1-c_2}{m_2-m_1}, \frac{c_1m_2-m_1c_2}{m_2-m_1}\right)$.

Hence the area of the triangle is

$$\frac{1}{2}\begin{vmatrix} 0 & c_1 & 1 \\ 0 & c_2 & 1 \\ \frac{c_1-c_2}{m_2-m_1} & \frac{c_1m_2-c_2m_1}{m_2-m_1} & 1 \end{vmatrix}$$

$$=\frac{1}{2}\frac{c_1-c_2}{m_2-m_1}(c_1-c_2)$$

$$=\frac{1}{2}\frac{(c_1-c_2)^2}{m_2-m_1}.$$

18. Find the equations to the diagonals of the parallelogram formed by the lines

$$lx+my+n=0,\ lx+my+n'=0,$$

$$l'x+m'y+n=0,\ l'x+m'y+n'=0$$

and shew that they are at right angles if

$$l^2+m^2=l'^2+m'^2.$$

19. Shew that the equations to the lines through the origin making an angle α with the line $y=mx+c$ are

$$\frac{y}{x}=\frac{m\pm\tan\alpha}{1\mp m\tan\alpha}.$$

20. The angular points of a triangle are $(a\cos\alpha, a\sin\alpha)$, $(a\cos\beta, a\sin\beta)$ and $(a\cos\gamma, a\sin\gamma)$. Prove that the coordinates of the orthocentre are

$$\{a(\cos\alpha+\cos\beta+\cos\gamma),\ a(\sin\alpha+\sin\beta+\sin\gamma)\}.$$

(Andhra 1948)

2·18. Polar equation to a straight line.

(a) Let OX be the initial line and let a line OP make an angle α with the initial line. Let the current coordinates of P be (r, θ). For every point P on the line OP, $\theta=\alpha$. Hence $\theta=\alpha$ is the equation to the straight line OP through the origin, making an angle α with the initial line.

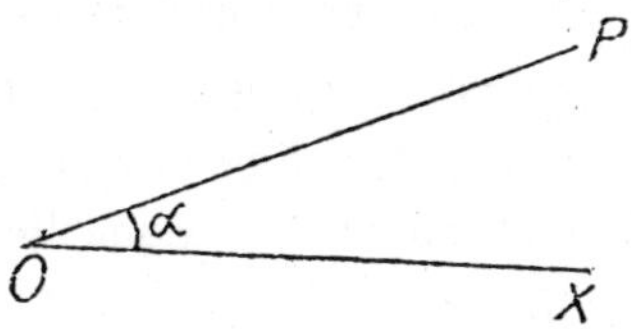

Cor. The equation to the initial line is $\theta=0$.

(b) Let AB be a line, the perpendicular on which from the origin is of length p and makes an angle α with the initial line OX.

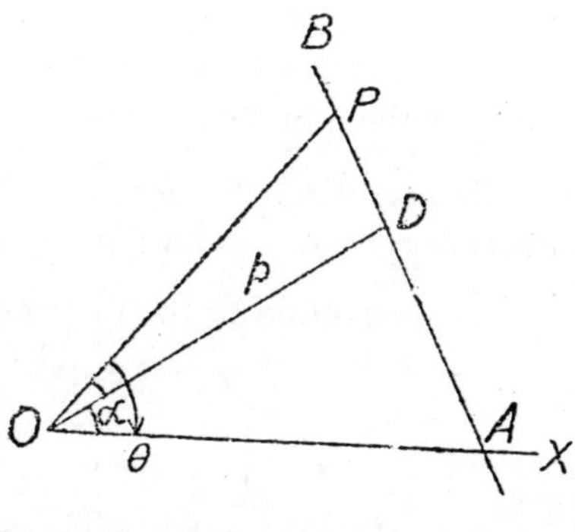

We have to find the equation to AB.

Let $P\ (r, \theta)$ be any point on the line AB, so that $OP=r$ and $\angle POD=\theta-\alpha$.

From $\triangle BOD$, $r \cos(\theta-\alpha)=p$, ...(1)
which is a relation between r and θ and is therefore the equation to the straight line AB.

Cor. The cartesian equation to a straight line is

$$ax+by+c=0.$$

Changing into polars, we get

$$ar\cos\theta+br\sin\theta=-c.$$

Putting $a=k\cos\alpha$ and $b=k\sin\alpha$, we get

$$kr(\cos\theta\cos\alpha+\sin\theta\sin\alpha)=-c$$

or

$$r\cos(\theta-\alpha)=-c/k,$$

which is of the same form as equation (1).

(c) Let AB be a straight line passing through two points $P_1(r_1, \theta_1)$ and $P_2(r_2, \theta_2)$. Let P be any point (r, θ) on the straight line AB.

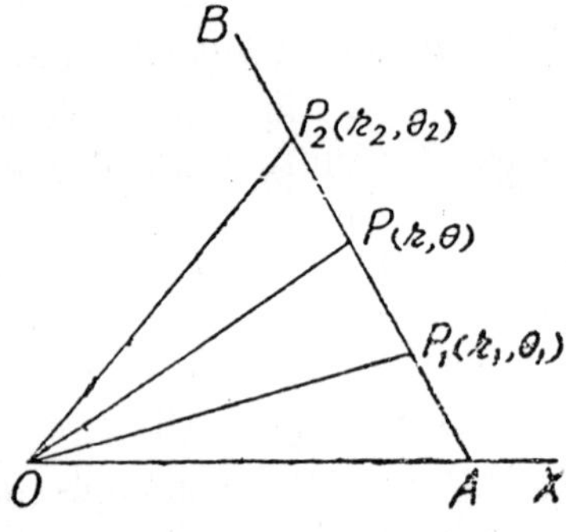

Now,

$$\triangle POP_1+\triangle POP_2=\triangle P_1OP_2$$

or

$$\tfrac{1}{2}rr_2\sin(\theta_2-\theta)+\tfrac{1}{2}rr_1\sin(\theta-\theta_1)=\tfrac{1}{2}r_1r_2\sin(\theta_2-\theta_1)$$

or

$$r_1r_2\sin(\theta_2-\theta_1)+rr_1\sin(\theta_1-\theta)+rr_2\sin(\theta-\theta_2)=0,$$

which is the equation to the straight line AB.

Ex. 1. *Find the length and direction of the perpendicular drawn from the pole on the line $1/r=\cos\theta-\sqrt{3}\sin\theta$.*

The equation to the line can be written as

$$r[\cos\theta-\sqrt{3}\sin\theta]=1$$

or

$$r\left[\frac{1}{2}\cos\theta-\frac{\sqrt{3}}{2}\sin\theta\right]=\frac{1}{2}$$

or

$$r\cos(\theta+60)=\tfrac{1}{2}.$$

Comparing the above equation with

$$r \cos (\theta - \alpha) = p,$$

we get $\theta = -60$ and $p = \frac{1}{2}$.

Hence the perpendicular on the line should be drawn in the direction 60° with the initial line on the negative side or if the angle is measured in the anticlockwise direction from the initial line, the angle is 300° and the length of the perpendicular is $\frac{1}{2}$.

Ex. 2. *Find the equation to any straight line perpendicular to the straight line*

$$\frac{l}{r} = A \cos \theta + B \sin \theta.$$

On changing to cartesian coordinates, the equation becomes

$$l = Ax + By \qquad \ldots \quad \ldots (1)$$

Any straight line perpendicular to (1) is given by

$$c = Bx - Ay$$

or

$$\frac{c}{r} = B \cos \theta - A \sin \theta$$

$$= A \cos \left(\theta + \frac{\pi}{2}\right) + B \sin \left(\theta + \frac{\pi}{2}\right).$$

Rule :—To change the equation $\frac{l}{r} = A \cos \theta + B \sin \theta$ into the equation to a straight line perpendicular to it, change l to another constant c and θ to $\theta + \pi/2$.

Exercise IV

1. Find the equation to a straight line perpendicular to $l = r \sin (\theta + \alpha)$ and at the same perpendicular distance from the pole.
2. Find the equation to the line through the pole and at right angles to the line
$$k = r \cos (\theta - \alpha).$$
3. Show that the equation
$$\frac{l}{r} = f(\theta), \text{ where } f(\theta) \equiv A \cos (\theta + \alpha) + B \cos (\theta + \beta),$$

represents a straight line, and that any straight line perpendicular to it is

$$\frac{\lambda}{r}=f\left(\theta+\frac{\pi}{2}\right),$$

where λ is a parameter.

(**Hint.** See solved example 2).

4. Shew that the four right lines

$$r(\cos\theta+\sin\theta)=1,$$
$$r(\cos\theta+\sin\theta)=-1,$$
$$r(-\cos\theta+\sin\theta)=1,$$
$$r(\cos\theta-\sin\theta)=1.$$

make up a square, and calculate the length of any side.

5. Prove that any straight line passing through the point of intersection of the lines $\frac{1}{r}=A\cos\theta+B\sin\theta$ and $\frac{1}{r}=A'\cos\theta+B'\sin\theta$ is

$$\frac{1+\lambda}{r}=(A+\lambda A')\cos\theta+(B+\lambda B')\sin\theta.$$

Hence obtain the equation of the line passing through the pole and the point of intersection of the given lines.

6. Find the polar coordinates of the foot of the perpendicular from the pole on the line joining the points (r_1, θ_1) and (r_2, θ_2). *(Punjab 1946)*

7. Show that the condition that the chord cut off by the curve $\frac{l}{r}=1+e\cos\theta$ from the line $\frac{1}{r}=A\cos\theta+B\sin\theta$ may subtend a right angle at the pole is

$$(lA-e)^2+l^2B^2=2.$$

CHAPTER III

CHANGE OF AXES : OBLIQUE AXES

3·1. Change of origin, the direction of the axes remaining the same.

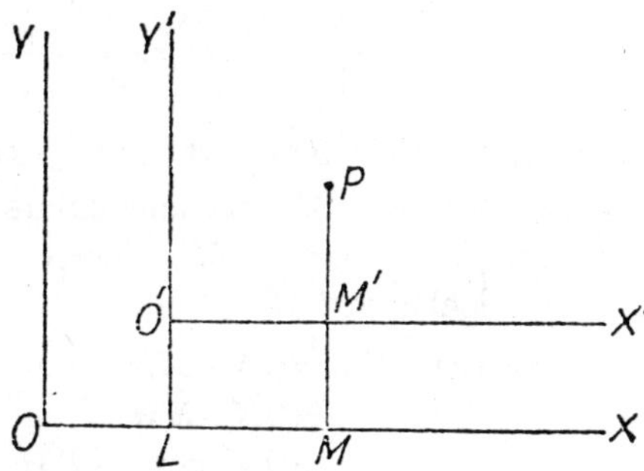

Let the point P have coordinates (x, y) with reference to axes OX and OY and coordinates (x_1, y_1) with reference to axes $O'X'$ and $O'Y'$. Let the coordinates of O' with reference to OX and OY be (h, k).

Then $$x = OM = OL + LM = OL + O'M'$$
$$= x_1 + h$$
and $$y = PM = PM' + M'M$$
$$= PM' + O'L$$
$$= y_1 + k.$$

Hence, if we change the origin from O (0, 0) to O' (h, k) an equation in (x, y) can be changed to one in (x_1, y_1) by putting $x = x_1 + h$ and $y = y_1 + k$.

3·2. Changing the direction of the axes, the origin remaining the same.

Let the coordinates of a point be (x, y) and (x_1, y_1) with reference to (OX, OY) and $(OX'$ and $OY')$ respectively. Let

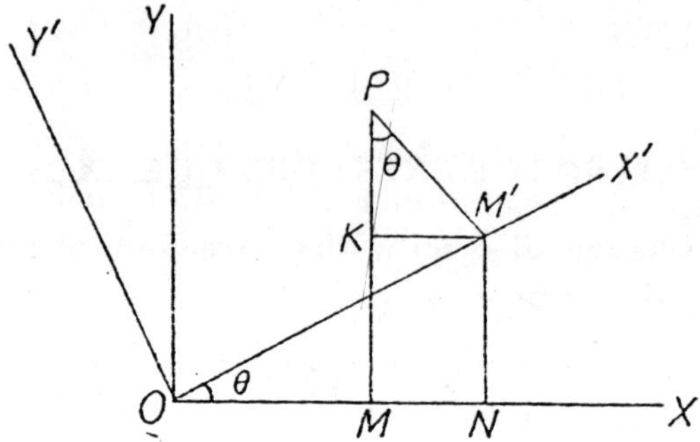

the angle $X'OX$ be θ, so that $\angle YOY'$ is also equal to θ. Draw PM and PM' perpendiculars from P on OX and OX' respectively. Draw $M'N$ perpendicular to OX and $M'K$ perpendicular to PM.

Now, $x = OM = ON - MN$
$= ON - KM'$
$= OM' \cos\theta - PM' \sin\theta$
$= x_1 \cos\theta - y_1 \sin\theta$

and $y = PM = MK + KP$
$= NM' + KP$
$= x_1 \sin\theta + y_1 \cos\theta.$

Hence an equation in (x, y) is converted to another in (x_1, y_1) by putting $x = x_1 \cos\theta - y_1 \sin\theta$,

$$y = x_1 \sin\theta + y_1 \cos\theta.$$

Cor. If the origin is transferred to (h, k) and the axes are simultaneously rotated through an angle θ, then

$$x = h + x_1 \cos\theta - y_1 \sin\theta,$$
$$y = k + x_1 \sin\theta + y_1 \cos\theta.$$

Ex. 1. *The equation $3x^2 + 2xy + 3y^2 - 18x - 22y + 50 = 0$ is transferred to $4x^2 + 2y^2 = 1$ when referred to rectangular axes through the point (2, 3). Find the inclination of the latter axes to the former.* (Calcutta 1945)

Transferring the origin to the point (2, 3) and rotating the axes through θ, we get

$$x=2+x_1\cos\theta-y_1\sin\theta,$$
$$y=3+x_1\sin\theta+y_1\cos\theta.$$

Substituting these values in the given equation, we get

$$3(2+x\cos\theta-y\sin\theta)^2+2(2+x\cos\theta-y\sin\theta)$$
$$\times(3+x\sin\theta+y\cos\theta)+3(3+x\sin\theta+y\cos\theta)^2$$
$$-18(2+x\cos\theta-y\sin\theta)-22(3+x\sin\theta+y\cos\theta)+50=0$$

or $x^2[3+\sin 2\theta]+y^2[3-\sin 2\theta]+xy[2\cos^2\theta-2\sin^2\theta]-1=0.$

As the transformed equation does not contain any term in xy, the coefficient of this term in the above equation should be zero, that is

$$2\cos^2\theta-2\sin^2\theta=0$$

or $$\tan^2\theta=1$$

or $$\tan\theta=\pm 1 \quad \text{or} \quad \theta=\pi/4.$$

On substituting $\theta=\pi/4$, the above equation becomes

$$4x^2+2y^2=1.$$

Hence the axes should be turned through $\pi/4$.

Ex. 2. *Show that of $ax^2+2hxy+by^2=1$ and $a'x^2+2h'xy+b'y^2=1$ represent the same conic, and the axes are rectangular, then*

(i) $(a-b)^2+4h^2=(a'-b')^2+4h'^2.$ *(Agra 1943)*

(ii) $a+b=a'+b'.$ *(Punjab 1955)*

(iii) $ab-h^2=a'b'-h'^2.$ *(Punjab 1955)*

(i) Let the axes be rotated through an angle θ to convert the equation

$$ax^2+2hxy+by^2=1 \quad \ldots \quad \ldots(A)$$

into

$$a'x^2+2h'xy+b'y^2=1 \quad \ldots \quad \ldots(B)$$

Putting $x=x_1\cos\theta-y_1\sin\theta$

and $y=x_1\sin\theta+y_1\cos\theta$ in (A),

we get $a(x_1\cos\theta-y_1\sin\theta)^2+2h(x_1\cos\theta-y_1\sin\theta)$

$$\times(x_1\sin\theta+y_1\cos\theta)+b(x_1\sin\theta+y_1\cos\theta)^2=1$$

or $x_1^2(a\cos^2\theta+2h\sin\theta\cos\theta+b\sin^2\theta)$

$$+y_1^2(a\sin^2\theta-2h\sin\theta\cos\theta+b\cos^2\theta)$$

$$+2x_1y_1(h\sin\theta\cos\theta-a\sin\theta\cos\theta-h\sin^2\theta+h\cos^2\theta)=1.$$

As (x_1, y_1) are current coordinates, we can change them to (x, y) so that

$$x^2(a\cos^2\theta+b\sin^2\theta+h\sin 2\theta)+y^2(a\sin^2\theta+b\cos^2\theta-h\sin 2\theta)$$

$$+2xy\,[h\cos 2\theta+\tfrac{1}{2}(b-a)\sin 2\theta]=1.$$

Comparing the above equation with (B), we get

$$a'=a\cos^2\theta+b\sin^2\theta+h\sin 2\theta.$$

$$b'=a\sin^2\theta+b\cos^2\theta-h\sin 2\theta.$$

$$h'=h\cos 2\theta+\tfrac{1}{2}(b-a)\sin 2\theta.$$

Now $(a'-b')^2+4h'^2$

$=[a\cos 2\theta-b\cos 2\theta+2h\sin 2\theta]^2+4\,[h\cos 2\theta+\tfrac{1}{2}(b-a)\sin 2\theta]^2$

$=[(a-b)\cos 2\theta+2h\sin 2\theta]^2+[2h\cos 2\theta-(a-b)\sin 2\theta]^2$

$=(a-b)^2+4h^2$

(ii) $a'+b'=a(\cos^2\theta+\sin^2\theta)+b(\sin^2\theta+\cos^2\theta)$

$=a+b.$

(iii) $a'b'-h'^2$

$=(a\cos^2\theta+b\sin^2\theta+h\sin 2\theta)(a\sin^2\theta+b\cos^2\theta-h\sin 2\theta)$

$-\tfrac{1}{4}[2h\cos 2\theta-(a-b)\sin 2\theta]^2$

$=ab[\sin^4\theta+\cos^4\theta+2\sin^2\theta\cos^2\theta]-h^2(\cos^2 2\theta+\sin^2 2\theta)$

$=ab(\sin^2\theta+\cos^2\theta)^2-h^2$

$=ab-h^2$

3·3. Change of oblique axes to rectangular axes with the same origin and same abscissa.

Let the coordinates of a point P be (x, y) with reference to rectangular axes OX and OY and (x', y') with reference to oblique axes OX and OY'. Draw PM perp. to OX and PN parallel to OY'.

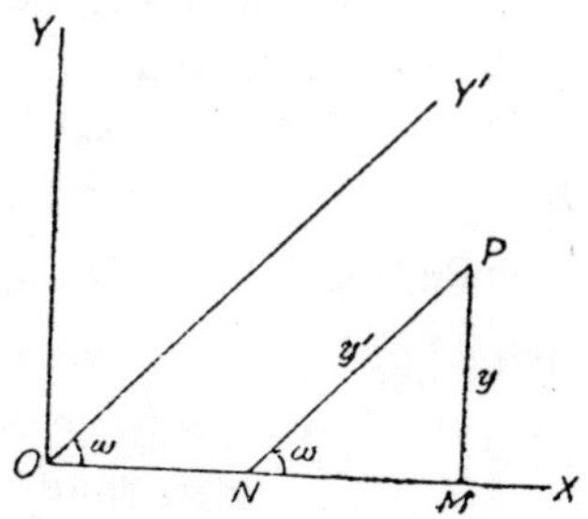

Now $OM = x = ON + NM$

$$= x' + y' \cos \omega.$$

and $$y = PM = y' \sin \omega.$$

3·4. Change from one set of oblique axes to another, without changing the origin.

Let OX and OY, the orignal axes be inclined to each other at angle ω. Let OX' and OY' be the new axes inclined to each other at angle ω'. Let angle XOX' be θ. Let P be a point whose coordinates with reference to (OX, OY) and (OX', OY') are (x, y) and (x', y') respectively, so that

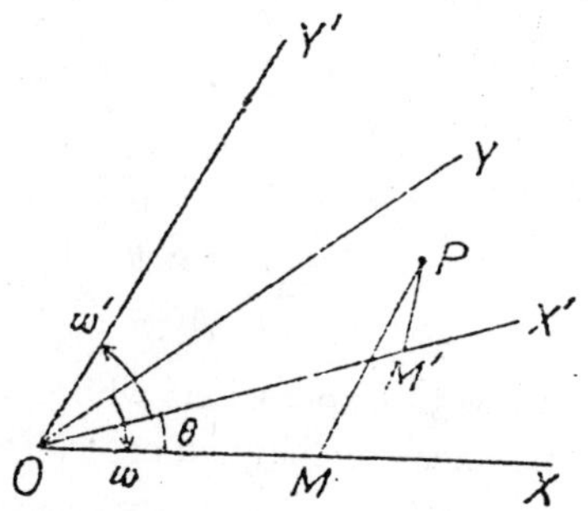

$$OM = x,\ MP = y,\ OM' = x' \text{ and } M'P = y'.$$

The sum of the projections of OM and MP on any line is equal to the sum of projections of OM' and $M'P$ on that line.

Angle between OM and a line perp. to $OX = \frac{\pi}{2}$

" " MP " " " " $= \frac{\pi}{2} - \omega.$

" " OM' " " " " $= \frac{\pi}{2} - \theta.$

" " PM' " " " " $= \omega' + \theta - \frac{\pi}{2}$

Hence projecting on a line perpendicular to OX, we get

$$y \cos\left(\frac{\pi}{2} - \omega\right) = x' \cos\left(\frac{\pi}{2} - \theta\right) + y' \cos\left(\omega' + \theta - \frac{\pi}{2}\right),$$

or $$y \sin \omega = x' \sin \theta + y' \sin (\omega' + \theta). \quad \ldots \quad \ldots (1)$$

Similarly, projecting on a line perpendicular to OY, we get

$$x \cos\left(\omega+\frac{\pi}{2}\right)=x' \cos\left(\omega+\frac{\pi}{2}-\theta\right)+y' \cos\left(\omega+\frac{\pi}{2}-\theta-\omega'\right)$$

or $$x \sin \omega=x' \sin(\omega-\theta)+y' \sin(\omega-\omega'-\theta) \quad \ldots \quad \ldots(2)$$

The relations (1) and (2) can be written as

$$x=\lambda x'+\mu y',$$
$$y=\lambda' x'+\mu' y',$$

where $$\lambda=\frac{\sin(\omega-\theta)}{\sin\omega}, \quad \mu=\frac{\sin(\omega-\omega'-\theta)}{\sin\omega},$$

$$\lambda'=\frac{\sin\theta}{\sin\omega}, \quad \mu'=\frac{\sin(\omega'+\theta)}{\sin\omega}.$$

3·5. Expressions whose values remain unchanged when the axes are changed are called **Invariants.**

We have seen in Example 2 that the expressions $a+b$ and $ab-h^2$ do not change when the axes are changed. Hence they are invariants.

Presently we are going to prove a more general proposition which runs as follows :—

If, by any change of axes, without change of origin, the expression $ax^2+2hxy+by^2$ becomes $a'x^2+2h'xy+b'y^2$,

then $$\frac{a+b-2h\cos\omega}{\sin^2\omega}=\frac{a'+b'-2h'\cos\omega'}{\sin^2\omega'}$$

and $$\frac{ab-h^2}{\sin^2\omega}=\frac{a'b'-h'^2}{\sin^2\omega'},$$

ω and ω' being the angles between the original and final pairs of axes.

Let the coordinates of any point P be (x, y) and (x', y') referred to the original and final axes respectively.

Because the origin is fixed, the distance of the point P from the origin with reference to both the sets of axes must be the same, that is

$$x^2+2xy\cos\omega+y^2=x'^2+2x'y'\cos\omega+y'^2 \quad \ldots(1)$$

Also by hypothesis,

$$ax^2+2hxy+by^2=a'x'^2+2h'x'y'+b'y'^2 \qquad \ldots(2)$$

Multiplying (1) by λ and adding it to (2), we have

$$x^2(a+\lambda)+2xy(h+\lambda\cos\omega)+y^2(b+\lambda)$$
$$=x'^2(a'+\lambda)+2x'y'(h'+\lambda\cos\omega')+y'^2(b'+\lambda) \qquad \ldots(3)$$

Then, if any value of λ makes the left hand side of (3) a perfect square, it must make the right hand side also a perfect square.

But the value of λ that makes the left-hand side a perfect square is given by

$$(h+\lambda\cos\omega)^2=(a+\lambda)(b+\lambda),$$

or $$\lambda^2(1-\cos^2\omega)+\lambda(a+b-2h\cos\omega)+ab-h^2=0,$$

or $$\lambda^2+\lambda\frac{(a+b-2h\cos\omega)}{\sin^2\omega}+\frac{ab-h^2}{\sin^2\omega}=0 \quad \ldots \quad \ldots(4)$$

Similarly, the value of λ that makes the right-hand side of (3) a perfect square is given by

$$\lambda^2+\lambda\frac{(a'+b'-2h'\cos\omega)}{\sin^2\omega'}+\frac{a'b'-h'^2}{\sin^2\omega'}=0 \quad \ldots \quad \ldots(5)$$

Since the values of λ from the above equations (4) and (5) are the same, comparing coefficients, we get

$$\frac{a+b-2h\cos\omega}{\sin^2\omega}=\frac{a'+b'-2h'\cos\omega'}{\sin^2\omega'}$$

and $$\frac{ab-h^2}{\sin^2\omega}=\frac{a'b'-h'^2}{\sin^2\omega'}.$$

Cor. If both the axes are rectangular, so that $\omega=\omega'=90°$, the above results become

$$a+b=a'+b' \text{ and } ab-h^2=a'b'-h'^2.$$

Ex. 1. *If one system of oblique axes be transformed to another system according to the formula*

$$x=mx'+ny',\ y=m'x'+n'y',$$

the origin being the same, then prove that

$$nn'\ (m^2+m'^2-1)=mm'\ (n^2+n'^2-1).$$

The distance between the origin and the point P remains the same with reference to both the sets of axes.

$$\therefore\quad x^2+y^2+2xy\cos\omega\equiv x'^2+y'^2+2x'y'\cos\omega'.$$

Using the given formula, we get

$$(mx'+ny')^2+(m'x'+n'y')^2+2\,(mx'+ny')\,(m'x'+n'y')\cos\omega$$
$$\equiv x'^2+y'^2+2x'y'\cos\omega'.$$

Comparing the coefficients of x'^2 and y'^2, we get

$$m^2+m'^2+2mm'\cos\omega=1 \qquad \ldots\quad \ldots(1)$$
$$n^2+n'^2+2nn'\cos\omega=1 \qquad \ldots\quad \ldots(2)$$

From (1) and (2), eliminating $\cos\omega$, we get

$$nn'\ (m^2+m'^2-1)=mm'\ (n^2+n'^2-1)$$

or
$$\frac{m^2+m'^2-1}{n^2+n'^2-1}=\frac{mm'}{nn'}.$$

Ex. 2. *Prove that the transformation of rectangular axes which converts* $\dfrac{X^2}{p}+\dfrac{Y^2}{q}$ *into* $ax^2+2hxy+by^2$ *will convert* $\dfrac{X^2}{p-\lambda}+\dfrac{Y^2}{q-\lambda}$ *into* $\dfrac{ax^2+2hxy+by^2-\lambda\,(ab-h^2)\,(x^2+y^2)}{1-(a+b)\,\lambda+(ab-h^2)\,\lambda^2}$. *(Lucknow 1944)*

Since the distance of a point P from the origin with reference to both the axes is the same,

$$X^2+Y^2\equiv x^2+y^2. \qquad \ldots\quad \ldots\quad \ldots(1)$$

Also, because

$$\frac{X^2}{p}+\frac{Y^2}{q}\equiv ax^2+2hxy+by^2,$$

the invariant relations give us

$$\frac{1}{p}+\frac{1}{q}=a+b \qquad \ldots\quad \ldots\quad \ldots(2)$$

and
$$\frac{1}{pq}=ab-h^2 \qquad \ldots\quad \ldots\quad \ldots(3)$$

Now,

$$\frac{X^2}{p-\lambda}+\frac{Y^2}{q-\lambda}=\frac{qX^2+pY^2-\lambda(X^2+Y^2)}{pq-\lambda(p+q)+\lambda^2}$$

$$=\frac{\dfrac{X^2}{p}+\dfrac{Y^2}{q}-\lambda\left(\dfrac{X^2+Y^2}{pq}\right)}{1-\lambda\left(\dfrac{1}{p}+\dfrac{1}{q}\right)+\dfrac{\lambda^2}{pq}}$$

$$=\frac{ax^2+2hxy+by^2-\lambda(ab-h^2)(x^2+y^2)}{1-\lambda(a+b)+(ab-h^2)\lambda^2}$$

3·6. Distance between two points when the axes are oblique.

Let the coordinates of P and Q be (x_1, y_1) and (x_2, y_2) respectively with reference to oblique axes OX and OY inclined at an angle ω. Draw PM and QN parallel to OY, and PR parallel to OX.

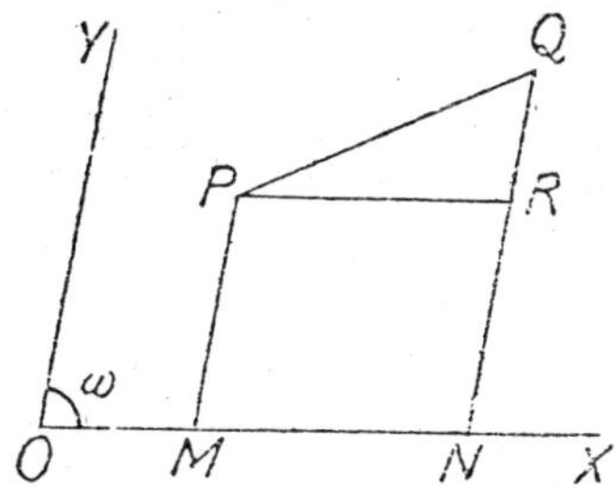

$$\angle QRP=\angle RNM=\angle PMO=\pi-\omega,$$

$$PR=ON-OM=x_2-x_1,$$

$$QR=QN-RN=QN-PM$$

$$=y_2-y_1.$$

From $\triangle PQR$,

$$PQ^2=PR^2+RQ^2-2PR.RQ.\cos(\pi-\omega)$$

$$=(x_2-x_1)^2+(y_2-y_1)^2+2(x_2-x_1)(y_2-y_1)\cos\omega$$

or $$PQ=\sqrt{\{(x_1-x_2)^2+(y_1-y_2)^2+2(x_1-x_2)(y_1-y_2)\cos\omega\}}.$$

3·7. Equation of a straight line in oblique coordinates when its inclination to the x-axis and the intercept it cuts off on y-axis are given.

Let a straight line AB cut off an intercept c from the y-axis (positive direction) and make an angle θ with the x-axis. Let P be a point (x, y) on AB. Draw PM parallel to OY and CR parallel to OX, cutting PM in R.

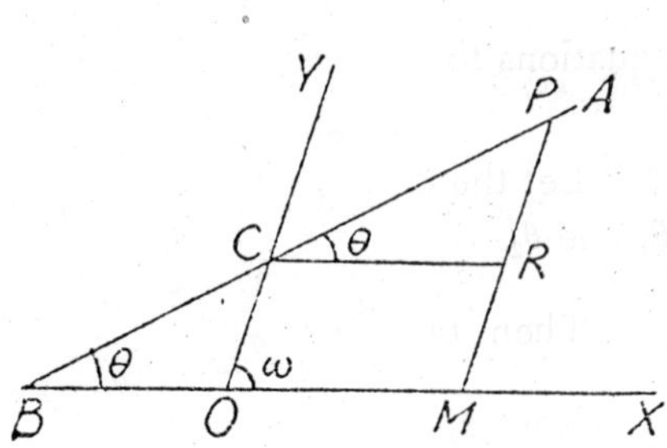

Then $$\angle PCR=\theta,\ \angle PRC=\pi-\omega,$$
$$\angle RPC=\angle PCY=\omega-\theta,$$
$$CR=OM=x,\ PR=PM-RM$$
$$=PM-OC=y-c.$$

From $\triangle PCR$,

$$\frac{y-c}{\sin\theta}=\frac{x}{\sin(\omega-\theta)},$$

or $$y-c=\frac{\sin\theta}{\sin(\omega-\theta)}.x,$$

or $$y=\frac{\sin\theta}{\sin(\omega-\theta)}.x+c.$$

This is of the same form as

$$y=mx+c$$

where $$m=\frac{\sin\theta}{\sin(\omega-\theta)}$$

or $$m\sin(\omega-\theta)=\sin\theta$$

or $$m\sin\omega\cos\theta-m\cos\omega\sin\theta=\sin\theta$$

or $$\sin\theta(1+m\cos\omega)=m\sin\omega\cos\theta$$

or $$\tan\theta=\frac{m\sin\omega}{1+m\cos\omega}.$$

3·8. Angle between two straight lines when the axes are oblique.

Let the axes be inclined at an angle ω and let the

equations to the two straight lines be

$$y=m_1x+c_1 \text{ and } y=m_2x+c_2.$$

Let the inclinations of the given lines to the x-axis be θ_1 and θ_2.

Then $\tan\theta_1=\dfrac{m_1\sin\omega}{1+m_1\cos\omega}$ and $\tan\theta_2=\dfrac{m_2\sin\omega}{1+m_2\cos\omega}$.

Now, $\theta_1-\theta_2$ is the angle between the lines and

$$\tan(\theta_1-\theta_2)=\frac{\tan\theta_1-\tan\theta_2}{1+\tan\theta_1\tan\theta_2}$$

$$=\frac{\dfrac{m_1\sin\omega}{1+m_1\cos\omega}-\dfrac{m_2\sin\omega}{1+m_2\cos\omega}}{1+\dfrac{m_1m_2\sin^2\omega}{(1+m_1\cos\omega)(1+m_2\cos\omega)}}$$

$$=\frac{(m_1-m_2)\sin\omega}{1+(m_1+m_2)\cos\omega+m_1m_2}.$$

Cor. 1. The lines are parallel if $m_1=m_2$.

Cor. 2. The lines are at right angles if

$$1+(m_1+m_2)\cos\omega+m_1m_2=0.$$

3·9. Length of the perpendicular from a given point on a given straight line when the axes are oblique.

Let the straight line be $ax+by+c=0$ and let the axes be inclined at an angle ω.

Transforming to rectangular axes with the same origin and same abscissa, (x', y') being the new coordinates, we get

$$\left.\begin{aligned} x'&=x+y\cos\omega \\ \text{and}\quad y'&=y\sin\omega \end{aligned}\right\} \qquad \ldots \quad \ldots(1)$$

$$\left.\begin{aligned} \text{or}\quad x&=x'-y'\cot\omega \\ \text{and}\quad y&=y'\operatorname{cosec}\omega \end{aligned}\right\} \qquad \ldots \quad \ldots(2)$$

Hence the equation to the given line becomes

$$a(x'-y'\cot\omega)+by'\operatorname{cosec}\omega+c=0 \qquad \text{from (2)}$$

or $$ax'+y'(b\operatorname{cosec}\omega-a\cot\omega)+c=0$$

or $$ax' \sin \omega + y' (b - a \cos \omega) + c \sin \omega = 0.$$

Length of perpendicular from (x', y') on the above line is

$$\frac{a \sin \omega . x' + (b - a \cos \omega) y' + c \sin \omega}{\sqrt{\{a^2 \sin^2 \omega + (b - a \cos \omega)^2\}}}$$

or $$\frac{a \sin \omega\, x' + (b - a \cos \omega) y' + c \sin \omega}{\sqrt{(a^2 + b^2 - 2ab \cos \omega)}}$$

or $$\frac{a \sin \omega (x + y \cos \omega) + (b - a \cos \omega)(y \sin \omega) + c \sin \omega}{\sqrt{(a^2 + b^2 - 2ab \cos \omega)}}$$

from (1)

or $$\frac{(ax + by + c) \sin \omega}{\sqrt{(a^2 + b^2 - 2ab \cos \omega)}}.$$

Ex. 1. *Find the equation to any line perpendicular to $ax + by + c = 0$, the axes being inclined at an angle ω.*

If the two lines $y = m_1 x + c_1$ and $y = m_2 x + c_2$ are at right angles, then we have deduced that

$$1 + (m_1 + m_2) \cos \omega + m_1 m_2 = 0.$$

Let the straight lines be

$$ax + by + c = 0 \text{ and } a'x + b'y + c' = 0.$$

Then the above condition becomes

$$1 + \left(-\frac{a}{b} - \frac{a'}{b'}\right) \cos \omega + \left(-\frac{a}{b}\right)\left(-\frac{a'}{b'}\right) = 0$$

or $$\frac{a'}{b'}\left(\frac{a}{b} - \cos \omega\right) = \frac{a}{b} \cos \omega - 1$$

or $$\frac{a'}{b'}(a - b \cos \omega) = (a \cos \omega - b)$$

or $$\frac{a'}{a \cos \omega - b} = \frac{b'}{a - b \cos \omega}.$$

Hence the line $a'x + b'y + c' = 0$ becomes

$$(a \cos \omega - b) x + (a - b \cos \omega) y + k = 0,$$

where k is an arbitrary constant.

Ex. 2. *The axes being inclined at an angle of 45°, obtain the equations to the straight lines which pass through the origin and are inclined at an angle of 30° to the straight line $x + y = 1$.*

Let one of the lines be $y=mx$.

The m' of the given line is (-1),

Hence $$\frac{(m-m')\sin\omega}{1+(m+m')\cos\omega+mm'}=\tan(\pm 30°),$$

where $$m'=-1 \text{ and } \omega=45°$$

or $$\frac{m+1}{\sqrt{2}+(m-1)-\sqrt{2}m}=\pm\frac{1}{\sqrt{3}}.$$

Taking the upper sign, $m=\dfrac{\sqrt{2}-\sqrt{3}-1}{\sqrt{2}+\sqrt{3}-1}$.

Taking the lower sign, $m=\dfrac{\sqrt{2}+\sqrt{3}-1}{\sqrt{2}-\sqrt{3}-1}$.

The required equations are therefore

$$y=\frac{\sqrt{2}-\sqrt{3}-1}{\sqrt{2}+\sqrt{3}-1}.x,\ y=\frac{\sqrt{2}+\sqrt{3}-1}{\sqrt{2}-\sqrt{3}-1}.x.$$

Exercise V

1. Prove that the straight lines $x+y=c$ and $x-y=d$ are at right angles, whatever be the angle between the axes.

2. Transform the equation $y^2+4y\cot\alpha-4x=0$ from rectangular axes to oblique axes meeting at an angle α, the axis of x being kept the same.

3. If (x, y) and (x', y') be the coordinates of the same point referred to two sets of rectangular axes with the same origin and if $ux+vy$, where u and v are independent of x, y, becomes $u'x'+v'y'$, show that
$$u^2+v^2=u'^2+v'^2.$$

4. Find the angle through which the axes must be turned so that the expression $ax^2+2hxy+by^2$ may become independent of the term containing xy.

5. Through what angle should the axes be turned in order to transform the equation $x^2+2\sqrt{3}xy-y^2=2a^2$ into $x'^2-y'^2=a^2$?

6. Transform to axes inclined at 45° to the original axes the equation $x^2-y^2=a^2$.

7. Prove that the equation to the straight line which passes through the point (h, k) and is perpendicular to the axis of x is $x+y \cos \omega=h+k \cos \omega$.

8. If the straight lines $y=m_1x+c_1$ and $y=m_2x+c_2$ make equal angles with the axis of X and are not parallel to one another, $m_1+m_2+2m_1m_2 \cos \omega=0$.

9. The coordinates of a point P referred to axes inclined to each other at an angle ω are (l, m); prove that the length of the straight line joining the feet of the perpendiculars from P upon the axes is

$$\sin \omega \sqrt{(l^2+m^2+2lm \cos \omega)}.$$

10. From a given point (h, k) perpendiculars are drawn to the axes, whose inclination is ω, and their feet are joined. Prove that the length of the perpendicular drawn from (h, k) on this line is

$$\frac{hk \sin^2 \omega}{\sqrt{(h^2+k^2+2hk \cos \omega)}}$$

and that its equation is $hx-ky=h^2-k^2$.

CHAPTER IV

THE STRAIGHT LINE : HIGHER DEGREE EQUATIONS.

4·1. Let us consider the equation

$$(ax+by+c)\,(a'x+b'y+c')=0 \qquad \ldots \qquad \ldots (1)$$

Any point on the straight line

$$ax+by+c=0 \qquad \ldots \qquad \ldots (2)$$

will satisfy equation (1). Also, any point on

$$a'x+b'y+c'=0 \qquad \ldots \qquad \ldots (3)$$

will satisfy it.

Hence any point on either of the straight lines (2) and (3) will satisfy (1) and no other point will satisfy it.

Therefore, equation (1) represents a pair of straight lines given by (2) and (3).

If we multiply out the factors in (1), we get an equation of the form

$$Ax^2+2Hxy+By^2+2Gx+2Fy+C=0,$$

which is the general equation of the second degree. It appears from the above analysis that the general equation of the second degree can represent a pair of straight lines, if it can be resolved into two linear factors.

4·2. Pair of lines through the origin.

The equation of a straight line through the origin is

$$ax+by=0.$$

Hence the equation to a pair of lines through the origin is given by $(ax+by)\,(a'x+b'y)=0$, which is of the form

$$Ax^2+2Hxy+By^2=0,$$

which is a homogeneous equation of the second degree.

4·3. The homogeneous equation of the second degree represents a pair of straight lines.

Let the equation be $ax^2+2hxy+by^2=0$.

It will represent a pair of straight lines, if it can be resolved into two linear factors.

Now, the above equation can be written as

$$x^2+\frac{2h}{a}xy+\frac{b}{a}y^2=0$$

or $$\left(x+\frac{h}{a}y\right)^2=y^2\left(\frac{h^2}{a^2}-\frac{b}{a}\right),$$

or $$\left(x+\frac{h}{a}y\right)^2-y^2\left(\frac{h^2}{a^2}-\frac{b}{a}\right)=0,$$

or $$\left[\left(x+\frac{h}{a}\right)y+\frac{y}{a}\sqrt{(h^2-ab)}\right]\left[\left(x+\frac{h}{a}\right)y-\frac{y}{a}\sqrt{(h^2-ab)}\right]=0.$$

The linear factors are

$$ax+[h+\sqrt{(h^2-ab)}]\,y=0$$

and $$ax+[h-\sqrt{(h^2-ab)}]\,y=0,$$

each of which represents a line passing through the origin.

4·4. Angle between the lines represented by $ax^2+2hxy+by^2=0$.

The equations to the lines represented by the above equation can be written in the form

$$ax+[h+\sqrt{(h^2-ab)}]\,y=0,$$
$$ax+[h-\sqrt{(h^2-ab)}]\,y=0.$$

If θ is the angle between these lines,

$$\tan\theta=\frac{m_1-m_2}{1+m_1m_2}.$$

Here $$m_1=-\frac{a}{h+\sqrt{(h^2-ab)}}$$

$$m_2=-\frac{a}{h-\sqrt{(h^2-ab)}}.$$

$$\therefore \quad \tan\theta = \frac{a[h+\sqrt{(h^2-ab)}-h+\sqrt{(h^2-ab)}]}{[h+\sqrt{(h^2-ab)}][h-\sqrt{(h^2-ab)}]+a^2}$$

$$= \frac{2a\sqrt{(h^2-ab)}}{ab+a^2} = \frac{2\sqrt{(h^2-ab)}}{b+a}.$$

Aliter :—Let the straight lines represented by

$$ax^2+2hxy+by^2=0$$

be

$$y=m_1x, \ y=m_2x.$$

Then

$$(y-m_1x)(y-m_2x)=0 \qquad \ldots \quad \ldots (1)$$

must represent the same straight lines as the equation

$$ax^2+2hxy+by^2=0 \qquad \ldots \quad \ldots (2)$$

The coefficients in both the equations must be proportional.

Comparing coefficients, we get

$$m_1+m_2=-\frac{2h}{b} \qquad \ldots \quad \ldots (3)$$

and

$$m_1m_2=\frac{a}{b} \qquad \ldots \quad \ldots (4)$$

If θ is the angle between the lines,

$$\tan\theta=\frac{m_1-m_2}{1+m_1m_2}$$

$$=\frac{\sqrt{[(m_1+m_2)^2-4m_1m_2]}}{1+m_1m_2}$$

$$=\frac{\sqrt{\left(\frac{4h^2}{b^2}-\frac{4a}{b}\right)}}{1+\frac{a}{b}} \qquad \text{[from (3) and (4)]}$$

$$=\frac{2\sqrt{(h^2-ab)}}{b+a} \qquad \ldots \quad \ldots (5)$$

Cor. 1. If the lines are perpendicular to each other, $\theta=90°$, and substituting this value in (5), we get $a+b=0$.

Cor. 2. If the lines are coincident, $\theta=0$ and therefore

$$h^2=ab.$$

4·5. Equation to the pair of bisectors of angles between the pair of straight lines given by

$$ax^2+2hxy+by^2=0.$$

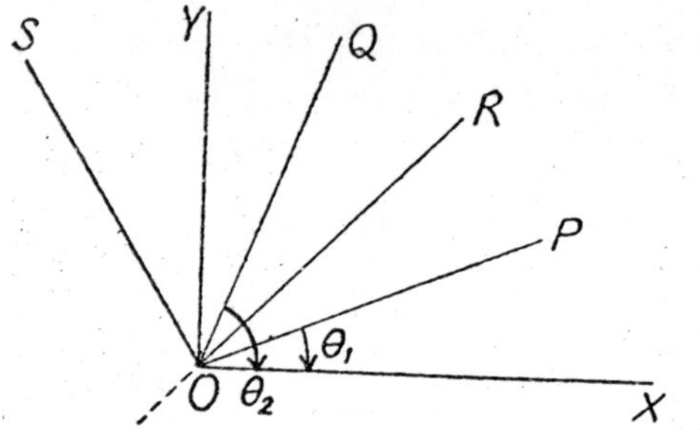

Let the lines represented by

$$ax^2+2hxy+by^2=0, \qquad \ldots \quad \ldots(1)$$

OP and OQ, make angles θ_1 and θ_2 with the x-axis.

Then (1) can be represented by

$$(y-x \tan \theta_1)(y-x \tan \theta_2)=0 \qquad \ldots \quad \ldots(2)$$

so that

$$\tan \theta_1+\tan \theta_2=2h/b \qquad \ldots \quad \ldots(3)$$

$$\tan \theta_1 \tan \theta_2=a/b \qquad \ldots \quad \ldots(4)$$

Let the bisector OR make an angle θ with the positive direction of the x-axis. Then the other bisector makes an angle $\pi/2+\theta$ with the positive direction of the x-axis.

From the figure,

$$\theta-\theta_1=\theta_2-\theta,$$

or

$$2\theta=\theta_1+\theta_2. \qquad \ldots \quad \ldots(5)$$

The equation of the bisector OR is

$$y-x \tan \theta=0 \qquad \ldots \quad \ldots(6)$$

and the equation of the bisector OS is

$$y-x \tan (\pi/2+\theta)=0 \qquad \ldots \quad \ldots(7)$$

Hence their combined equation is

$$(y-x \tan \theta)\{y-x \tan (\pi/2+\theta)\}=0,$$

or

$$(y-x \tan \theta)(y+x \cot \theta)=0,$$

or $$y^2+xy(\cot\theta-\tan\theta)-x^2=0 \quad \ldots \quad \ldots(8)$$

But $\cot\theta-\tan\theta=\dfrac{\cos^2\theta-\sin^2\theta}{\sin\theta\cos\theta}$

$$=2.\frac{\cos 2\theta}{\sin 2\theta}$$

$$=2\cot 2\theta$$

$$=2\cot(\theta_1+\theta_2) \qquad \text{from (5),}$$

$$=2\,\frac{1-\tan\theta_1\tan\theta_2}{\tan\theta_1+\tan\theta_2}$$

$$=2.\frac{1-\frac{a}{b}}{-\frac{2h}{b}}$$

$$=\frac{a-b}{h}.$$

Substituting for $\cot\theta-\tan\theta$ in (8), we get

$$y^2+\frac{a-b}{h}xy-x^2=0,$$

or $$\frac{x^2-y^2}{a-b}=\frac{xy}{h},$$

which is the equation of the bisectors.

Aliter. Let the lines represented by the pair

$$ax^2+2hxy+by^2=0 \quad \ldots \quad \ldots(1)$$

be $$y=m_1x \text{ and } y=m_2x \quad \ldots \quad \ldots(2)$$

so that $$m_1+m_2=-\frac{2h}{b} \quad \ldots \quad \ldots(3)$$

and $$m_1m_2=\frac{a}{b}. \quad \ldots \quad \ldots(4)$$

The equation to the bisectors of the angles between (2) is given by

$$\frac{y-m_1x}{\sqrt{(1+m_1^2)}}=\pm\frac{y-m_2x}{\sqrt{(1+m_2^2)}}$$

This is equivalent to

$$\left\{\frac{y-m_1x}{\sqrt{(1+m_1^2)}}+\frac{y-m_2x}{\sqrt{(1+m_2^2)}}\right\}\left\{\frac{y-m_1x}{\sqrt{(1+m_1^2)}}-\frac{y-m_2x}{\sqrt{(1+m_2^2)}}\right\}=0,$$

or $$\frac{(y-m_1x)^2}{1+m_1^2}-\frac{(y-m_2x)^2}{1+m_2^2}=0,$$

or $$(1+m_2^2)(y-m_1x)^2-(1+m_1^2)(y-m_2x)^2=0,$$

or $$y^2(1+m_2^2-1-m_1^2)+x^2(m_1^2+m_1^2m_2^2-m_2^2-m_1^2m_2^2)$$
$$+2xy(-m_1-m_1m_2^2+m_2+m_2m_1^2)=0,$$

or $$x^2(m_1^2-m_2^2)-y^2(m_1^2-m_2^2)+2(m_1-m_2)(m_1m_2-1)xy=0,$$

or $$(x^2-y^2)(m_1+m_2)+2(m_1m_2-1)xy=0$$

or $$\frac{x^2-y^2}{2(m_1m_2-1)}=-\frac{xy}{m_1+m_2}$$

or $$\frac{x^2-y^2}{a-b}=\frac{xy}{h} \text{ from (3) and (4).}$$

4·6. Angle between the lines $ax^2+2hxy+by^2=0$ if the axes be oblique.

Let the axes be inclined at an angle ω. If θ is the angle between the lines $y=m_1x$ and $y=m_2x$, it is given by

$$\tan\theta=\frac{(m_1-m_2)\sin\omega}{1+(m_1+m_2)\cos\omega+m_1m_2}.$$

Again, substituting

$$m_1+m_2=-\frac{2h}{b}$$

and $$m_1m_2=\frac{a}{b},$$

we get $$\tan\theta=\frac{2\sqrt{(h^2-ab)}.\sin\omega}{a+b-2h\cos\omega}.$$

Cor. 1. The lines are perpendicular to each other if

$$a+b-2h\cos\omega=0.$$

Cor. 2. The lines are coincident if $h^2=ab$.

4·7. Homogeneous equation of the nth degree.

A homogeneous equation of the nth degreee in x and y represents n straight lines through the origin.

Let the equation be

$$a_0y^n+a_1y^{n-1}x+a_2y^{n-2}x^2+\ldots+a_nx^n=0.$$

Dividing the equation by a_0x^n, we get

$$\left(\frac{y}{x}\right)^n+\frac{a_1}{a_0}\left(\frac{y}{x}\right)^{n-1}+\frac{a_2}{a_0}\left(\frac{y}{x}\right)^{n-2}+\ldots+\frac{a_n}{a_0}=0.$$

This is an equation of the nth degree in $\frac{y}{x}$ and must have n roots. Let these roots be $m_1, m_2 \ldots m_n$. Then the equation can be written as

$$\left(\frac{y}{x}-m_1\right)\left(\frac{y}{x}-m_2\right)\left(\frac{y}{x}-m_3\right)\ldots\left(\frac{y}{x}-m_n\right)=0,$$

which represents n straight lines

$$y=m_1x,\ y=m_2x,\ \ldots\ y=m_nx,$$

all of which pass through the origin.

Some of the m's in the above case may be imaginary or equal in which case the corresponding straight lines will be imaginary or coincident.

Ex. 1. *Find the condition that one of the lines*

$$ax^2+2hxy+by^2=0$$

may coincide with one of the lines $a'x^2+2h'xy+b'y^2=0$.

(Allahabad 1948)

Let the line common to both the equations be $y=mx$ and let the other lines representing the above equations be $y=m_1x$ and $y=m_2x$ respectively.

Then $$m+m_1=-\frac{2h}{b} \qquad \ldots\ \ldots\ \ldots(1)$$

$$mm_1=\frac{a}{b}. \qquad \ldots\ \ldots\ \ldots(2)$$

Also $$m+m_2=-\frac{2h'}{b'}. \qquad \ldots\ \ldots\ \ldots(3)$$

$$mm_2=\frac{a'}{b'}. \qquad \ldots\ \ldots\ \ldots(4)$$

From (1) and (2), eliminating m_1, we get

$$m+\frac{a}{b}.\frac{1}{m}=-\frac{2h}{b}$$

or $$a+2hm+bm^2=0. \qquad \ldots \quad \ldots \quad \ldots(5)$$

Similarly, from (3) and (4), eliminating m_2, we get

$$a'+2h'm+b'm^2=0. \qquad \ldots \quad \ldots \quad \ldots(6)$$

From (5) and (6),

$$\frac{1}{2(hb'-h'b)}=\frac{m}{a'b-ab'}=\frac{m^2}{2(ah'-a'h)},$$

which give

$$m=\frac{a'b-ab'}{2(hb'-h'b)} \text{ and } m^2=\frac{ah'-a'h}{hb'-h'b}.$$

Eliminating m from these, we get

$$\frac{ah'-a'h}{hb'-h'b}=\tfrac{1}{4}\left(\frac{a'b-ab'}{hb'-h'b}\right)^2,$$

or $$(a'b-ab')^2=4(ah'-a'h)(hb'-h'b).$$

Ex. 2. *Show that the equation $bx^2-2hxy+ay^2=0$ represents a pair of straight lines which are at right angles to the pair given by the equation*

$$ax^2+2hxy+by^2=0. \qquad (C.\ U.\ 1951)$$

Let the lines represented by

$$ax^2+2hxy+by^2=0$$

be $$y=m_1x \text{ and } y=m_2x.$$

Then $$m_1+m_2=-\frac{2h}{b}, \qquad \ldots \quad \ldots \quad \ldots(1)$$

$$m_1m_2=\frac{a}{b}. \qquad \ldots \quad \ldots \quad \ldots(2)$$

Equations to the straight lines perpendicular to $y=m_1x$ and $y=m_2x$ and passing through the origin are

$$y=-\frac{1}{m_1}x \text{ and } y=-\frac{1}{m_2}x \text{ respectively.}$$

Their combined equation is given

$$\left(y+\frac{1}{m_1}x\right)\left(y+\frac{1}{m_2}x\right)=0,$$

or $$(m_1 y+x)(m_2 y+x)=0,$$

or $$m_1 m_2 y^2+xy(m_1+m_2)+x^2=0$$

or $$\frac{a}{b}y^2-\frac{2h}{b}xy+x^2=0 \qquad \text{[from (1) and (2)]}$$

or $$bx^2-2hxy+ay^2=0.$$

Ex. 3. *Prove that the two straight lines*

$$(x^2+y^2)(\cos^2\theta\sin^2\alpha+\sin^2\theta)=(x\tan\alpha-y\sin\theta)^2$$

include an angle 2α.

The equation can be written as

$$x^2(\cos^2\theta\sin^2\alpha+\sin^2\theta-\tan^2\alpha)+y^2\cos^2\theta\sin^2\alpha$$
$$+2xy\tan\alpha\sin\theta=0.$$

In this equation

$$a=\cos^2\theta\sin^2\alpha+\sin^2\theta-\tan^2\alpha,$$
$$b=\cos^2\theta\sin^2\alpha,$$
$$h=\tan\alpha\sin\theta.$$

If ϕ is the angle between the lines, then

$$\tan\phi=\frac{2\sqrt{(h^2-ab)}}{a+b}$$

and $h^2-ab=\tan^2\alpha\sin^2\theta-\cos^4\theta\sin^4\alpha-\cos^2\theta\sin^2\theta\sin^2\alpha$
$$+\cos^2\theta\sin^2\alpha\tan^2\alpha$$

$$=\frac{\sin^2\alpha\sin^2\theta-\cos^4\theta\sin^4\alpha\cos^2\alpha-\cos^2\theta\sin^2\theta\sin^2\alpha\cos^2\alpha+\cos^2\theta\sin^4\alpha}{\cos^2\alpha}$$

$$=\frac{\sin^2\alpha\sin^2\theta(1-\cos^2\theta\cos^2\alpha)+\sin^4\alpha\cos^2\theta(1-\cos^2\theta\cos^2\alpha)}{\cos^2\alpha}$$

$$=\frac{\sin^2\alpha(\sin^2\theta+\sin^2\alpha\cos^2\theta)(1-\cos^2\theta\cos^2\alpha)}{\cos^2\alpha}$$

$$=\frac{\sin^2\alpha(1-\cos^2\theta\cos^2\alpha)^2}{\cos^2\alpha}.$$

Also, $a+b=\cos^2\theta\sin^2\alpha+\sin^2\theta-\tan^2\alpha+\cos^2\theta\sin^2\alpha$

$$=\frac{2\cos^2\theta\sin^2\alpha\cos^2\alpha+\sin^2\theta\cos^2\alpha-\sin^2\alpha}{\cos^2\alpha}$$

$$=\frac{\cos^2\alpha-\sin^2\alpha-\cos^2\theta\cos^2\alpha+2\cos^2\theta\cos^2\alpha\sin^2\alpha}{\cos^2\alpha}$$

$$=\frac{\cos 2\alpha-\cos^2\theta\cos^2\alpha\,(1-2\sin^2\alpha)}{\cos^2\alpha}$$

$$=\frac{\cos 2\alpha\,(1-\cos^2\theta\cos^2\alpha)}{\cos^2\alpha}.$$

Hence $\tan\phi=\dfrac{2\sin\alpha\cos\alpha}{\cos 2\alpha}$

$$=\frac{\sin 2\alpha}{\cos 2\alpha}=\tan 2\alpha.$$

$$\therefore\quad \phi=2\alpha.$$

Ex. 4. *If the pairs of straight lines $x^2-2pxy-y^2=0$ and $x^2-2qxy-y^2=0$ be such that each pair bisects the angles between the other pair, prove that $pq=-1$.*

Equation of the bisector of the angle between the lines

$$x^2-2pxy-y^2=0$$

is given by

$$\frac{x^2-y^2}{2}=\frac{xy}{-p},$$

or

$$-px^2+py^2-2xy=0.$$

This is the same as

$$x^2-2qxy-y^2=0.$$

Comparing coefficients, we get

$$\frac{-p}{1}=\frac{p}{-1}=\frac{1}{q},$$

or

$$pq=-1.$$

Exercise VI

1. Shew that the two straight lines

$$x^2 (\tan^2 \alpha+\cos^2 \alpha)-2xy \tan \alpha+y^2 \sin^2 \alpha=0$$

make with the x-axis angles such that the difference of their tangents is 2.

2. Prove that the angles between the lines

$$ax^2+2hxy+by^2+\lambda (x^2+y^2)=0$$

have the same pair of bisectors as the lines

$$ax^2+2hxy+by^2=0.$$

Interpret the case $\lambda=-(a+b)$. *(Agra 1943)*

3. Shew that the straight lines

$$(A^2-3B^2) x^2+8ABxy+(B^2-3A^2) y^2=0$$

form with the line $Ax+By+C=0$ an equilateral triangle whose area is $\dfrac{c^2}{\sqrt{\{3 (A^2+B^2)\}}}$.

(Agra 1941, Allahabad 44, Punjab 47)

4. Show that the condition that two of the lines represented by the equation $ax^3+3bx^2y+3cxy^2+dy^3=0$ may be at right angles is $a^2+3ac+3bd+d^2=0$,

(Agra 1947, 51, Aligarh 45)

5. Show that the equation

$$a (x^4+y^4)-4bxy (x^2-y^2)+6cx^2y^2=0$$

represents two pairs of straight lines at right angles and that if $2b^2=a^2+3ac$, the two pairs will coincide.

(Agra 1949)

6. Find the condition that one of the two lines

$$ax^2+2hxy+by^2=0$$

may be perpendicular to one of the lines

$$a'x^2+2h'xy+b'y^2=0.$$ *(Delhi 1947)*

7. If one of the lines $ax^2+2hxy+by^2=0$ coincides with one of the lines $a'x^2+2h'xy+b'y^2=0$ and the remaining two are at right angles, prove that

$$h\left(\frac{1}{b}-\frac{1}{a}\right)=h'\left(\frac{1}{b'}-\frac{1}{a'}\right)=\frac{(a-b)(a'-b')}{2\sqrt{(-aa'bb')}}.$$

8. If the lines $ax^2+2hxy+by^2=0$ be two sides of a parallelgram and the line $lx+my=1$ be one of its diagonals, show that the equation of the other diagonal is $y\,(bl-hm)=x\,(am-hl)$. *(Allahabad 1942, Punjab 1955)*

9. A parallelogram is formed by the lines $ax^2+2hxy+by^2=0$ and the lines through (p, q) parallel to them. Prove that the equation of the diagonal not passing through the origin is $(2x-p)\,(ap+hq)+(2y-p)\,(hp+bq)=0$.

(Rajputana 1950)

10. Show that the product of the perpendiculars from the point (x', y') on the lines $ax^2+2hxy+by^2=0$ is

$$\frac{ax'^2+2hx'y'+by'^2}{\sqrt{[(a-b)^2+4h^2]}}.$$ *(Agra 1950, Allahabad 1943)*

11. Show that the area of the triangle formed by the lines

$$ax^2+2hxy+by^2=0 \text{ and } lx+my+n=0$$

is
$$\frac{x^2\sqrt{(h^2-ab)}}{am^2-2hlm+bl^2}.$$ *(Agra 1955)*

12. A point moves so that the distance between the feet of the perpendiculars from it on the lines

$$ax^2+2hxy+by^2=0$$

is a constant $2k$. Show that the equation to its locus is $(x^2+y^2)\,(h^2-ab)=k^3\,[(a-b)^2+4h^2]$. *(Agra 1952)*

13. If the axes be inclined at an angle ω, shew that the equation $x^2+2xy \cos \omega+y^2 \cos 2\omega=0$ represents a pair of straight lines at right angles.

14. Prove that the straight lines $ax^2+2hxy+by^2=0$ make equal angles with the axis of X if $h=a \cos \omega$, the axes being inclined at an angle ω.

15. Prove that the equation to the bisector of the angle between the straight lines $ax^2+2hxy+by^2=0$ is

$$h\,(x^2-y^2)+(b-a)\,xy=(ax^2-by^2) \cos \omega,$$

the axes being inclined at an angle ω.

16. Show that the ortho-centre of the triangle formed by the straight lines $ax^2+2hxy+by^2=0$ and $lx+my=1$ is the point (kl, km) where $k=\dfrac{a+b}{am^2-2hlm+bl^2}$.

(*Rajputana 1949*)

17. Prove that the equation

$$(a+2h+b)\,x^2-2\,(a-b)\,xy+(a-2h+b)\,y^2=0$$

denotes a pair of straight lines each inclined at an angle of 45° to one or the other of the lines given by

$$ax^2+2hxy+by^2=0. \qquad (\textit{Dacca 1945})$$

18. Prove that two of the lines

$$ax^4+4bx^3y+6cx^2y^2+4dxy^3+ey^4=0$$

will bisect the angle between the other two if $a+c=0$ and $b+d=0$.

19. If one of the lines $ax^3+bx^2y+cxy^2+dy^3=0$ bisects the angle between the other two, then

$(3a+c)^2\,(bc+2cd-3ad)=(b+3d)^2\,(bc+2ab-3ad)$.

20. Shew that the distance between the points of intersection of the straight line $x\cos\alpha+y\sin\alpha-p=0$ with the straight line $ax^2+2hxy+by^2=0$ is

$$\frac{2p\sqrt{(h^2-ab)}}{b\cos^2\alpha-2h\cos\alpha\sin\alpha+a\sin^2\alpha}.$$

4·8. Condition that the general equation of the second degree represents a pair of straight lines.

The general equation of the second degree is represented by

$$ax^2+2hxy+by^2+2gx+2fy+c=0.$$

It will represent two straight lines if the left-hand side can be solved into two linear factors. Multiplying the above expression by a, we get

$a^2x^2+2ahxy+aby^2+2agx+2afy+ac$

$$=a^2x^2+2ax\,(hy+g)+(hy+g)^2-(hy+g)^2+aby^2+2afy+ac$$
$$=(ax+hy+g)^2+y^2\,(ab-h^2)+2y\,(af-hg)-g^2+ac$$
$$=(ax+hy+g)^2-[\sqrt{\{(h^2-ab)\,y^2-2y\,(af-hg)+g^2-ac\}}]^2.$$

The above expression can be resolved into two linear factors if

$$(h^2-ab)\,y^2-2y\,(af-hg)+g^2-ac$$

is a perfect square, the condition for which is that

$$(af-hg)^2=(h^2-ab)\,(g^2-ac),$$

or $$a^2f^2-2afhg+h^2g^2=h^2g^2-abg^2-ach^2+a^2bc,$$

or $$a^2bc+2afgh-a^2f^2-abg^2-ach^2=0$$

or $$abc+2fgh-af^2-bg^2-ch^2=0.$$

Aliter. Let the straight lines representing

$$ax^2+2hxy+by^2+2gx+2fy+c=0$$

be represented by $lx+my+n=0$

and $l_1x+m_1y+n_1=0.$

Then

$$ax^2+2hxy+by^2+2gx+2fy+c\equiv(lx+my+n)\,(l_1x+m_1y+n_1).$$

Comparing coefficients, we get

$$\frac{ll_1}{a}=\frac{mm_1}{b}=\frac{nn_1}{c}=\frac{(l_1m+lm_1)}{2h}=\frac{ln_1+l_1n}{2g}=\frac{mn_1+m_1n}{2f}$$
$$=k\text{ (say)} \quad \ldots \quad \ldots(1)$$

Now consider the determinants

$$\begin{vmatrix} l & l_1 & 0 \\ m & m_1 & 0 \\ n & n_1 & 0 \end{vmatrix} \text{ and } \begin{vmatrix} l_1 & l & 0 \\ m_1 & m & 0 \\ n_2 & n & 0 \end{vmatrix},$$

each of which is equal to zero. Their product will also be zero, so that

$$\begin{vmatrix} l & l_1 & 0 \\ m & m_1 & 0 \\ n & n_1 & 0 \end{vmatrix} \times \begin{vmatrix} l_1 & l & 0 \\ m_1 & m & 0 \\ n_1 & n & 0 \end{vmatrix}=0,$$

or
$$\begin{vmatrix} 2ll_1 & lm_1+l_1m & ln_1+l_1n \\ l_1m+lm_1 & 2mm_1 & mn_1+m_1n \\ l_1n+ln_1 & m_1n+mn_1 & 2nn_1 \end{vmatrix}=0,$$

or
$$\begin{vmatrix} 2ak & 2hk & 2gk \\ 2hk & 2bk & 2fk \\ 2gh & 2fk & 2ck \end{vmatrix}=0, \text{ substituting from (1),}$$

or
$$\begin{vmatrix} a & h & g \\ h & b & f \\ g & f & c \end{vmatrix}=0$$

or $\quad abc+2fgh-af^2-bg^2-ch^2=0.$

4·9. Angle between the lines

$$\mathbf{ax^2+2hxy+by^2+2gx+2fy+c=0.}$$

The lines represented by the above equation are parallel to the lines represented by $ax^2+2hxy+by^2=0$.

Hence the angle between the pair

$$ax^2+2hxy+by^2+2gx+2fy+c=0$$

is equal to that between the pair

$$ax^2+2hxy+by^2=0,$$

so that the angle θ is given by

$$\tan\theta=\frac{2\sqrt{(h^2-ab)}}{a+b}.$$

4·10. Equation to the lines joining the origin to the points of intersection of a given straight line with a given curve.

Let the curve be

$$ax^2+2hxy+by^2+2gx+2fy+c=0\ldots \quad \ldots \quad \ldots(1)$$

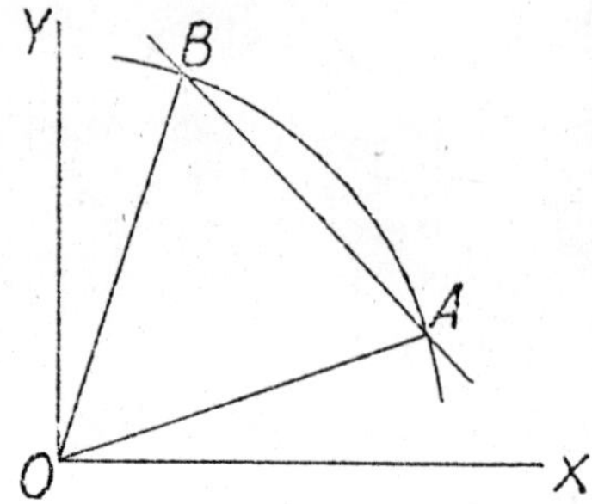

and let the given straight line be

$$lx+my+n=0 \quad .. \quad ...(1)$$

As equation (1) is of the second degree, it is cut by the straight line AB in two points A and B. We have to find the equation to the straight lines OA and OB.

As OA and OB pass through the origin, their equation must be homogeneous and of the second degree. Moreover, the lines pass through the points of intersection of (1) and (2), so that the required equation must be satisfied by all points where (1) and (2) are simultaneously satisfied.

Consider the equation

$$ax^2+2hxy+by^2+2gx\left(\frac{lx+my}{n}\right)+2fy\left(\frac{lx+my}{n}\right)$$
$$+c\left(\frac{lx+my}{n}\right)^2=0. \quad ... \quad ...(3)$$

The above equation is homogeneous in x and y and is satisfied by all points which satisfy (1) and (2) simultaneously. Hence the required equation is (3).

Rule. The equation to the straight line is $lx+my=n$

or $$\left(\frac{lx+my}{n}\right)=1$$

$\frac{lx+my}{n}$ is the multiplying factor.

To make the equation

$$ax^2+2hxy+by^2+2gx+2fy+c=0$$

homogeneous with the help of the above multiplying factor, we multiply the first degree terms by it and the constant terms by its square.

Ex. 1. *Find λ in order that the equation*

$$x^2+\lambda xy+y^2-5x-7y+6=0$$

may represent two straight lines, and find the lines.

Here $a=1$, $b=1$, $h=\frac{\lambda}{2}$, $g=-\frac{5}{2}$, $f=-\frac{7}{2}$ and $c=6$.

Hence $abc+2fgh-af^2-bg^2-ch^2=0$ gives

$$6+\frac{35\lambda}{4}-\frac{49}{4}-\frac{25}{4}-\frac{3\lambda^2}{2}=0,$$

or
$$6\lambda^2-35\lambda+50=0,$$

or
$$\lambda=\frac{35\pm\sqrt{(35\times 35-4\times 6\times 50)}}{12}$$

$$=\frac{35\pm 5}{12}$$

$$=\frac{10}{3} \quad \text{or} \quad \frac{5}{2}.$$

On substituting these values in the above equation, we get

$$2x^2+5xy+2y^2-10x-14y+12=0$$

and
$$3x^2+10xy+3y^2-15x-21y+18=0$$

for the two pairs of lines.

Ex. 2. *Find the equation to the pair of straight lines through the origin perpendicular to the pair of straight lines given by*

$$2x^2+5xy+2y^2+10x+5y=0.$$

Resolving the homogeneous expression $2x^2+5xy+2y^2$ into factors, we get

$$2x^2+5xy+2y^2=(2x+y)(x+2y).$$

Let the lines be

$$2x+y+c_1=0, \quad x+2y+c_2=0.$$

Hence $(2x+y+c_1)(x+2y+c_2)$ is equivalent to

$$2x^2+5xy+2y^2+10x+5y.$$

Comparing coefficients, we get $c_1+2c_2=10$ ... ...(1)

and $2c_1+c_2=5$ ... ...(2)

Solving (1) and (2) for c_1 and c_2, we get $c_2=5$, $c_1=0$.

Hence the straight lines representing the above equation are $2x+y=0$ and $x+2y+5=0$.

The lines passing through the origin and perpendicular to the above lines are $x-2y=0$ and $2x-y=0$.

Their combined equation is given by

$$(x-2y)(2x-y)=0$$

or

$$2x^2-5xy+2y^2=0.$$

Ex. 3. *Prove that the general equation*

$$ax^2+2hxy+by^2+2gx+2fy+c=0$$

represents two parallel straight lines if $h^2=ab$ and $bg^2=af^2$.

Prove also that the distance between them is

$$2\sqrt{\left\{\frac{g^2-ac}{a(a+b)}\right\}}$$

Let the parallel lines be

$$y=mx+c_1 \qquad \ldots \quad \ldots \quad \ldots(1)$$

$$y=mx+c_2. \qquad \ldots \quad \ldots \quad \ldots(2)$$

Then $(y-mx-c_1)(y-mx-c_2)$ is equivalent to the given equation. To make the coefficients of y^2 equal in both let us multiply the above expression by b.

Then $b(y-mx-c_1)(y-mx-c_2)$ is the same as

$$ax^2+2hxy+by^2+2gx+2fy+c.$$

Comparing coefficients, we get

$$bm^2=a \qquad \ldots \quad \ldots \quad \ldots(1)$$

$$bm=-h \qquad \ldots \quad \ldots \quad \ldots(2)$$

$$bm(c_1+c_2)=2g \qquad \ldots \quad \ldots \quad \ldots(3)$$

$$-b(c_1+c_2)=2f \qquad \ldots \quad \ldots \quad \ldots(4)$$

$$bc_1c_2=c \qquad \ldots \quad \ldots \quad \ldots(5)$$

From (1) and (2),

$$\frac{h^2}{b^2}=\frac{a}{b} \text{ or } h^2=ab. \qquad \ldots \quad \ldots \quad \ldots(6)$$

From (3) and (4),

$$-m=\frac{g}{f}$$

or $$m^2=\frac{g^2}{f^2}$$

or $$\frac{a}{b}=\frac{g^2}{f^2} \qquad \text{[from (1)]}$$

or $$af^2=bg^2. \qquad \ldots \quad \ldots \quad \ldots(7)$$

Let (x_1, y_1) be a point on $y=mx+c_1$,

then $$mx_1-y_1+c_1=0. \qquad \ldots \quad \ldots \quad \ldots(8)$$

Length of the perpendicular from (x_1, y_1) on the line $y=mx+c_2$ is given by

$$\frac{mx_1-y_1+c_2}{\sqrt{(1+m^2)}}=\frac{c_2-c_1}{\sqrt{(1+m^2)}}$$

substituting from (8) for mx_1-y_1

$$=\sqrt{\left\{\frac{(c_2+c_1)^2-4c_1c_2}{1+m^2}\right\}}$$

$$=\sqrt{\left[\left(\frac{4f^2}{b^2}-\frac{4c}{b}\right)\Big/\left(1+\frac{a}{b}\right)\right]}$$

substituting from (4) and (1),

$$=2\sqrt{\left[\left(\frac{g^2}{ab}-\frac{c}{b}\right)\Big/\left(1+\frac{a}{b}\right)\right]}$$

substituting from (7) for f.

$$=2\sqrt{\left[\frac{g^2-ac}{a\,(a+b)}\right]}.$$

Ex. 4. *If $ax^2+2hxy+by^2+2gx+2fy+c=0$ represent a pair of straight lines intersecting in (x_1, y_1), then the equation to the lines bisecting the angles between them will be*

$$\frac{(x-x_1)^2-(y-y_1)^2}{a-b}=\frac{(x-x_1)\,(y-y_1)}{h}.$$

On transferring the origin to the point (x_1, y_1) the above equation will become homogeneous, as it will then represent two straight lines passing through the origin. The transformed equation is therefore

$$ax^2+2hxy+by^2=0.$$

The equation to the lines bisecting the angles between them is

$$\frac{x^2-y^2}{a-b}=\frac{xy}{h}.$$

On transferring the origin back to the original point, the above equation becomes

$$\frac{(x-x_1)^2-(y-y_1)^2}{a-b}=\frac{(x-x_1)(y-y_1)}{h}.$$

Ex. 5: *Prove that the straight lines joining the origin to the points of intersection of the straight line $kx+hy=2hk$ with the circle $(x-h)^2+(y-k)^2=c^2$ are at right angles if $h^2+k^2=c^2$.*

The equation to the circle can be written as

$$x^2+y^2-2xh-2yk+h^2+k^2-c^2=0 \quad \ldots \quad \ldots \quad \ldots(1)$$

The equation to the straight line is

$$kx+hy=2hk. \quad \ldots \quad \ldots \quad \ldots(2)$$

Making (1) homogeneous with the help of (2), we get the required equation as

$$x^2+y^2-2xh\left(\frac{kx+hy}{2hk}\right)-2yk\left(\frac{kx+hy}{2hk}\right)$$
$$+(h^2+k^2-c^2)\left(\frac{kx+hy}{2hk}\right)^2=0.$$

If these lines are at right angles, the sum of the coefficients of x^2 and y^2 should be zero, which gives

$$1-1+1-1+(h^2+k^2-c^2)\left(\frac{1}{4h^2}+\frac{1}{4k^2}\right)=0,$$

or $$(h^2+k^2-c^2)\,\frac{(h^2+k^2)}{4h^2k^2}=0$$

or $$h^2+k^2-c^2=0$$

or $$h^2+k^2=c^2.$$

Ex. 6. *Show that the straight lines joining the origin to the other two points of intersection of the curves whose equations are*

$$ax^2+2hxy+by^2+2gx=0$$

and

$$a'x^2+2h'xy+b'y^2+2g'x=0$$

will be at right angles if

$$g(a'+b')-g(a+b)=0.$$

A curve passing through the intersections of

$$ax^2+2hxy+by^2+2gx=0$$

and

$$a'x^2+2h'xy+b'y^2+2g'x=0$$

is given by

$$ax^2+2hxy+by^2+2gx+\lambda(a'x^2+2h'xy+b'y^2+2g'x)=0$$

for all values of λ.

The above equation will represent two straight lines passing through the origin and through the intersection of the given curves if the equation is homogeneous, so that λ must be such that the terms of the first degree in x and y vanish, that is

$$2g+2\lambda g'=0,$$

or

$$\lambda=-\frac{g}{g'}.$$

The equation to the straight lines become

$$\left(a-\frac{g}{g'}a'\right)x^2+2xy\left(h-\frac{g}{g'}h'\right)+y^2\left(b-\frac{g}{g'}b'\right)=0.$$

If these straight lines are at right angles,

$$a-\frac{g}{g'}a'+b-\frac{g}{g'}b'=0,$$

or

$$g'(a+b)-g(a'+b')=0.$$

Ex. 7. *If the equation* $ax^2+2hxy+by^2+2gx+2fy+c=0$ *represents two straight lines, prove that the square of the distance of their point of intersection from the origin is*

$$\frac{c(a-b)-f^2-g^2}{ab-h^2}:$$

Let the point of intersection be (x_1, y_1). On transferring the origin to (x_1, y_1) the above equation becomes

$$a(x+x_1)^2+2h(x+x_1)(y+y_1)+b(y+y_1)^2+2g(x+x_1)$$
$$+2f(y+y_1)+c=0,$$

or $$ax^2+2hxy+by^2+2x(ax_1+hy_1+g)+2y(hx_1+by_1+f)$$
$$+ax_1^2+2hx_1y_1+by_1^2+2gx_1+2fy_1+c=0.$$

The above equation must be homogeneous as it represents two straight lines passing through the origin. Hence the terms of the first degree in x and y and the absolute term in the above equation should vanish.

Hence $$ax_1+hy_1+g=0, \qquad \ldots \quad \ldots \quad \ldots(1)$$
$$hx_1+by_1+f=0 \qquad \ldots \quad \ldots \quad \ldots(2)$$
and $$ax_1^2+2hx_1y_1+by_1^2+2gx_1+2fy_1+c=0 \ldots \quad \ldots \quad \ldots(3)$$

Solving (1) and (2) for x_1 and y_1, we get

$$\frac{x_1}{hf-bg}=\frac{y_1}{gh-af}=\frac{1}{ab-h^2} \qquad \ldots \quad \ldots \quad \ldots(4)$$

(3) can be rewritten as

$$x_1(ax_1+hy_1+g)+y_1(hx_1+by_1+f)+gx_1+fy_1+c=0$$

or $$gx_1+fy_1+c=0,$$

or $$g\frac{(hf-bg)}{ab-h^2}+f\frac{(gh-af)}{ab-h^2}+c=0,$$

substituting for x_1 and y_1 from (4),

or $$abc+2fgh-af^2-bg^2-ch^2=0 \ldots \quad \ldots \quad \ldots(5)$$

The square of the distance from the origin to the point $(x_1, y_1)=x_1^2+y_1^2$

$$=\frac{(hf-bg)^2}{(ab-h^2)^2}+\frac{(gh-af)^2}{(ab-h^2)^2} \qquad \text{[from (4)]}$$

$$=\frac{h^2f^2+b^2g^2-2bfgh+g^2h^2+a^2f^2-2ghfa}{(ab-h^2)^2}$$

$$=\frac{h^2f^2+b^2g^2+g^2h^2+a^2f^2-2fgh(a+h)}{(ah-h^2)^2}$$

$$=\frac{h^2f^2+b^2g^2+g^2h^2+a^2f^2-(a+b)(af^2+bg^2-ch^2-abc)}{(ab-h^2)^2}$$

substituting for $2fgh$ from (5),

$$=\frac{h^2f^2+b^2g^2+g^2h^2+a^2f^2-a^2f^2-abf^2-abg^2-b^2g^2-ach^2-bch^2+a^2bc+ab^2c}{(ab-h^2)^2}$$

$$=\frac{-f^2(ab-h^2)-g^2(ah-h^2)+ac(ab-h^2)+bc(ab-h^2)}{(ab-h^2)}$$

$$=\frac{bc+ac-f^2-g^2}{ab-h^2}.$$

$$=\frac{c(a+b)-f^2-g^2}{ab-b^2}.$$

Exercise VII.

1. Find the value of λ for which

$$x^2-\lambda xy+2y^2+3x-5y+2=0$$

represents a pair of straight lines and find the lines.

2. Show that the equation

$\cos 3\alpha\,(x^3-3xy^2)+\sin 3\alpha\,(y^3-3x^2y)+3a\,(x^2+y^2)-4a^3=0$

represents three straight lines forming an equilateral triangle.

Prove also that its area is $3\sqrt{3}a^2$.

3. If the equation $ax^2+2hxy+by^2+2gx+2fy+c=0$ represents two straight lines, prove that the product of the perpendiculars from the origin on these lines is

$$\frac{c}{\sqrt{\{(a-b^2)+4h^2\}}}.$$

(Agra 1957)

4. If $ax^2+2hxy+by^2+2gx+2fy+c=0$ represents a pair of straight lines, prove that the area of the triangle formed by the x-axis and the bisectors of the above pair is

$$\sqrt{\left\{\frac{(a-b)^2+4h^2}{2h}\right\}}\cdot\frac{ca-g^2}{ab-h^2}.$$

(Agra 1953)

5. The equation $ax^2+2hxy+by^2+2gx+2fy+c=0$ repre-

sents two straight lines at right angles. Prove that the square of the distance of their point of intersection from the origin is

$$\frac{f^2+g^2}{b^2+h^2}.$$ (Lucknow 1946)

6. If $f(x, y) \equiv ax^2+2hxy+by^2+2gx+2fy+c=0$ represents a pair of straight lines, prove that the equation of the third pair of straight lines passing through the points where these meet the axes is

$$c\, f(x, y)+4\,(fg-ch)\, xy=0.$$

7. If the equation $ax^2+2hxy+by^2+2gx+2fy+c=0$ represents a pair of straight lines equidistant from the origin show that

$$f^4-g^4=c\,(bf^2-ag^2)$$

and $$h\,(g^2-f^2)=fg\,(a-h).$$

8. If each of the equations $ax^2+2hxy+by^2+2gx+2fy+c=0$, $ax^2+2hxy+by^2-2gx-2fy+c=0$ represents a pair of straight lines, prove that they enclose a parallelogram of area

$$\frac{2c}{\sqrt{(h^2-ab)}}.$$

9. Perpendiculars are drawn from a point (x', y') on the lines $ax^2+2hxy+by^2=0$. Shew that the area of the triangle formed by their feet and the origin is

$$\frac{(bx'^2-2hx'y'+ay'^2)}{(a-b)^3+4h^2}.$$

10. Show that the length of the straight line joining the feet of the perpendiculars let fall from the point (f, g) on the pair of lines $ax^2+2hxy+by^2=0$ is

$$2\sqrt{\left\{\frac{(h^2-ab)\,f^2+g^2)}{(a-b)^2+4h^2}\right\}}$$

11, Find the equation to the pair of straight lines joining

the origin to the intersections of the straight line $y=mx+c$ and the curve $x^2+y^2=a^2$.

Prove that they are at right angles if $2c^2=a^2(1+m^2)$.

12. Prove that the angle between the straight lines joining the origin to the intersection of the straight line $y=3x+2$ with the curve $x^2+2xy+3y^2+4x+8y-11=0$ is $\tan^{-1}\frac{2\sqrt{2}}{3}$.

13. If the circle $x^2+y^2+2gx+2fy+c=0$ intercepts on the line $lx+my=1$ a length which subtends a right angle at the origin, shew that $c(l^2+m^2)+2(gl+fm+1)=0$.

(Andhra 1947)

14. Find the angle which the chord $xx'+yy'=a^2$ of the circle $x^2+y^2=a^2$ subtends at the origin. Also obtain the condition that this angle may be a right angle.

15. The line $px+qy=1$ is a chord of $ax^2+2hxy+by=c$ and subtends a right angle at the origin. Prove that

$$c(p^2+q^2)=a+b.$$ *(Lucknow 1951)*

CHAPTER V

THE CIRCLE

5·1. The locus of a point which moves so that its distance from a fixed point always remains the same is a **circle.**

The fixed point is called the **centre** and the fixed distance the **radius** of the circle.

5·2. Equation of a circle whose centre and radius are given.

Let the coordinates of the centre of the circle be (h, k) with reference to OX and OY as axes and let its radius be a.

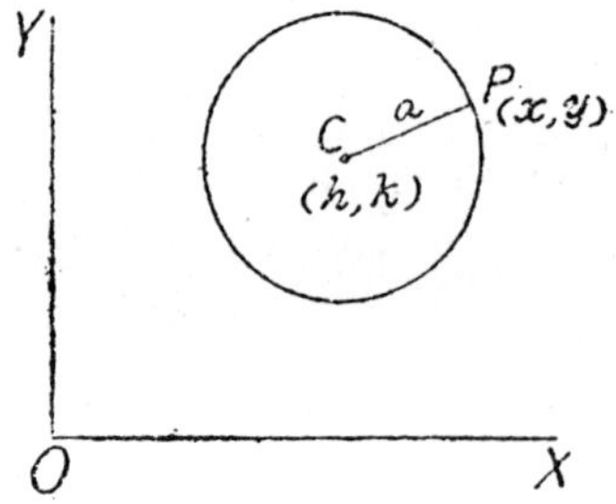

Let $P(x, y)$ be a point on the circle.

Then $CP = a = \sqrt{[(x-h)^2+(y-k)^2]}$.

$$\therefore \quad (x-h)^2+(y-k)^2=a^2 \qquad \dots \qquad \dots(1)$$

which is a relation between the coordinates of any point on the circle and is therefore the equation to the circle.

Cor. 1. If the centre is the origin, then equation to the circle becomes

$$x^2+y^2=a^2.$$

Cor. 2. If the axes are inclined to each other at an angle ω, the equation to the circle becomes

$$(x-h)^2+(y-k)^2+2(x-h)(y-k)\cos\omega=a^2.$$

Ex. 1. *Find the equation to a circle whose centre is (2, 3) and radius 5.*

The equation to the circle is

$$(x-2)^2+(y-3)^2=5^2,$$

or
$$x^2+y^2-4x-6y-12=0.$$

Ex. 2. *The equation to a curve is* $x^2+y^2-4y=0$. *Find its centre and radius.*

$x^2+y^2-4y=0$ can be written as $x^2+(y-2)^2=4$.
Its centre is $(0, 2)$ and the radius is 2.

5·3. General equation to a circle.

The equation to a circle whose centre is (h, k) and radius a is given by $(x-h)^2+(y-k)^2=a^2$.

or
$$x^2+y^2-2hx-2ky+h^2+k^2-a^2=0.$$

The above equation can be written in the more general form $x^2+y^2+2gx+2fy+c=0$, where g, f and c are arbitrary constants.

Rewriting the above equation as

$$(x+g)^2+(y+f)^2=g^2+f^2-c,$$

we find that the centre of this circle is $(-g, -f)$ and its radius is $\sqrt{(g^2+f^2-c)}$.

If $g^2+f^2-c=0$, the radius is zero and the circle reduces to a point circle.

If $g^2+f^2-c > 0$, it is a real circle.

If $g^2+f^2-c < 0$, the circle is imaginary.

5·4. Condition that the general equation of the second degree may represent a circle.

The general equation of the second degree is

$$ax^2+2hxy+by^2+2gx+2fy+c=0 \quad \ldots \quad \ldots (1)$$

Comparing this equation with the general equation to a circle

$$x^2+y^2+2g_1x+2f_1y+c_1=0 \quad \ldots \quad \ldots (2)$$

we find that $a=b$ and $h=0$, because by making these substitutions in (1), the equation becomes

$$ax^2+ay^2+2gx+2fy+c=0$$

or
$$x^2+y^2+\frac{2g}{a}x+\frac{2f}{a}y+\frac{c}{a}=0$$
which is of the same form as (2).

Cor. If the axes are inclined at an angle ω, the general equation will be of the form
$$x^2+y^2+2xy\cos\omega+2gx+2fy+c=0.$$

Hence the condition that the general equation of the second degree may represent a circle is that $a=b$ and $h=a\cos\omega$.

5·5. Condition that the straight line $y=mx+c$ may intersect the circle $x^2+y^2=a^2$.

Eliminating y from $x^2+y^2=a^2$ and $y=mx+c$, we get
$$x^2+(mx+c)^2=a^2.$$
or
$$x^2(1+m^2)+2mxc+c^2-a^2=0. \qquad \ldots \quad \ldots(1)$$

This equation gives the abscissae of the points of intersection. The roots of the above equation are real and different, real and coincident, or imaginary according as $4m^2c^2-4(1+m^2)(c^2-a^2)>,=$ or <0,

i. e. according as $-c^2+a^2+a^2m^2>,=$ or <0,

" " $a^2(1+m^2)>,=$ or $<c^2$.

" " $c<,=$ or $a\sqrt{(1+m^2)}$.

Cor. 1. The straight line $y=mx+c$ will cut the circle $x^2+y^2=a^2$ in coincident points if $c=\pm a\sqrt{(1+m^2)}$.

Hence $y=mx\pm a\sqrt{(1+m^2)}$ is the equation to any tangent to the circle.

Cor. 2. If the roots of equation (1) above are x_1 and x_2, we have
$$x_1+x_2=\frac{-2mc}{1+m^2}, \qquad \ldots \quad \ldots \quad \ldots(2)$$
$$x_1x_2=\frac{c^2-a^2}{1+m^2}. \qquad \ldots \quad \ldots \quad \ldots(3)$$

The length of the chord intercepted on the line $y=mx+c$

is $\sqrt{\{(x_1-x_2)^2+(y_1-y_2)^2\}}$.

But (x_1, y_1) and (x_2, y_2) are points on the line $y=mx+c$.

$$\therefore \quad y_1=mx_1+c,$$

$$y_2=mx_2+c,$$

which give $$y_1-y_2=m(x_1-x_2).$$

Hence the length of the chord

$$=\sqrt{\{(x_1-x_2)^2+m^2(x_1-x_2)^2\}}$$
$$=(x_1-x_2)\sqrt{(1+m^2)}.$$

But $(x_1-x_2)=\sqrt{\{(x_1+x_2)^2-4x_1x_2\}}$

$$=\sqrt{\left\{\frac{4m^2c^2}{(1+m^2)^2}-\frac{4(c^2-a^2)}{1+m^2}\right\}}$$

substituting from (2) and (3),

Hence $$x_1-x_2=2\sqrt{\left\{\frac{m^2c^2-(c^2-a^2)(1+m^2)}{(1+m^2)^2}\right\}}$$
$$=2\sqrt{\left\{\frac{a^2-c^2+a^2m^2}{(1+m^2)^2}\right\}}$$
$$=2\sqrt{\left\{\frac{a^2(1+m^2)-c^2}{(1+m^2)^2}\right\}}.$$

$\therefore$ the length of the chord

$$=2\sqrt{\left\{\frac{a^2(1+m^2)-c^2}{1+m^2}\right\}}.$$

5·6. Equation of the circle on a given diameter.

Let A and B, the extremities of the diameter of the circle, be (x_1, y_1) and (x_2, y_2) respectively.

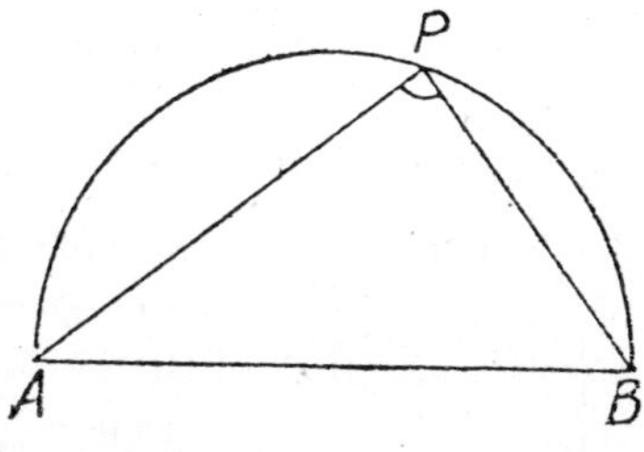

Let $P(x, y)$ be a point on the circle.

Now $\angle APB$ is a right angle, so that the straight lines AP and PB are at right angles. Hence the product of their direction tangents should be equal to -1, or $$\frac{y-y_1}{x-x_1} \times \frac{y-y_2}{x-x_2} = -1$$

or $$(x-x_1)(x-x_2)+(y-y_2)(y-y_2)=0.$$

This is a relation between the coordinates of any point on the circle and is therefore the equation to the circle.

5·7. The equation to the circle passing through three given points.

Let the given points be (x_1, y_1), (x_2, y_2) and (x_3, y_3).

Suppose the equation to the circle is

$$x^2+y^2+2gx+2fy+c=0 \quad \ldots \quad \ldots \quad \ldots(1)$$

As the points (x_1, y_1), (x_2, y_2) and (x_3, y_3) lie on the circle, we have

$$x_1^2+y_1^2+2gx_1+2fy_1+c=0, \quad \ldots \quad \ldots \quad \ldots(2)$$

$$x_2^2+y_2^2+2gx_2+2fy_2+c=0, \quad \ldots \quad \ldots \quad \ldots(3)$$

$$x_3^2+y_3^2+2gx_3+2fy_3+c=0. \quad \ldots \quad \ldots \quad \ldots(4)$$

Eliminating g, f and c from (1), (2), (3) and (4), we get the equation to the circle as

$$\begin{vmatrix} x^2+y^2 & x & y & 1 \\ x_1^2+y_1^2 & x_1 & y_1 & 1 \\ x_2^2+y_2^2 & x_2 & y_2 & 1 \\ x_3^2+y_3^2 & x_3 & y_3 & 1 \end{vmatrix}=0.$$

Ex. 1. *Obtain the equation of the circle which passes through the two points on the axis of x which are at a distance 2 from the origin and whose radius is 5.*

Let the equation to the circle be

$$x^2+y^2+2gx+2fy+c=0. \quad \ldots \quad \ldots \quad \ldots(1)$$

As the radius of the circle is 5,

$$g^2+f^2-c=25. \quad \ldots \quad \ldots \quad \ldots(2)$$

The absissae of the points of intersection of the circle (1) with the x-axis are given by

$$x^2+2gx+c=0.$$

By hypothesis, the roots of this equation are $+2$ and -2.

$\therefore$ the above equation must be equivalent to $x^2=4$.

$$\therefore \quad g=0 \text{ and } c=-4.$$

Substituting these values of g and c in (2), we get

$$f^2=21.$$

$$\therefore \quad f=\pm\sqrt{21}.$$

Hence the equation to the circle is

$$x^2+y^2\pm 2\sqrt{21}y-4=0.$$

Ex. 2. *Find the length of the common chord of the circles*

$$(x-a)^2+y^2=a^2 \qquad \ldots \quad \ldots (1)$$

and

$$x^2+(y-b)^2=b^2 \qquad \ldots \quad \ldots (2)$$

and prove that the equation of the circle whose diameter is this chord is $(a^2+b^2)(x^2+y^2)=2ab(bx+ay)$. *(Rajputana 1954)*

For all points common to (1) and (2), we get, by subtraction,

$$ax-by=0 \qquad \ldots \quad \ldots (3)$$

Hence the equation of the common chord is (3).

The abscissae of the points of intersection of (3) with (1) are given by $x^2-2ax+\frac{a^2}{b^2}x^2=0$,

or

$$x^2(a^2+b^2)=2ab^2x.$$

$$\therefore \quad x=0 \quad \text{or} \quad \frac{2ab^2}{a^2+b^2}.$$

Corresponding to these values of x_1, x_2 the values of y_1 and y_2 are 0 and $\frac{2a^2h}{a^2+b^2}$ respectively.

From the equation of the chord $y=\frac{a}{b}x$.

$\therefore$ the length of the chord

$$=\sqrt{[(x_1-x_2)^2+(y_1-y_2)^2]}$$

$$=\sqrt{\left\{\frac{4a^2b^4}{(a^2+b^2)^2}+\frac{4a^4b^2}{(a^2+b^2)^2}\right\}}$$

$$=\frac{2ab}{\sqrt{(a^2+b^2)}}.$$

The equation to the circle on (x_1, y_1) and (x_2, y_2) as diameter is

$$(x-x_1)(x-x_2)+(y-y_1)(y-y_2)=0,$$

which in this case becomes

$$x\left(x-\frac{2ab^2}{a^2+b^2}\right)+y\left(y-\frac{2a^2b}{a^2+b^2}\right)=0,$$

or
$$(a^2+b^2)(x^2+y^2)=2ab(bx+ay).$$

Note. The equation $S+\lambda S'=0$ is a conic passing through the intersection of $S=0$ and $S'=0$ for all values of λ.

If $S=0$ and $S'=0$ are circles, then $S+\lambda S'=0$ is also a circle because the coefficient of xy is even now zero and the coefficients of x^2 and y^2 are equal. If $\lambda=-1$ the coefficients of x^2 and y^2 in $S-S'=0$ vanish and the equation becomes the equation to the common chord it $S=0$ and $S'=0$.

If $u=0$ is the equation of a straight line then $S+\lambda u=0$ is the equation to a conic passing through the intersection of $S=0$ and $u=0$.

If $S=0$ is a circle, $S+\lambda u=0$ is also a circle because it satisfies the conditions of being a circle.

If $L=0$, $M=0$, $N=0$ are straight lines, the equation $LM+\lambda MN+\mu NL=0$ is satisfied by the points of intersection of $L=0$ and $M=0$ or $M=0$ and $N=0$ or $N=0$ and $L=0$. Also, this equation is of the second degree; it therefore represents a conic passing through the vertices of the triangle formed by the lines $L=0$, $M=0$ and $N=0$. If λ and μ are so chosen that the coefficients of x^2 and y^2 are equal and the coefficient of xy vanishes, it represents the circle circumscribing the triangle.

Ex. 3. *Find the equation to the circumcircle of the triangle formed by the lines*

$$a_1x+b_1y+c_1=0,$$
$$a_2x+b_2y+c_2=0,$$
$$a_3x+b_3y+c_3=0.$$

Any curve passing through the vertices of the triangle is

$$(a_1x+b_1y+c_1)\,(a_2x+b_2y+c_2)+\lambda\,(a_2x+b_2y+c_2)\,(a_3x+b_3y+c_3)$$
$$+\mu\,(a_3x+b_3y+c_3)\,(a_1x+b_1y+c_1)=0 \quad \ldots(1)$$

It represents a circle if the coefficients of x^2 and y^2 are equal and the coefficient of xy vanishes, which give

$$a_1a_2+\lambda a_2a_3+\mu a_3a_1=b_1b_2+\lambda b_2b_3+\mu b_3b_1 \quad \ldots(1)$$

or

$$a_1a_2-b_1b_2+\lambda\,(a_2a_3-b_2b_3)+\mu\,(a_3a_1-b_3b_1)=0 \quad \ldots(2)$$

and

$$a_1b_2+a_2b_1+\lambda\,(a_2b_3+a_3b_2)+\mu\,(a_3b_1+b_3a_1)=0 \quad \ldots(3)$$

(1) can be written as

$$\frac{1}{a_3x+b_3y+c_3}+\frac{\lambda}{a_1x+b_1y+c_1}+\frac{\mu}{a_2x+b_2y+c_2}=0 \quad \ldots(4)$$

Eliminating λ and μ from (2), (3) and (4), we get

$$\begin{vmatrix} \dfrac{1}{a_3x+b_3y+c_3} & \dfrac{1}{a_1x+b_1y+c_1} & \dfrac{1}{a_2x+b_2y+c_2} \\ a_1a_2-b_1b_2 & a_2a_3-b_2b_3 & a_3a_1-b_3b_1 \\ a_1b_2+b_1a_2 & a_2b_3+b_2a_3 & a_3b_1+b_3a_1 \end{vmatrix}=0$$

as the equation to the required circle.

Exercise VIII

1. Find the equation to a circle which passes through the points (1, 0) and $(-1, 0)$ and has its centre on the line
$$3x-y=1.$$
2. Find the equation to the circle which passes through the origin and the points (h, k) and (k, h).
3. Find the equation to a circle which passes through the origin and cuts off intercepts equal to a and b from the positive parts of the axes.

4. Find the equation to a circle touching the axis of x at a distance 2 from the origin and cutting off an intercept of length 3 from the axis of y.

5. A circle passes through the three points $(5, -8)$, $(-2, 9)$ and $(2, 1)$. Find its centre and radius. *(Delhi 1945)*

6. If $y=mx$ is a chord of the circle $x^2+y^2-2ax=0$, prove that the equation to the circle on this chord as diameter is $(1+m^2)(x^2+y^2)-2a(x+my)=0$.

7. Find the equation of the circle whose diameter is the common chord of the circles $x^2+y^2+2x+3y+1=0$ and
$$x^2+y^2+4x+3y+2=0. \quad \textit{(Agra 1950)}$$

8. Show that the circle on the chord
$$x \cos \alpha+y \sin \alpha-p=0$$
of the circle $x^2+y^2-a^2=0$ as diameter is
$$x^2+y^2-a^2-2p(x \cos \alpha+y \sin \alpha-p)=0. \quad \textit{(Agra 1943)}$$

9. Prove that the length of the common chord of the circles $(x-h)^2+(y-k)^2=a^2$ and $(x-k)^2+(y-h)^2=a^2$ is $\sqrt{[4a^2-2(h-k)^2]}$.

10. Prove that the equation to the circle described on the line joining the points of intersection of $lx+my=1$ with the lines $ax^2+2hxy+by^2=0$ as diameter is
$$(am^2-2hlm+bl^2)(x^2+y^2)+2x(mh-bl)$$
$$+2y(hl-am)+a+b=0$$
Deduce the condition that the lines
$$ax^2+2hxy+by^2=0$$
may be at right angles. *(Travancore 1941)*

11. Find the equation of the circle circumscribing the triangle formed by the straight lines
$$x+y=6, \; 2x+y=4 \text{ and } x+2y=5.$$
(Rajputana 1954, Agra 1947)

12. Show that the equation to the circle circumscribing the triangle formed by the lines $ax^2+2hxy+by^2=0$ and

$lx+my=n$ is

$$(am^2-2hlm+bl^2)(x^2+y^2)+x(al+2hm-bl)n$$
$$+y(bm+2lh-am)n=0.$$

(Rangoon 1939)

13. Prove that the equation of the circle circumscribing the triangle formed by the lines $a_r x+b_r y+c_r=0$, $r=1, 2, 3$ is

$$\begin{vmatrix} \dfrac{a_1^2+b_1^2}{a_1x+b_1y+c_1} & a_1 & b_1 \\ \dfrac{a_2^2+b_2^2}{a_2x+b_2y+c_2} & a_2 & b_2 \\ \dfrac{a_3^2+b_3^2}{a_3x+b_3y+c_3} & a_3 & b_3 \end{vmatrix}=0$$

(Agra 1952)

14. Show that the circumcircle of the triangle formed by the lines $ax+by+c=0$, $bx+cy+a=0$ and $cx+ay+b=0$ passes through the origin if

$$(b^2+c^2)(c^2+a^2)(a^2+b^2)=abc(a+b)(b+c)(c+a).$$

(Punjab 1955)

15. Find the equation to the circle circumscribing the quadrilateral formed by the straight lines

$$2x+3y=2,\ 3x-2y=4,\ x+2y=3 \text{ and } 2x-y=3.$$

5·8. Tangent to a circle.

Let the equation to the circle be

$$x^2+y^2+2gx+2fy+c=0 \qquad \ldots \quad \ldots \quad \ldots(1)$$

Let $P(x_1, y_1)$ and $Q, (x_2, y_2)$ be two points on the above circle.

The equation to the straight line joining the points (x_1, y_1) and (x_2, y_2) is

$$y-y_1=\frac{y_2-y_1}{x_2-x_1}(x-x_1) \qquad \ldots \quad \ldots \quad \ldots(2)$$

But (x_1, y_1) and (x_2, y_2) lie on the circle.

$$\therefore\ x_1^2+y_1^2+2gx_1+2fy_1+c=0 \ldots \quad \ldots \quad \ldots(3)$$

and $$x_2^2+y_2^2+2gx_2+2fy_2+c=0 \quad \ldots \quad \ldots \quad \ldots(4)$$

Subtracting (4) from (3), we get

$$(x_1^2-x_2^2)+(y_1^2-y_2^2)+2g(x_1-x_2)+2f(y_1-y_2)=0$$

or $$(x_1-x_2)(x_1+x_2+2g)+(y_1-y_2)(y_1+y_2+2f)=0$$

or $$\frac{y_2-y_1}{x_2-x_1}=-\frac{x_1+x_2+2g}{y_1+y_2+2f} \quad \ldots \quad \ldots \quad \ldots(5)$$

Hence the equation to line PQ becomes

$$y-y_1=-\frac{x_1+x_2+2g}{y_1+y_2+2f}(x-x_1) \ldots \quad \ldots \quad \ldots(6)$$

The line PQ becomes a tangent at P if Q approaches P, and ultimately coincides with it, that is, when $x_2=x_1$ and $y_2=y_1$. Therefore, the equation to the tangent at P is

$$y-y_1=-\frac{x_1+g}{y_1+f}(x-x_1) \quad \ldots \quad \ldots \quad \ldots(7)$$

or $$y(y_1+f)+x(x_1+g)=y_1(y_1+f)+x_1(x_1+g)$$
$$=y_1^2+x_1^2+fy_1+gx_1$$
$$=-gx_1-fy_1-c \qquad \text{by (3)},$$

or $$xx_1+yy_1+g(x+x_1)+f(y+y_1)+c=0.$$

Aliter. The direction tangent of the tangent will be given by $\frac{dy}{ax}$ at the point (x_1, y_1).

Finding $\frac{dy}{dx}$ from $x^2+y^2+2gx+2fy+c=0$,

$$\frac{dy}{dx}=-\frac{x_1+g}{y_1+g} \text{ at } (x_1, y_1).$$

Hence the equation to the tangent at (x_1, y_1) is

$$y-y_1=-\frac{x_1+g}{y_1+f}(x-x_1)$$

which is the same as (7) above.

Aliter. Tangent is the straight line through (x_1, y_1) perpendicular to the radius at P.

The direction tangent of the radius is $\frac{y_1+f}{x_1+g}$ as it joins the point (x_1, y_1) and the centre $(-g, -f)$.

Hence the direction tangent of the tangent is

$$-\frac{x_1+g}{y_1+f}.$$

$\therefore$ the equation of the tangent is

$$y-y_1=-\frac{x_1+g}{y_1+f}(x-x_1)$$

which is again the same as (7) above.

Cor. If the circle is $x^2+y^2=a^2$, the equation to the tangent at (x_1, y_1) is

$$xx_1+yy_1=a^2.$$

5·9. We have seen that the line $y=mx+c$ becomes a tangent to $\qquad x^2+y^2=a^2$

if $\qquad c=\pm a\sqrt{(1+m^2)}.$

Hence $y=mx\pm a\sqrt{(1+m^2)}$ are any tangents to the circle.

Let the point of contact of $y=mx+a\sqrt{(1+m^2)}$ with the circle $x^2+y^2=a^2$ be (x_1, y_1).

Then the equation to the tangent at (x_1, y_1) is

$$xx_1+yy_1=a^2,$$

which is the same as $y=mx+a\sqrt{(1+m^2)}$.

Compairing coefficients in the two equations, we get

$$\frac{x_1}{m}=\frac{y_1}{-1}=\frac{-a^2}{a\sqrt{(1+m^2)}},$$

which give

$$x_1=\frac{-am}{\sqrt{(1+m^2)}} \text{ and } y_1=\frac{a}{\sqrt{(1+m^2)}}.$$

5·10. Normal at a point of a circle.

The normal is the line passing through the point of contact perpendicular to the tangent.

Let the circle be $x^2+y^2+2gx+2fy+c=0$. Let the

point of contact be (x_1, y_1). The equation to the tangent at (x_1, y_1) is

$$xx_1+yy_1+g(x+x_1)+f(y+y_1)+c=0.$$

The direction tangent of the line is

$$-\frac{x_1+g}{y_1+f}.$$

Hence the direction tangent of the normal is

$$\frac{y_1+f}{x_1+g}.$$

The equation to the normal is therefore

$$y-y_1=\frac{y_1+f}{x_1+g}(x-x_1),$$

or $$y(x_1+g)-x(y_1+f)=y_1(x_1+g)-x_1(y_1+f)$$
$$=y_1 g-x_1 f,$$

or $$x(y_1+f)-y(x_1+g)-x_1 f+y_1 g=0.$$

Cor. If the circle is $x^2+y^2=a^2$, the equation to the normal is $xy_1-yx_1=0$.

5·11. Tangents from a point to a circle.

Let the circle be $x^2+y^2=a^2$ and let the given point be (x_1, y_1).

The equation to any tangent to the circle,

$$y=mx+a\sqrt{(1+m^2)}.$$

As it passing through (x_1, y_1)

$$y_1=mx_1+a\sqrt{(1+m)^2}$$

or $$(y_1-mx_1)^2=a^3(1+m^2)$$

or $$m^2(x_1^2-a^2)-2mx_1 y_1+y_1^2-a^2=0,$$

which is a quadratic equation in m and gives two values of m corresponding to the two tangents which can be drawn from the point (x_1, y_1) to the circle.

These two tangents will be real and distinct, real and coincident or imaginary according as the two values of m in the above equation are real and distinct, real and equal or imaginary, that is if

$$x_1^2y_1^2 >, = \text{ or } < (x_1^2-a^2)(y_1^2-a^2),$$

i. e. if $\quad a^2(x_1^2+y_1^2) >, = \text{ or } < a^4$

i. e. if $\quad x_1^2+y_1^2 >, = \text{ or } < a^2,$

which means that the point (x_1, y_1) lies outside the circle, on the circle, or in the circle.

5·12. Length of the tangent from an external point.

Let the equation to the circle be $x^2+y^2+2gx+2fy+c=0$ and let the external point P be (x_1, y_1).

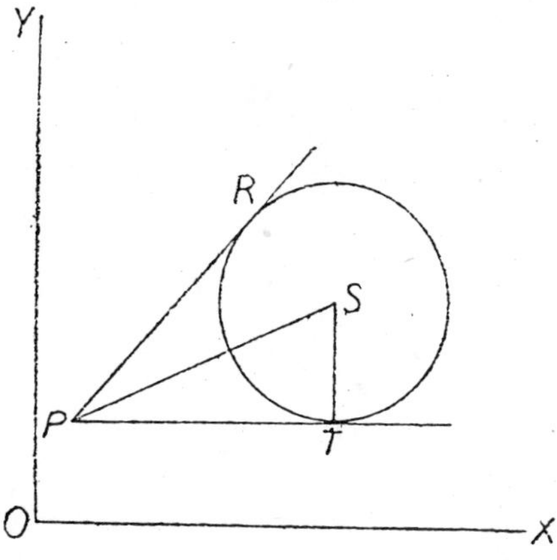

The coordinates of the centre of the circle are $(-g, -f)$ and its radius $ST=\sqrt{(g^2+f^2-c)}$.

Length $PS=\sqrt{[(x_1+g)^2+(y_1+f)^2]}$.

From $\triangle PST$,

$$PT^2=PS^2-ST^2$$
$$=(x_1+g)^2+(y_1+f)^2-(g^2+f^2-c)$$
$$=x_1^2+y_1^2+2gx_1+2fy_1+c.$$

Note. The square of the length of the tangent is obtained by substituting x_1, y_1 for x and y respectively in the left hand side of the equation to the circle when the coefficients of x^2 and y^2 are each equal to unity.

If the equation to the circle is

$$ax^2+ay^2+2gx+2fy+c=0,$$

the square of the length of the tangent is

$$x^2+y^2+\frac{2g}{a}x+\frac{2f}{a}y+\frac{c}{a}.$$

Cor. If the equation to the circle is $x^2+y^2=a^2$, the square of the length of the tangent is $x_1^2+y_1^2-a^2$.

5·13. Chord of contact.

The straight line joining the points of contact of two tangents from an external point to a circle is called the chord of contact of that point.

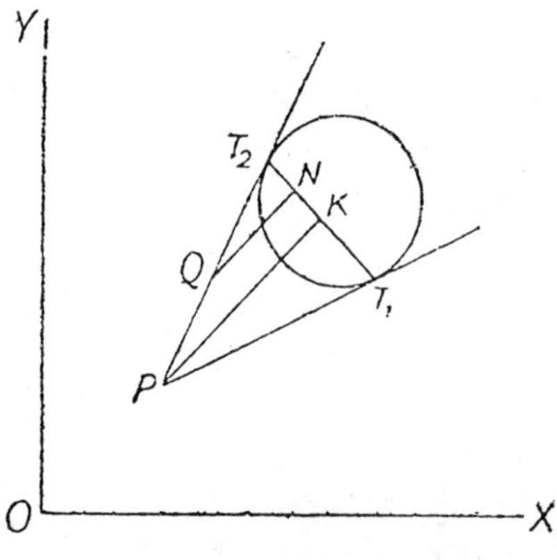

Let the circle be $x^2+y^2+2gx+2fy+c=0$ and let the external point P be (x_1, y_1).

Let the points of contact T_1 and T_2 be (p_1, q_1) and (p_2, q_2) respectively.

The equation to the tangent at (p_1, q_1) is

$$xp_1+yq_1+g(x+p_1)+f(y+q_1)+c=0.$$

It passes through (x_1, y_1). Therefore

$$x_1p_1+y_1q_1+g(x_1+p_1)+f(y_1+q_1)+c=0. \quad \ldots \quad \ldots (1)$$

Similarly, the tangent at T_2 (p_2, q_2) passes through the point (x_1, y_1).

$$\therefore \quad x_1p_2+y_1q_2+g(x_1+p_2)+f(y_1+q_2)+c=0. \quad \ldots \quad \ldots(2)$$

But (1) and (2) are the conditions that the points (p_1, q_1) and (p_2, q_2) should lie on the line

$$xx_1+yy_1+g(x+x_1)+f(y+y_1)+c=0. \quad \ldots \quad \ldots(3)$$

Hence (3) is the equation to the chord of contact which passes through the points T_1 and T_2, as it is of first degree in x and y.

If the point (x_1, y_1) is outside the circle, the chord of contact is real.

If the point (x_1, y_1) is on the circle the chord of contact becomes the tangent at the point (x_1, y_1).

If the point (x_1, y_1) is inside the circle the chord of contact is imaginary.

Cor. If the equation to the circle is $x^2+y^2=a^2$ the chord of contact of the point (x_1, y_1) is $xx_1+yy_1=a^2$.

5·14. Pair of tangents from an external point.

Let the circle be $x^2+y^2+2gx+2fy+c=0$ and let the external point be (x_1, y_1).

Let (h, k) be a point Q on one of the tangents PT_2 (see figure of the last article). From P and Q draw PK and QN perpendiculars respectively on the chord of contact T_1T_2. From similar ΔsPKT_2 and QNT_2,

$$\frac{QT_2}{PT_2}=\frac{QN}{PK} \quad \ldots \quad \ldots(1)$$

But PT_2=length of the tangent from (x_1, y_1)

$$=\sqrt{(x_1^2+y_1^2+2gx_1+2fy_1+c)}.$$

QT_2=length of the tangent from (h, k)

$$=\sqrt{(h^2+k^2+2gh+2fk+c)}.$$

QN=length of the perpendicular from (h, k) on the

chord of contact T_1T_2

$$=\frac{hx_1+ky_1+g(h+x_1)+f(k+y_1)+c}{\sqrt{[(x_1+g)^2+(y_1+f)^2]}}$$

PK=length of the perpendicular from (x_1, y_1) on the chord of contact T_1T_2

$$=\frac{x_1^2+y_1^2+2gx_1+2fy_1+c}{\sqrt{[(x_1+g)^2+(y_1+f)^2]}}$$

Substituting these values in (1), we get

$$\sqrt{(h^2+k^2+2gh+2fk+c)}.(x_1^2+y_1^2+2gx_1+2fy_1+c)$$
$$=\sqrt{(x_1^2+y_1^2+2gx_1+2fy_1+c)}\times[hx_1+ky_1+g(h+x_1)+f(k+y_1)+c]$$

or $\quad \sqrt{(h^2+k^2+2gh+2fk+c)}.\sqrt{(x_1^2+y_1^2+2gx_1+2fy_1+c)}$
$$=hx_1+ky_1+g(h+x_1)+f(k+y_1)+c.$$

Hence the locus of the point (h, k) is

$$(x^2+y^2+2gx+2fy+c)(x_1^2+y_1^2+2gx_1+2fy_1+c)$$
$$=[xx_1+yy_1+g(x+x_1)+f(y+y_1)+c]^2$$

which is the equation of the pair of tangents.

The above equation of the pair of tangents can be written as $\qquad SS_1=T^2,$

where $\qquad S\equiv x^2+y^2+2gx+2fy+c=0,$

$$S_1\equiv x_1^2+y_1^2+2gx_1+2fy_1+c=0,$$

$$T\equiv xx_1+yy_1+g(x+x_1)+f(y+y_1)+c=0.$$

Cor. If the equation to the circle is $x^2+y^2=a^2$, the equation to the pair of tangents from (x_1, y_1) is

$$(x^2+y^2-a^2)(x_1^2+y_1^2+a^2)=(xx_1+yy_1-a^2)^2.$$

5·15. Polar and Pole.

The locus of the point of intersection of tangents at the extremities of a chord through a fixed point, is called the polar of the point. The fixed point is called the pole of the polar.

Let the fixed point be (x_1, y_1). Let a line through P cut the circle in Q and R. Let the tangents at Q and R cut

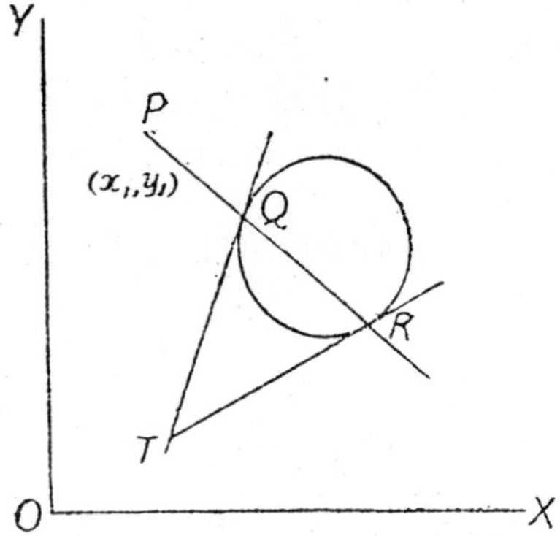

each other at T. We have to find the locus of T. Let its coordinates be (h, k). Then the chord of contact of T (h, k) passes through the point P (x_1, y_1).

$$\therefore \quad hx_1+ky_1+g(x_1+h)+f(y_1+k)+c=0.$$

Hence the locus of the point (h, k) is

$$xx_1+yy_1+g(x+x_1)+f(y+y_1)+c=0,$$

which is the equation to the polar of the point (x_1, y_1).

Cor. If the circle be $x^2+y^2=a^2$, the equation to the polar of (x_1, y_1) is $xx_1+yy_1-a^2=0$.

5·16. Pole of a given straight line.

Let the given line be $lx+my+n=0$ ... ...(1)

and let the equation of the circle be

$$x^2+y^2+2gx+2fy+c=0. \quad \ldots \quad \ldots(2)$$

Let the pole of line (1) be (h, k); then the polar of (h, k) is

$$xh+yk+g(x+h)+f(y+k)+c=0 \quad \ldots \quad \ldots(3)$$

which is the same as (1).

Comparing coefficients in (1) and (3), we get

$$\frac{g+h}{l}=\frac{f+k}{m}=\frac{gh+fk+c}{n}. \quad \ldots \quad \ldots(4)$$

From the first and third terms of these equations, we get

$$h(x-lg)-klf+ng-lc=0 \quad \ldots \quad \ldots(5)$$

and from the second and third, we get

$$-hmg+k(n-fm)+nf-mc=0. \qquad \ldots \qquad \ldots(6)$$

Solving (5) and (6) for h and k we get

$$\frac{h}{-lf(nf-mc)-(n-fm)(ng-lc)}=\frac{k}{-mg(ng-lc)-(n-lg)(nf-mc)}$$

$$=\frac{1}{(n-lg)(n-fm)-lfmg},$$

or
$$\frac{h}{-lnf^2-n^2g+fmg+nlc}=\frac{k}{-mng^2-n^2f+nlgf+mnc}$$

$$=\frac{1}{n^2-lgn-nfm},$$

from which h and k can be found.

Cor. If the circle be $x^2+y^2=a^2$, the above equations become

$$\frac{h}{-lna^2}=\frac{k}{-mna^2}=\frac{1}{n^2}$$

$$\therefore \quad h=-\frac{la^2}{n} \text{ and } k=-\frac{ma^2}{n}.$$

5·17. If the polar of P passes through Q, then the polar of Q passes through P.

Let the points P and Q be (x_1, y_1) and (x_2, y_2) respectively.

Let the circle be

$$x^2+y^2+2gx+2fy+c=0 \qquad \ldots \qquad \ldots(1)$$

The polar of (x_1, y_1) with respect to this circle is

$$xx_1+yy_1+g(x+x_1)+f(y+y_1)+c=0.$$

If the point (x_2, y_2) lies on it,

$$x_1x_2+y_1y_2+g(x_1+x_2)+f(y_1+y_2)+c=0 \ldots \qquad \ldots(2)$$

which is also the condition for the point (x_1, y_1) to lie on the polar of (x_2, y_2).

Two points P and Q which are such that each lies on the polar of the other are called **Conjugate Points.**

5·18. If the pole of a line AB lies on another line CD, then the pole of the line CD lies on AB.

Let the pole of the line AB be (x_1, y_1). Then the equation to AB is

$$xx_1+yy_1+g(x+x_1)+f(y+y_1)+c=0 \quad \dots \quad \dots(1)$$

Let the pole of line CD be (x_2, y_2). Then the equation to CD is

$$xx_2+yy_2+g(x+x_2)+f(y+y_2)+c=0 \quad \dots \quad \dots(2)$$

If (x_1, y_1) lies on (2)

$$x_1x_2+y_1y_2+g(x_1+x_2)+f(y_1+y_2)+c=0,$$

which is also satisfied when (x_2, y_2) lies on (1).

Two lines AB and CD which are such that the pole of one lies on the other are called **Conjugate lines.**

5·19. Let the point P be (x_1, y_1) and let the circle be

$$x^2+y^2+2gx+2fy+c=0.$$

The polar of the point P (x_1, y_1) with respect to the circle is

$$xx_1+yy_1+g(x+x_1)$$
$$+f(y+y_1)+c=0 \quad \dots(1)$$

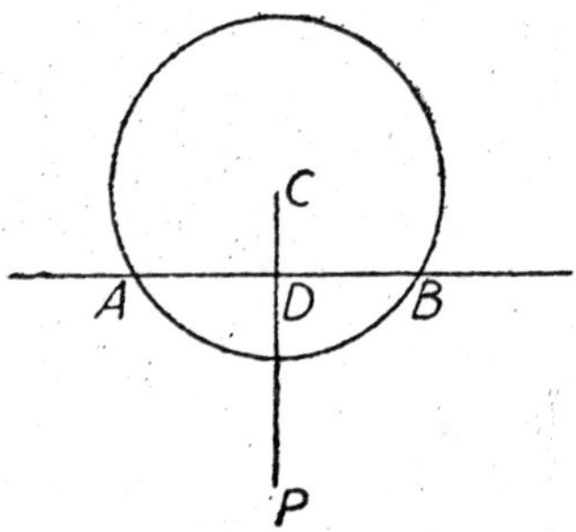

Let it be represented by AB in the figure. Length of the perpendicular CD from the centre $(-g, -f)$ on (1)

$$=\frac{-gx_1-fy_1+g(-g+x_1)+f(-f+y_1)+c}{\sqrt{[(x_1+g)^2+(y_1+f)^2]}}$$

$$=\frac{g^2+f^2-c}{\sqrt{[(x_1+g)^2+(y_1+f)^2]}}.$$

Also $CP=\sqrt{[(x_1+g)^2+(y_1+f)^2]}$

$$\therefore \quad CP.CD=g^2+f^2-c$$
$$=(\text{radius})^2.$$

Two such points P and D are called **Inverse points** with respect to the circle.

5·20. Self-Conjugate Triangle and Polar Circle.

A triangle in which the vertices are the poles of the opposite side with respect to a circle is called **self-conjugate** or **self-polar** with respect to the circle. The circle is called the **polar circle** of the triangle.

Let (x_1, y_1), (x_2, y_2) and (x_3, y_3) be the vertices A, B and C respectively of a triangle ABC and let the given circle be

$$x^2+y^2+2gx+2fy+c=0 \qquad \ldots \quad \ldots(1)$$

Then the polar of A with respect to the circle is

$$xx_1+yy_1+g(x+x_1)+f(y+y_1)+c=0.$$

As this passes through the points B and C,

$$x_1x_2+y_1y_2+g(x_1+x_2)+f(y_1+y_2)+c=0 \ldots \quad \ldots(2)$$

and $$x_1x_3+y_1y_3+g(x_1+x_3)+f(y_1+y_3)+c=0 \ldots \quad \ldots(3)$$

Similarly, because the polar of B passes through C,

we have $$x_2x_3+y_2y_3+g(x_2+x_3)+f(y_2+y_3)+c=0 \ldots \quad \ldots(4)$$

Eliminating g, f and c from (1), (2), (3) and (4), we get the equation to the polar circle in the form

$$\begin{vmatrix} x^2+y^2 & 2x & 2y & 1 \\ x_1x_2 & x_1+x_2 & y_1+y_2 & 1 \\ x_2x_3 & x_2+x_3 & y_2+y_3 & 1 \\ x_3x_1 & x_3+x_1 & y_3+y_1 & 1 \end{vmatrix}=0.$$

5·21. Conjugate Lines.

Condition that the lines

$$p_1x+q_1y+r_1=0 \qquad \ldots \quad \ldots(1)$$

$$p_2x+q_2y+r_2=0 \qquad \ldots \quad \ldots(2)$$

may be conjugate with respect to the circle

$$x^2+y^2+2gx+2fy+c=0.$$

Let the pole of (1) be (x_1, y_1).

Then the polar of (x_1, y_1) with respect to the circle is

$$xx_1+yy_1+g(x+x_1)+f(y+y_1)+c=0,$$

or $$x(x_1+g)+y(y_1+f)+gx_1+fy_1+c=0. \quad \ldots \quad \ldots(3)$$

As (1) and (3) represent the same straight line, comparing coefficients, we get

$$\frac{x_1+g}{p_1}=\frac{y_1+f}{q_1}=\frac{gx_1+fy_1+c}{r_1}=\lambda \text{ (suppose)}$$

The above equations give

$$x_1+g-p_1\lambda=0 \quad \ldots \quad \ldots(4)$$

$$y_1+f-q_1\lambda=0 \quad \ldots \quad \ldots(5)$$

$$yx_1+fy_1+c-r_1\lambda=0 \quad \ldots \quad \ldots(6)$$

Also, by hypothesis, (x_1, y_1) lies on (2).

$$\therefore \quad p_2x_1+q_2y_1+r_2=0 \quad \ldots \quad \ldots(7)$$

Eliminating λ, x_1 and y_1 from (4), (5), (6) and (7), we get

$$\begin{vmatrix} 1 & 0 & g & p_1 \\ 0 & 1 & f & q_1 \\ g & f & c & r_1 \\ p_2 & q_2 & r_2 & 0 \end{vmatrix}=0,$$

which is the required condition.

Ex. 1. *Find the equation of the two tangents to $x^2+y^2=3$ which make an angle of 60° with the axis of X.*

(Calcutta University, 1955)

The direction tangent of the tangent is tan 60°.

Let the line be $$y=x\tan 60+c,$$

or $$y=\sqrt{3}x+c.$$

This line will be a tangent to the circle $x^2+y^2=3$, if

$$c=\pm 3\sqrt{(1+3)}$$
$$=\pm 6.$$

The two tangents are therefore

$$y=\sqrt{3}x+6,$$
$$y=\sqrt{3}x-6.$$

Ex. 2. *A circle passes through $(-1, 1)$, $(0, 6)$ and $(5, 5)$.*

Find the equations to the tangents which are parallel to the line obtained by joining the origin to the centre. (Dacca 1945)

Let the circle be

$$x^2+y^2+2gx+2fy+c=0 \qquad \ldots \quad \ldots(1)$$

It passes through $(-1, 1)$, $(0, 6)$ and $(5, 5)$

$$\therefore \quad -2g+2f+c+2=0 \qquad \ldots \quad \ldots(2)$$

$$12f+c+36=0, \qquad \ldots \quad \ldots(3)$$

$$10g+10f+c+50=0 \qquad \ldots \quad \ldots(4)$$

Solving (2), and (3) and (4) for g, f and c we get

$$f=-3,\ g=-2 \text{ and } c=0.$$

Hence the equation to the circle is

$$x^2+y^2-4x-6y=0$$

The centre of this circle is $(2, 3)$ and its radius is $\sqrt{13}$.

The direction tangent of the line joining the origin and $(2, 3)$ is $\frac{3}{2}$.

The equation to a line parallel to this line is $y=\frac{3}{2}x+c$.

If will be a tangent to the circle

$$x^2+y^2-4x-6y=0$$

if the perpendicular from the centre on this line is equal to the radius of the circle, that is to say,

$$\frac{\frac{3}{2}.2-3+c}{\sqrt{(1+\frac{9}{4})}}=\pm\sqrt{13},$$

or

$$c=\pm\tfrac{13}{2}.$$

$\therefore$ the equation to the tangents is

$$y=\tfrac{3}{2}x\pm\tfrac{13}{2}$$

or

$$2y=3x\pm 13.$$

Ex. 3. *Find the locus of the middle points of all chords of the circle $x^2+y^2=a^2$ which pass through the point (h, k).*

(Calcutta 1952)

Let the coordinates of the middle point of the chord be (l, m).

The line joining the centre $(0, 0)$ to the middle point of

the chord (l, m) has the slope $\frac{m}{l}$. Hence the slope of the chord is $-\frac{l}{m}$. Its equation is therefore

$$y-m=-\frac{l}{m}(x-l).$$

It also passes through the point (h, k).

$$\therefore \quad k-m=-\frac{l}{m}(h-l),$$

or $$km-m^2=-lh+l^2,$$

or $$l^2+m^2-lh-km=0.$$

Hence the locus of (l, m) is

$$x^2+y^2-xh-ym=0.$$

Ex. 4. *Tangents are drawn from the point (h, k) to the circle $x^2+y^2=a^2$; prove that the area of the triangle formed by them and their chord of contact is* $\dfrac{a\,(h^2+k^2-a^2)^{3/2}}{(h^2+k^2)}$.

The equation to the chord of contact of (h, k), PQ is

$$xh+yk-a^2=0 \quad \ldots(1)$$

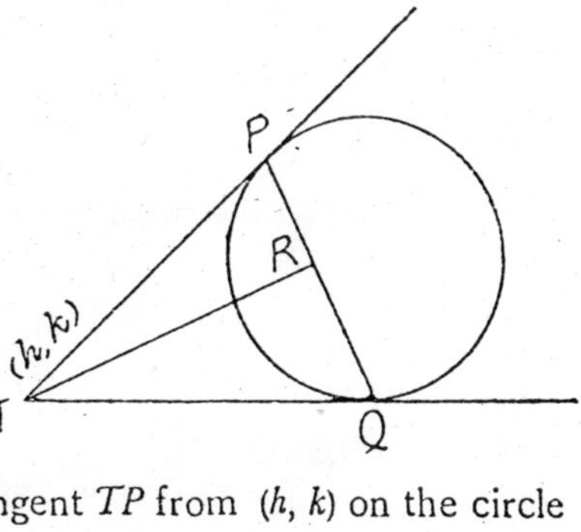

Perpendicular TR from T (h, k) on this chord is given by

$$\frac{h^2+k^2-a^2}{\sqrt{(h^2+k^2)}}.$$

The length of the tangent TP from (h, k) on the circle is given by $\sqrt{(h^2+k^2-a^2)}$.

Now $PR=\sqrt{(TP^2-TR^2)}$

$$=\sqrt{\left\{h^2+k^2-a^2-\frac{(h^2+k^2-a^2)^2}{(h^2+k^2)}\right\}}$$

$$=\sqrt{\left\{\frac{a^2(h^2+k^2-a^2)}{h^2+k^2}\right\}}.$$

The area of the $\triangle TPQ$

$$=TR.RP$$

$$=\frac{\sqrt{(h^2+k^2-a^2)}}{\sqrt{(h^2+k^2)}}.a\sqrt{\left\{\frac{(h^2+k^2-a^2)}{h^2+k^2}\right\}}$$

$$=\frac{a(h^2+k^2-a^2)^{3/2}}{h^2+k^2}.$$

Ex. 5. *Shew that the locus of the poles of the line* $\frac{x}{h}+\frac{y}{k}=1$ *with respect to the circles which touch the rectangular axes is given by the equations*

$$(hx-ky)(hy-kx)+hk(h\pm k)(x\pm y)=0.$$

(*Lucknow 1940*)

As the circle touches both the axes, its centre can be (a, a), $(-a, a)$, $(-a, -a)$ or $(a, -a)$ and its radius a. Hence there will be four circles touching the axes and their equations will be $(x\mp a)^2+(y\mp a)^2=a^2$.

Let the pole of the line $\frac{x}{h}+\frac{y}{k}=1$ be (x_1, y_1).

The polar of (x_1, y_1) with respect to the above circles is

$$xx_1+yy_1\mp a(x+x_1)\mp a(y+y_1)+a^2=0 \quad \ldots \quad \ldots(1)$$

This must be the same as

$$\frac{x}{h}+\frac{y}{k}=1 \quad \ldots \quad \ldots(2)$$

Comparing coefficients in (1) and (2),

$$\frac{x_1\mp a}{1/h}=\frac{y_1\mp a}{1/k}=\pm\frac{(ax_1+ay_1-a^2)}{1} \quad \ldots \quad \ldots(3)$$

Considering one sign only,

$$h(x_1-a)=k(y_1-a)=a(x_1+y_1-a) \quad \ldots \quad \ldots(4)$$

which gives $\qquad h(x_1-a)=k(y_1-a)$

or $\qquad a=\frac{hx_1-ky_1}{h-k}.$

Substituting this value of a in $k\ (y_1-a)=a\ (x_1+y_1-a)$, we get

$$k\left[y_1-\frac{hx_1-ky_1}{h-k}\right]=\frac{hx_1-ky_1}{h-k}\left[x_1+y_1-\frac{hx_1-ky_1}{h-k}\right]$$

or $$k\ [h\ (y_1-x_1)]=\frac{hx_1-ky_1}{h-k}\ (y_1h-x_1k]$$

or $$kh\ (h-k)\ (y_1-x_1)=(hx_1-ky_1)\ (y_1h-x_1k)$$

The locus of (x_1, y_1) is therefore

$$(hx-ky)\ (hy-kx)+hk\ (h-k)\ (x-y)=0.$$

Similarly considering the other signs, the locus is

$$(hx-ky)\ (hy-kx)+hk\ (h+k)\ (x+y)=0.$$

Hence the locus is

$$(hx-ky)\ (hy-kx)+hk\ (h\mp k)\ (x\mp g)=0.$$

Examples IX

1. Find the condition that the straight line $lx+my+n=0$ should touch the circle
$$x^2+y^2+2gx+2fy+c=0. \qquad \textit{(Utkal 1946)}$$

2. Find the equation of the normal to the circle $x^2+y^2-2ax=0$ at the point $\{a\ (1+\cos\alpha),\ a\sin\alpha\}$.

3. Find the equation of the circle whose centre is at the point (h, k) and which passes through the origin, and prove that the equation of the tangent at the origin is $hx+ky=0$.

4. Show that the locus of the feet of the perpendiculars from the point $(a, 0)$ on tangents to the circle $x^2+y^2=a^2$ is $$(x^2+y^2-ax)^2=a^2\ \{y^2+(x-a)^2\}.$$

5. Prove that the polar of the point (p, q) with respect to the circle $x^2+y^2=a^2$ touches $(x-c)^2+(y-d)^2=b^2$, if
$$b^2\ (p^2+q^2)=(a^2-cp-dq)^2.$$

6. The straight line $hx+my=1$ makes with $ax^2+2hxy+by^2=c$ a chord which subtends a right angle at the origin. Shew that $$c\ (h^2+m^2)=a+b.$$

7. Find the value of p so that the straight line $x \cos \alpha + y \sin \alpha = p$ may touch the circle

$$x^2+y^2-2ax \cos \alpha - 2by \sin \alpha = a^2 \sin^2 \alpha.$$

8. The distances from the origin of the centres of three circles $x^2+y^2-2\lambda x=c^2$ (where c is a constant and λ a variable) are in geometrical progression; prove that the lengths of the tangents drawn to them from any point on the circle $x^2+y^2=c^2$ are also in geometrical progression.

9. Shew that the equation to the pair of tangents drawn from the origin to the circle $x^2+y^2+2gx+2fy+c=0$ is

$$(gx+fy)^2=c(x^2+y^2).$$

10. From any point on the circle $x^2+y^2+2gx+2fy+c=0$ tangents are drawn to the circle

$$x^2+y^2+2gx+2fy+c+(g^2+f^2-c)\cos^2 \alpha=0.$$

Prove that the angle between them is 2α.

11. Prove that the locus of the middle points of the chords of contact of tangents from points lying on a straight line is another circle.

12. A variable circle is drawn to touch the x-axis at the origin. Find the locus of the pole of the straight line $lx+my+n=0$ with respect to it.

13. Find the locus of the feet of the normals from (h, k) to the circles $x^2+y^2-2\mu x=0$ where μ is a variable.

14. Shew that if two points are conjugate with respect to a circle, the square of the distance between them is equal to the sum of the squares of tangents from them to the circle.

15. Shew that the lines

$$a_1x+b_1y+c_1=0,$$
$$a_2x+b_2y+c_2=0$$

are conjugate with respect to the circle $x^2+y^2=a^2$ if

$$a^2(a_1a_2+b_1b_2)=c_1c_2.$$

16. A point P is taken on the circle $x^2+y^2=a^2$ and PA and PB are drawn perpendicular to the axes. If (h, k) is the pole of AB, shew that $a^2(h^2+k^2)=h^2k^2$.

17. Shew that the equation of the straight line meeting the circle $x^2+y^2=a^2$ in two points at equal distances d from a point (x_1, y_1) on the circumference is

$$xx_1+yy_1-a^2+\frac{d^2}{2}=0.$$

Apply this method to find the equation of the tangent at (x_1, y_1).

18. A point moves so that the sum of the squares of its distances from the angular points of a triangle is constant; prove that its locus is a circle. *(Calcutta 1952)*

19. Circles are drawn through the point $(0, h)$ touching the circle $x^2+y^2=a^2$. Shew that the locus of the pole of the axis of y with respect to these circles is

$$4a^2(y-h)^4=(a^2-h^2)\{a^2-(h-2y)^2\}x^2.$$

20. Tangents are drawn from any point on the circle $x^2+y^2=R^2$ to the circle $x^2+y^2=r^2$. Shew that if the line joining the points of intersection of these tangents with the first circle also touches the second, then

$$R=2r.$$

5·22. Angle of intersection of two circles.

The angle of intersection of two curves at a point is the angle between the tangents to the curves at that point.

Let the circles be

$$x^2+y^2+2g_1x+2f_1y+c_1=0$$

and

$$x^2+y^2+2g_2x+2f_2y+c_2=0.$$

Let O_1 and O_2 be the centres of the circles and let T be their point of intersection. Let the tangents at T be TT_1 and TT_2. Then the angle between the tangents is

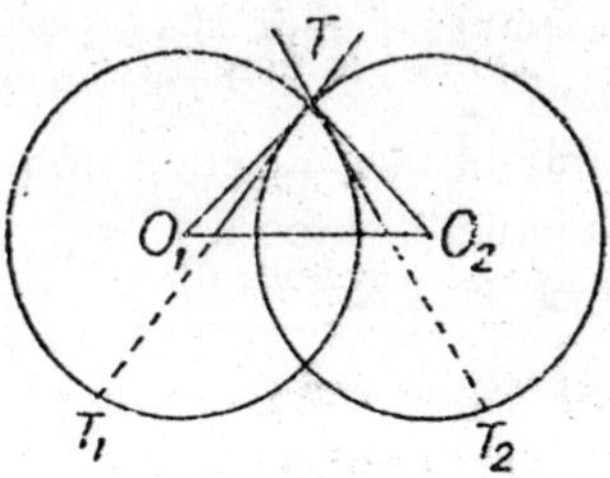

Case I.

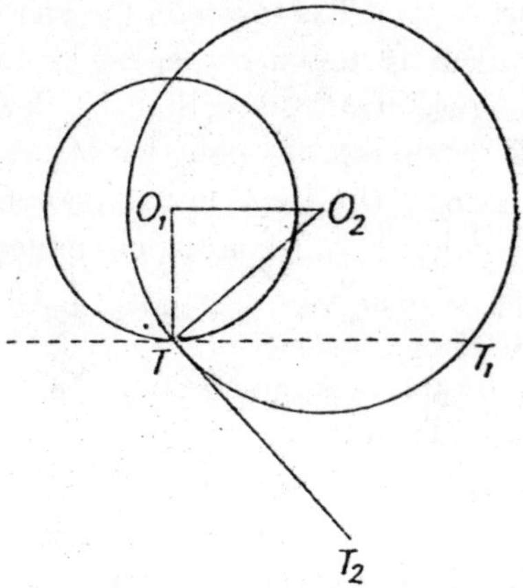

Case II.

supplementary to $\angle O_1TO_2$ as in case I or is equal to $\angle O_1TO_2$ as in case II. In any case,

$$O_1O_2^2=O_1T^2+O_2T^2-2O_1T_1.O_2T \cos O_1TO_2$$

or
$$\cos O_1TO_2=\frac{O_1T^2+O_2T^2-O_1O_2^2}{2O_1T.O_2T}.$$

But
$$O_1T^2=g_1^2+f_1^2-c_1=\text{radius of the circle.}$$

$$O_2T^2=g_2^2+f_2^2-c_2$$

and
$$O_1O_2^2=(g_1-g_2)^2+(f_1-f_2)^2$$

$$\therefore \cos O_1TO_2=\frac{g_1^2+f_1^2-c_1+g_2^2+f_2^2-c_2-(g_1-g_2)^2-(f_1-f_2)^2}{2\sqrt{[(g_1^2+f_1^2-c_1)(g_2^2+f_2^2-c_2)]}}$$

$$=\frac{2g_1g_2+2fg_1f_2-c_1-c_2}{2\sqrt{[(g_1^2+f_1^2-c_1)(g_2^2+f_2^2-c_2)]}}.$$

5·23. Orthogonal intersection.

Two circles intersect orthogonally if they intersect at right angles.

It is obvious from the previous article that if an le O_1TO_2 is a right angle,

$$2g_1g_2+2f_1f_2-c_1-c_2=0$$

which is the condition for orthogonal intersection of the circles.

The condition can also be found independently by taking $\angle O_1TO_2$ a right angle in the ΔO_1TO_2, and

$$\therefore \quad O_1T^2+O_2T^2=O_1O_2^2$$

or $$g_1^2+f_1^2-c_1+g_2^2+f_2^2-c_2=(g-g_2)^2+(f_1-f_2)^2$$

or $$2g_1g_2+2f_1f_2-c_1-c_2=0.$$

5·24. Radical axis of two circles.

The radical axis is the locus of a point which moves so that the lengths of tangents from it to two circles are equal.

Let the circles be

$$x^2+y^2+2g_1x+2f_1y+c_1=0$$

and $$x^2+y^2+2g_2x+2f_2y+c_2=0.$$

Let (h, k) be a point on the locus. Then the lengths of tangents from (h, k) to the circles are equal, that is

$$h^2+k^2+2g_1h+2f_1k+c_1=h^2+k^2+2g_2h+2f_2k+c_2$$

or $$2h\,(g_1-g_2)+2k\,(f_1-f_2)+c_1-c_2=0.$$

Hence the locus of (h, k) is the line

$$2x\,(g_1-g_2)+2y\,(f_1-f_2)+c_1-c_2=0.$$

This is the equation to the radical axis.

Cor. 1. If $S_1\equiv x^2+y^2+2g_1x+2f_1y+c_1=0$

and $$S_2\equiv x^2+y^2+2g_2x+2f_2y+c_2=0,$$

then $$S_1-S_2\equiv 2\,(g_1-g_2)\,x+2\,(f_1-f_2)\,y+c_1-c_2=0.$$

The radical axis is therefore given by $S_1-S_2=0$. But $S_1+\lambda S_2=0$ is a curve passing through the intersection of $S_1=0$ and $S_2=0$ for all values of λ. For $\lambda=-1$, this becomes the equation to the radical axis. Hence radical axis passes through the points of intersection (real or imaginary) of the circles $S_1=0$ and $S_2=0$ or $S_1-S_2=0$ is also the common chord of the two circles.

If the circles touch, the radical axis becomes the

common tangent to these circles.

Cor. 2. The slope of the radical axis is

$$-\frac{g_1-g_2}{f_1-f_2}.$$

Also the slope of the line joining the centres of the two circles is

$$\frac{f_1-f_2}{g_1-g_2}$$

The product of the slopes is equal to -1. Hence the radical axis is perpendicular to the line joining the centres.

Cor. 3. Suppose CD is the common tangent to two circles and the radical axis cuts it at T. Then $TC=TD$. Hence the radical axis bisects the common tangent to two circles.

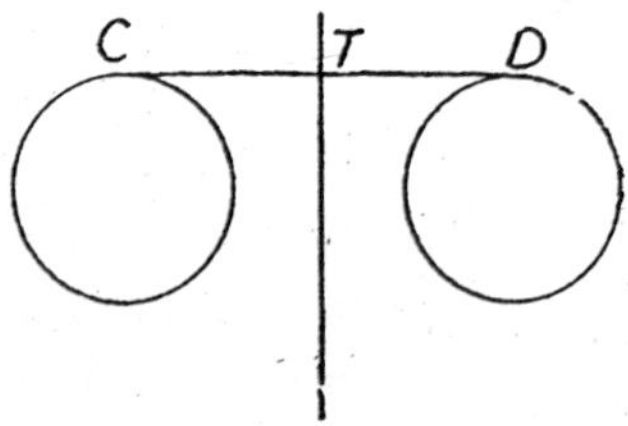

Cor. 4. The radical axes of three circles, taken in pairs, meet in a point, called the radical centre of three circles.

The radical axis of $S_1=0$ and $S_2=0$ is $S_1-S_2=0$ ...(1)

Similarly the radical axes of the other pairs are

$S_2-S_3=0$ ... ...(2)

$S_3-S_1=0$ ... ...(3)

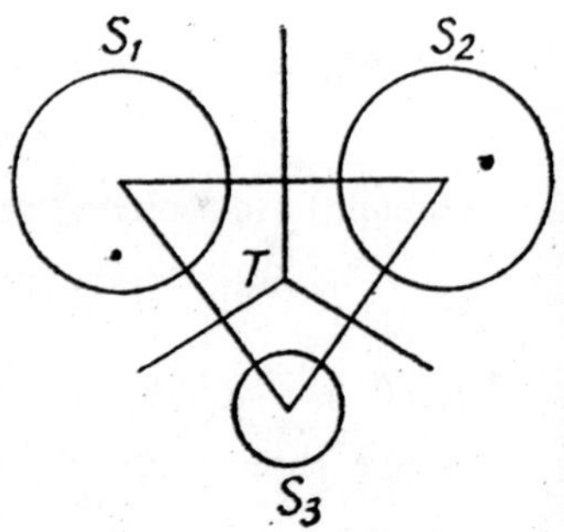

But the sum of (1), (2) and (3) is identically equal to zero. Hence they meet in a point T called the radical centre of the three circles. The lengths of the tangents from T to the circles will be all equal.

5·25. Coaxial circles.

(a) Let $S_1=0$ and $S_2=0$ be the equations to two circles such that

$$S_1 \equiv x^2+y^2+2g_1x+2f_1y+c_1=0,$$
$$S_2 \equiv x^2+y^2+2g_2x+2f_2y+c_2=0.$$

Then $S_1+\lambda S_2 = x^2(1+\lambda)+y^2(1+\lambda)+2(g_1+\lambda g_2)x$
$$+2(f_1+\lambda f_2)y+c_1+\lambda c_2=0.$$

The above equation has also got the coefficients of x^2 and y^2 equal and the coefficient of xy is also absent in it. Hence it represents a circle passing through the points of intersection of the given circles. As these circles pass through the same points of intersection for different values of λ, they represent a system of circles whose radical axis is the same.

Let two circles of this system be

$$S_1+\lambda_1S_2 \equiv x^2(1+\lambda_1)+y^2(1+\lambda_1)+2(g_1+\lambda_1g_2)x$$
$$+2(f_1+\lambda_1f_2)y+c_1+\lambda_1c_2=0 \quad \ldots \quad \ldots (1)$$

and $S_1+\lambda_2S_2 \equiv (1+\lambda_2)x^2+(1+\lambda_2)y^2+2(g_1+\lambda_2g_2)x$
$$+(f_1+\lambda_2f_2)y+c_1+\lambda_2c_2=0 \quad \ldots \quad \ldots (2)$$

Multiplying the first equation by $(1+\lambda_2)$ and the second by $(1+\lambda_1)$ and subtracting (2) from (1), we get

$$2x\{(g_1+\lambda_1g_2)(1+\lambda_2)-(1+\lambda_1)(g_1+\lambda_2g_2)\}$$
$$+2y\{(f_1+\lambda_1f_2)(1+\lambda_2)-(1+\lambda_1)(f_1+\lambda_2f_2)\}$$
$$+(c_1+\lambda_1c_2)(1+\lambda_2)-(c_1+\lambda_2c_2)(1+\lambda_1)=0$$

or $-2x(\lambda_1-\lambda_2)(g_1-g_2)-2y(\lambda_1-\lambda_2)(f_1-f_2)$
$$-(\lambda_1-\lambda_2)(c_1-c_2)=0.$$

or $$2x(g_1-g_2)+2y(f_1-f_2)+(c_1-c_2)=0,$$

which shows that every pair of the system $S_1+\lambda S_2=0$ has the same radical axis as $S_1=0$ and $S_2=0$ and therefore the system of circles coaxial with $S_1=0$ and $S_2=0$ is

$$S_1+\lambda S_2=0$$

(b) Let a circle be

$$S\equiv x^2+y^2+2gx+2fy+c=0$$

and let $P\equiv lx+my+n=0$ be a straight line.

Then $S+\lambda P=x^2+y^2+x\,(2g+\lambda l)+y\,(2f+\lambda m)+c+\lambda n=0$ also represents a circle, which passes through the points of intersection of $S=0$ and $P=0$.

Let $S+\lambda_1 P=0$ and $S+\lambda_1 P=0$ be two circles of this system. Then their radical axis is given by

$$S+\lambda_1 P-(S+\lambda_2 P)=0$$

or $$(\lambda_1-\lambda_2)\,P=0$$

or $$P=0 \qquad \text{as } \lambda_1\neq\lambda_2.$$

Hence the radical axis of this system is $P=0$.

We conclude that if $S=0$ is a circle, and $P=0$ is its radical axis with another circle then $S+\lambda P=0$ is a system of circles coaxial with $S=0$.

5·26. Equation of coaxial circles.

Let y-axis be the radical axis of a given system and let the circle with centre A be one of them. Let a line through the centre A perpendicular to the radical axis be the x-axis. Then the equation to this circle can be written as

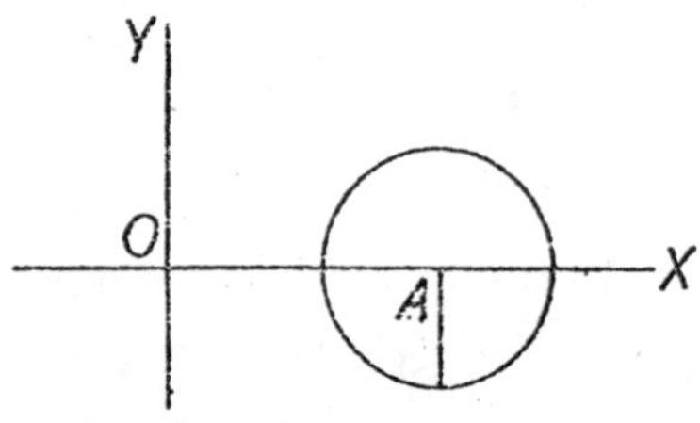

$$x^2+y^2+2gx+c=0.$$

Let the equation to another circle of the same system be

$$x^2+y^2+2g_1x+2f_1y+c_1=0.$$

Their radical axis is

$$2x\,(g_1-g)+2f_1y+c_1-c=0.$$

But this must be $x=0$.

$$\therefore\ f_1=0 \text{ and } c_1=c.$$

Hence the equation to another circle of the system is

$$x^2+y^2+2g_1x+c=0$$

By changing g only in the above equation we shall get all the circles of the same system whose radical axis is the y-axis.

5·26. Coaxial systems.

(a) Let the equation to a system whose radical axis is the y-axis be

$$x^2+y^2+2\lambda x+c=0.$$

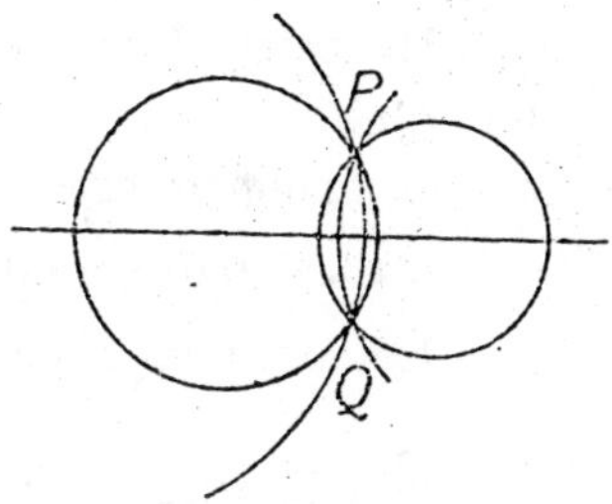

Here λ is the variable parameter and c a constant. The centres of these circles all lie on the x-axis. The coordinates of centre are $(-\lambda, 0)$ and the radius is $\sqrt{(\lambda^2-c)}$.

The points of intersection of this circle with the radical axis $x=0$ are given by $y^2+c=0$ or $y^2=-c$

or $$y=\pm\sqrt{(-c)}.$$

If c is negative, the ordinates of the points of intersection are $\pm\sqrt{c}$ for all circles of this system, that is, all the circles pass through the two real points of intersection P and Q as shown in the figure. This system of circles is called the intersecting system.

(b) When c is positive, $y=\pm i\sqrt{c}$ which means that the points of intersection of the circles with the y-axis are imaginary. The circles will still have real centres on the x-axis. Such a system of circles is called the non-intersecting system.

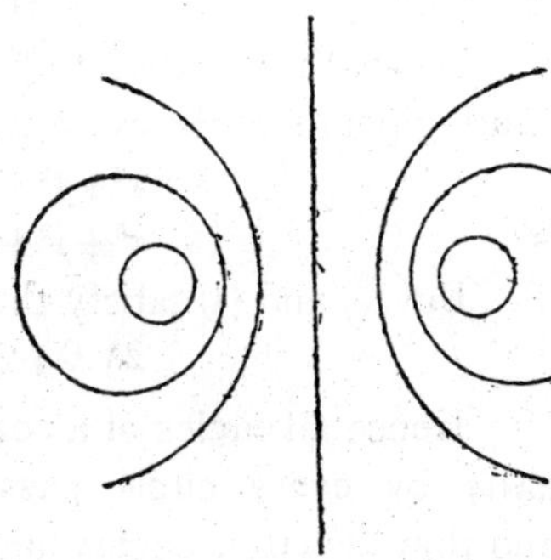

5·27. Limiting points of a coaxial system.

The circles of the system

$$x^2+y^2+2\lambda x+c=0$$

whose radius is zero are point circles and are called the limiting points of the coaxial system. In this case

$$\lambda^2-c=0. \quad \text{or} \quad \lambda=\pm\sqrt{c}.$$

Hence the limiting points are $(\pm\sqrt{c}, 0)$. They are two for each system on each side of the radical axis and are real or imaginary according as c is positive or negative.

5·28. Orthogonal circles of a coaxial system.

Let a system of circles be

$$x^2+y^2+2\lambda x+c=0.$$

The limiting points of this system are

$$\sqrt{c},\ 0 \text{ and } -\sqrt{c},\ 0.$$

Let $x^2+y^2+2gx+2fy+c_1=0$ be another circle.

As it passes through $\sqrt{c}, 0$ and $-\sqrt{c}, 0$,

$$c+2g\sqrt{c}+c_1=0 \qquad \ldots \quad \ldots(1)$$

$$c-2g\sqrt{c}+c_1=0 \qquad \ldots \quad \ldots(2)$$

Subtracting (2) from (1),

$$4g\sqrt{c}=0$$

or $$g=0 \qquad \text{as } c\neq 0.$$

Also $$c_1=-c.$$

Hence the equation to the circle passing through the limiting points of the system

$$x^2+y^2+2\lambda x+c=0 \qquad \ldots \quad \ldots(3)$$

is $$x^2+y^2+2fy-c=0 \qquad \ldots \quad \ldots(4)$$

But (3) and (4) satisfy the condition of orthogonality as

$$2\lambda.0+2f.0=c-c.$$

Hence all circles of a coaxial system are cut orthogonally by every circle passing through the limiting points and that the latter circles form a coaxial system orthogonal to the former system.

5·29. Circles cutting two circles orthogonally.

Let two circles of the same system be

$$x^2+y^2+2\lambda_1 x+c=0 \qquad \ldots \quad \ldots(1)$$

and

$$x^2+y^2+2\lambda_2 x+c=0 \qquad \ldots \quad \ldots(2)$$

whose centres are on the x-axis and the radical axis is the y-axis.

Let the equation of another circle be

$$x^2+y^2+2gx+2fy+c_1=0 \qquad \ldots \quad \ldots(3)$$

As (3) cuts (1) and (2) orthogonally,

$$2g.\lambda_1+2f.0=c+c_1 \qquad \ldots \quad \ldots(4)$$

and

$$2g.\lambda_2+2f.0=c+c_1 \qquad \ldots \quad \ldots(5)$$

The equations (4) and (5) give

$$g=0 \text{ and } c_1=-c.$$

Hence the circle cutting (1) and (2) orthogonally is given by

$$x^2+y^2+2fy-c=0$$

which is another system with parameter f.

5·30. Orthogonal coaxial systems.

(a) We have proved that the system orthogonal to

$$x^2+y^2+2\lambda x+c=0 \qquad \ldots \quad \ldots(1)$$

is given by

$$x^2+y^2+2fy-c=0. \qquad \ldots \quad \ldots(2)$$

The radical axis of (1) is the y-axis and the centres of (2) lie on the y-axis. Hence the radical axis of one is the line of centres of the other.

(b) The limiting points of the first system are $(\pm\sqrt{c}, 0)$ and it has been shown that the second system passes through them. Similarly it can be shown that the first system passes through the limiting points of the second system $(0, \pm\sqrt{-c})$.

If c is positive, the limiting points of the first system are real and of the second system imaginary and if c is negative, the limiting points of the first system are imaginary and of the second system real. So if one system is of intersecting type, the other is of the non-intersecting type.

(c) The centre of the first system circle (1) is $-\lambda, 0$.

The length of the tangent from it to a circle (2) of the other system is $\sqrt{(\lambda^2-c)}$ which is the same as the radius of the first circle.

Similarly it can be shown that the length of the tangent from the centre of the second circle to the first circle is equal to the radius of the second circle.

5·31. Common tangents to two circles.

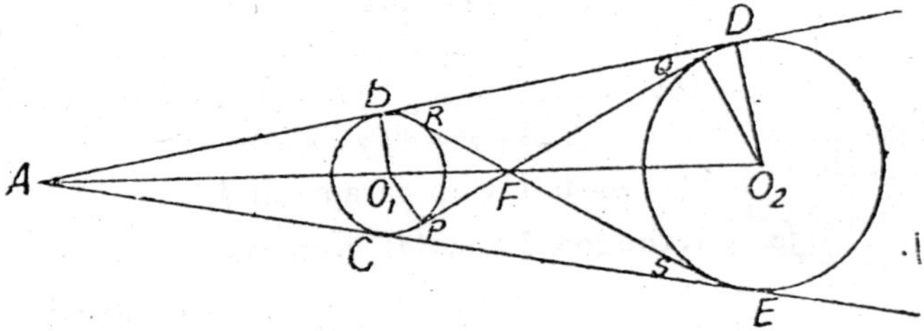

Let BD and CE be direct common tangents to two given circles whose centres are O_1 and O_2 and let PQ and RS be their transverse common tangents.

From similar Δs O_1PF and O_2QF,

$$\frac{PF}{FQ}=\frac{O_1P}{O_2Q}=\frac{r_1}{r_2}=\frac{O_1F}{O_2F}.$$

Hence F divides the lines O_1O_2 in the ratio of the radii of the given circles internally.

If the points O_1 and O_2 are given, the point F can be formed and the equation to a line passing through F and tangent to any one of the circles, gives the equation to the transverse common tangent to the circles.

Similarly A is the point of external division of O_1O_2 in the ratio of the radii of the given circles.

Hence point A can be found and the equation to a line through A,t angent to any one of the circles, gives the equation to the direct common tangent to the circles.

The points A and F are called the centres of similitude of the two circles and the circle on AF as diameter is called the circle of similitude.

5·33. Parametric Coordinates.

Let P be a point on the circle $x^2+y^2=a^2$.

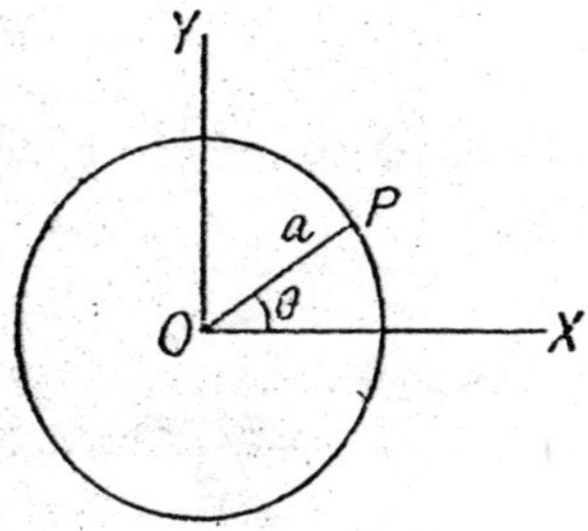

Let OP make an angle θ with the x-axis. If the coordinates of P are (x, y),

$$x=a \cos \theta,$$

$$y=a \sin \theta.$$

Hence any point on the circle $x^2+y^2=a^2$ can be taken as $(a \cos \theta, a \sin \theta)$. Here θ is the variable and the coordinates of two points on a circle can be taken as $(a \cos \theta_1, a \sin \theta_1)$ and $(a \cos \theta_2, a \sin \theta_2)$ by varying the parameter θ.

Sometimes the parametric coordinates are more useful in solving a question as they involve only one variable θ instead of two x and y.

5·33. Equation of a straight line joining two points on a circle.

Let the equation to the circle be $x^2+y^2=a^2$ and let the points P and Q on it be $(a \cos \alpha_1, a \sin \alpha_1)$ and $(a \cos \alpha_2, a \sin \alpha_2)$ respectively.

The equation to the line PQ is given by

$$y-a \sin \alpha_1=\frac{a \sin \alpha_2-a \sin \alpha_1}{a \cos \alpha_2-a \cos \alpha_1}(x-a \cos \alpha_1)$$

$$=\frac{2\cos\frac{\alpha_1+\alpha_2}{2}\sin\frac{\alpha_2-\alpha_1}{2}}{2\sin\frac{\alpha_1+\alpha_2}{2}\sin\frac{\alpha_1-\alpha_2}{2}}(x_1-a\cos\alpha_1)$$

$$=-\frac{\cos\frac{(\alpha_1+\alpha_2)}{2}}{\sin\frac{\alpha_1+\alpha_2}{2}}(x-a\cos\alpha_1)$$

or $$x\cos\frac{a_1+\alpha_2}{2}+y\sin\frac{\alpha_1+\alpha_2}{2}=a\sin\alpha_1\sin\frac{\alpha_1+\alpha_2}{2}$$
$$+a\cos\alpha_1\cos\frac{\alpha_1+\alpha_2}{2}$$

or $$x\cos\frac{\alpha_1+\alpha_2}{2}+y\sin\frac{\alpha_1+\alpha_2}{2}=a\cos\frac{\alpha_1-\alpha_2}{2}\quad\ldots\quad\ldots(1)$$

Cor. If $\alpha_2\to\alpha_1$ in (1) the above equation becomes the equation of the tangent at the point $(a\cos\alpha_1, a\sin\alpha_1)$. The equation of the tangent is therefore

$$x\cos\alpha_1+y\sin\alpha_1=a.$$

5·34. Polar equation of a circle.

(a) Let the origin be a point on the circle and let the diameter through O be the initial line. Let $P(r, \theta)$ be a point on the circle.

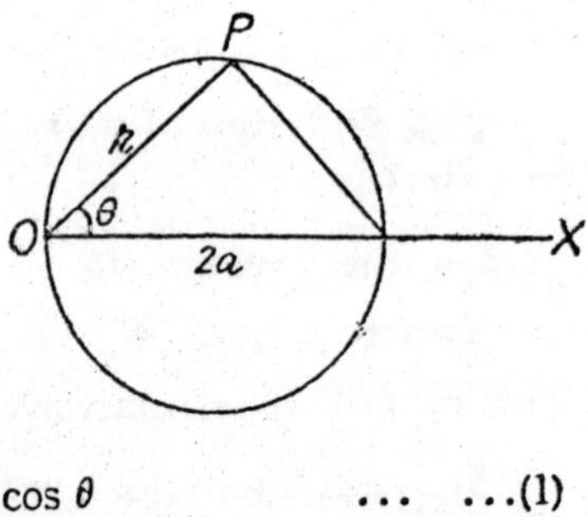

From the figure if $2a$ is the diameter of the circle,

$$r=2a\cos\theta\quad\ldots\quad\ldots(1)$$

which is a relation between r and θ, the coordinates of any point on the locus. Hence (1) is the equation to the circle.

(b) Let the origin O be taken more generally anywhere in the plane of the circle. Let a be the radius of the circle and let the line joining O to the centre C make an angle α with the initial line and let OC be equal to r_1 or let (r_1, α) be the coordinates of the centre C.

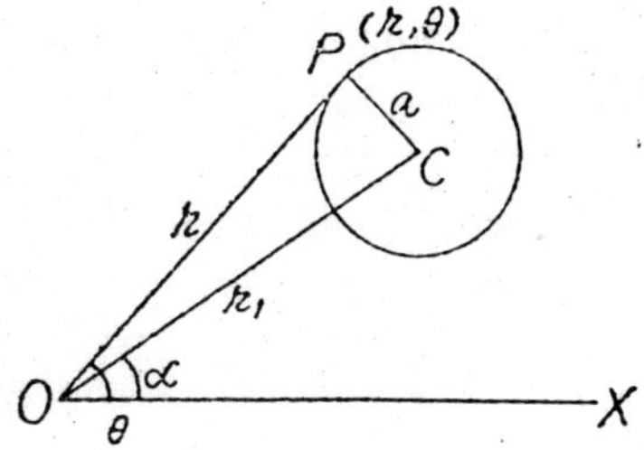

Let P be a point (r, θ) on the circle.

Then $\angle POC = \theta - \alpha$.

From $\triangle POC$,

$$PC^2 = OP^2 + OC^2 - 2OP.OC.\cos(\theta - \alpha)$$

or $$a^2 = r^2 + r_1^2 - 2rr_1 \cos(\theta - \alpha)$$

or $$r^2 + r_1^2 - 2rr_1 \cos(\theta - \alpha) - a^2 = 0$$

which is the equation to the circle.

Cor. 1. If the initial line passes through the centre C, $\alpha = 0$ and the equation to the circle becomes

$$r^2 + r_1^2 - 2rr_1 \cos\theta - a^2 = 0.$$

Cor. 2. If the origin is on the circle, $r_1 = a$ and the equation becomes

$$r = 2r_1 \cos(\theta - \alpha).$$

Cor. 3. If the origin lies on the circle and the centre lies on the initial line, $r_1 = a$ and $\alpha = 0$ and the equation becomes $$r = 2r_1 \cos\theta$$ which has been found before.

Cor. 4. If the origin be on the centre of the circle, its equation is $r = a$.

5·35. Equation of the line joining two points on the circle r = 2a cos θ.

Let the points P and Q be (r_1, θ_1) and (r_2, θ_2) respectively.

The equation to PQ is

$$\frac{\sin(\theta_2-\theta_1)}{r}=\frac{\sin(\theta_1-\theta)}{r_2}+\frac{\sin(\theta-\theta_2)}{r_1}=0 \quad \ldots \quad \ldots(1)$$

Since (r_1, θ_1) and (r_2, θ_2) lie on the circle

$$r=2a\cos\theta,$$

$$r_1=2a\cos\theta_1 \quad \ldots \quad \ldots(2)$$

and
$$r_2=2a\cos\theta_2. \quad \ldots \quad \ldots(3)$$

The equation (1) then becomes

$$\frac{\sin(\theta_2-\theta_1)}{r}+\frac{\sin(\theta_1-\theta)}{2a\cos\theta_2}+\frac{\sin(\theta-\theta_2)}{2a\cos\theta_1}=0$$

or
$$\frac{\sin(\theta_2-\theta_1)}{r}=\frac{\sin(\theta-\theta_1)\cos\theta_1+\sin(\theta_2-\theta)\cos\theta_2}{2a\cos\theta_1\cos\theta_2}$$

or
$$\frac{1}{r}=\frac{\cos(\theta_1+\theta_2-\theta)}{2a\cos\theta_1\cos\theta_2}$$

or
$$r\cos(\theta_1+\theta_2-\theta)=2a\cos\theta_1\cos\theta_2$$

Cor. If θ_1 coincides with θ_2, the equation to the tangent at θ_1 is $r\cos(2\theta_1-\theta)=2a\cos^2\theta_1$.

Ex. 1. *Prove that the locus of the point of intersection of the lines*

$$x\cos\theta+y\sin\theta=a$$

and
$$x\sin\theta-y\cos\theta=b$$

is a circle, θ being the parameter.

Let the point of intersection be (h, k). As (h, k) lies on both the lines,

$$h\cos\theta+k\sin\theta=a, \quad \ldots \quad \ldots(1)$$

$$h\sin\theta-k\cos\theta=b. \quad \ldots \quad \ldots(2)$$

Squaring and adding (1) and (2), we get

$$h^2+k^2=a^2+b^2.$$

Hence the locus of (h, k) is

$$x^2+y^2=a^2+b^2$$

which is a circle.

Ex. 2. *If the equations to two circles whose radii are a and a' be $S=0$, $S'=0$, show that the circles*

$$\frac{S}{a} \pm \frac{S'}{a'} = 0$$

intersect orthogonally. (*Agra 1955*)

Let the equations to the circles be

$$S \equiv x^2 + y^2 + 2g_1 x + 2f_1 y + c_1 = 0$$

and $$S' \equiv x^2 + y^2 + 2g_2 x + 2f_2 y + c_2 = 0$$

so that their radii are

$$a = \sqrt{(g_1^2 + f_1^2 - c_1)} \qquad \ldots \quad \ldots (1)$$

$$a' = \sqrt{(g_2^2 + f_2^2 - c_2)} \qquad \ldots \quad \ldots (2)$$

Now $$\frac{S}{a} + \frac{S'}{a'} \equiv x^2 \left(\frac{1}{a} + \frac{1}{a'}\right) + y^2 \left(\frac{1}{a} + \frac{1}{a'}\right)$$

$$+ 2x \left(\frac{g_1}{a} + \frac{g_2}{a'}\right) + 2y \left(\frac{g_2}{a} + \frac{g_2}{a'}\right) + \frac{c_1}{a'} + \frac{c_2}{a'} = 0.$$

Making the coefficients of x^2 and y^2 each equal to unity, the above equation becomes

$$x^2 + y^2 + 2x \left(\frac{g_1 a' + g_2 a}{a' + a}\right) + 2y \left(\frac{f_1 a' + f_2 a}{a' + a}\right)$$

$$+ \frac{c_1 a' + c_2 a}{a' + a} = 0 \qquad \ldots \quad \ldots (3)$$

Similarly the other equation $\frac{S}{a} - \frac{S'}{a'} = 0$ is equivalent to

$$x^2 + y^2 + 2x \left(\frac{g_1 a' - g_2 a}{a' - a}\right) + 2y \left(\frac{f_1 a' - f_2 a}{a' - a}\right)$$

$$+ \frac{c_1 a' - c_2 a}{a' - a} = 0 \qquad \ldots \quad \ldots (4)$$

(3) and (4) will be orthogonal to each other if

$$2 \frac{(g_1 a' + g_2 a)}{a' + a} \frac{(g_1 a' - g_2 a)}{a' - a} + \frac{2 (f_1 a' + f_2 a)}{(a' + a)} \cdot \frac{(f_1 a' - f_2 a)}{(a' - a)}$$

$$= \frac{c_1 a' + c_2 a}{a' + a} + \frac{c_1 a' - c_2 a}{a' - a}$$

or $$2 \frac{(g_1^2 a'^2 - g_2^2 a^2)}{a'^2 - a^2} + 2 \frac{(f_1^2 a'^2 - f_2 a^2)}{a'^2 - a^2} - \frac{2 (c_1^2 a'^2 - c_2^2 a^2)}{a'^2 - a^2} = 0$$

or $$a'^2(g_1^2+f_1^2-c_1^2)-a^2(g_2^2+f_2^2-c_2^2)=0$$

or $$a'^2a^2-a^2a'^2=0 \text{ from (1) and (2),}$$

which is true.

Ex. 3. *Find the radical axis of the circles*

$$x^2+y^2+2gx+2fy+c=0$$

and $$x^2+y^2+2g'x+2f'y+c'=0.$$

Show that the first of these circles will bisect the circumference of the other if $2g'(g-g')+2f'(f-f')=c-c'$.

(Aligarh 1953, Travancore 1946)

The equation to the radical axis of the circles is

$$2x(g-g')+2y(f-f')+c-c'=0.$$

If the first circle bisects the circumference of the second, the radical axis must pass through the centre of the second circle $(-g', -f')$. Hence

$$-2g'(g-g')-2f'(f-f')+c-c'=0$$

or $$2g'(g-g')+2f'(f-f')=c-c'.$$

Ex. 4. *If the four points in which the two circles*

$$S\equiv x^2+y^2+ax+by+c=0$$

and $$S_1\equiv x^2+y^2+a_1x+b_1y+c_1=0$$

are intersected by the straight lines

$$P\equiv Ax+By+c=0$$

and $$P_1\equiv A_1x+B_1y+C_1=0$$

respectively, lie on a circle, show that

$$\begin{vmatrix} a-a_1 & b-b_1 & c-c_1 \\ A & B & C \\ A_1 & B_1 & C_1 \end{vmatrix}=0.$$

(Agra 1940, 1951)

The equation to the circle passing through the intersec- of $S=0$ and $P=0$ is $S+\lambda P=0$.

or $$x^2+y^2+(a+\lambda A)x+(b+\lambda B)y+c+\lambda C=0 \quad \ldots(1)$$

Similarly the circle passing through the intersection

of $S_1=0$ and $P_1=0$ is given by

$$x^2+y^2+(a_1+\mu A_1)\,x+(b_1+\mu B_1)\,y+c_1+\mu C_1=0\ldots \qquad \ldots(2)$$

If (1) and (2) represent the same circle

$$\frac{a+\lambda A}{a_1+\mu A_1}=\frac{b+\lambda B}{b_1+\mu B_1}=\frac{c+\lambda C}{c_1+\mu C_1}=1.$$

Hence

$$a+\lambda A=a_1+\mu A_1$$

or
$$a-a_1+\lambda A-\mu A_1=0 \qquad \ldots \quad \ldots(3)$$

Similarly,
$$b-b_1+\lambda B-\mu B_1=0 \qquad \ldots \quad \ldots(4)$$

and
$$c-c_1+\lambda C-\mu C_1=0 \qquad \ldots \quad \ldots(5)$$

Eliminating λ and μ from (3), (4) and (5), we get

$$\begin{vmatrix} a-a_1 & A & A_1 \\ b-b_1 & B & B_1 \\ c-c_1 & C & C_1 \end{vmatrix}=0.$$

Changing rows into columns, we get

$$\begin{vmatrix} a-a_1 & b-b_1 & c-c_1 \\ A & B & C \\ A_1 & B_1 & C_1 \end{vmatrix}=0.$$

Ex. 5. *Find the radical centre of the three circles*

$$x^2+y^2=9,\ x^2+y^2-2x-2y=5$$

and
$$x^2+y^2+4x+6y=19.$$
(Utkal 1946)

The radical axis of the first two circles is

$$x+y=2. \qquad \ldots \quad \ldots(1)$$

The radical axis of the last two circles is

$$3x+4y=7. \qquad \ldots \quad \ldots(2)$$

The point of intersection of (1) and (2) is (1, 1) which is the radical centre of the circles.

Ex. 6. *Find the equation of the circle which cuts orthogonally* *each of the three circles*

$$x^2+y^2+2x+17y+4=0,$$
$$x^2+y^2+7x+6y+11=0,$$
$$x^2+y^2-x+22y+3=0. \quad \textit{(Allahabad 1948)}$$

Let the required circle be

$$x^2+y^2+2gx+2fy+c=0 \qquad \ldots \quad \ldots(1)$$

As it cuts the given circles orthogonally,

$$2g+2f.\tfrac{17}{2}=c+4$$

or
$$-4+2g+17f-c=0 \qquad \ldots \quad \ldots(2)$$

and
$$-11+7g+6f-c=0 \qquad \ldots \quad \ldots(3)$$

and
$$-3-g+22f-c=0. \qquad \ldots \quad \ldots(4)$$

From (2) and (3), we get

$$7-5g+11f=0. \qquad \ldots \quad \ldots(5)$$

From (3) and (4) similarly,

$$8-8g+16f=0$$

or
$$1-g+2f=0. \qquad \ldots \quad \ldots(6)$$

Solving (5) and (6), we get

$$g=-3, f=-2.$$

Substituting these values of g and f in (2), we get $c=-44$.

Hence the required equation to the circle is

$$x^2+y^2-6x-4y-44=0.$$

Another method.

The radical axis of the first two circles is

$$5x-11y+7=0 \qquad \ldots \quad \ldots(1)$$

The radical axis of the last two circles is

$$8x-16y+8=0$$

or
$$x-2y+1=0. \qquad \ldots \quad \ldots(2)$$

Solving (1) and (2), the radical centre is (3, 2).

The length of the tangent from (3, 2) on the first circle

$$=\sqrt{(9+4+6+34+4)}=\sqrt{(57)}.$$

Henee the equation to the required circle is

$$(x-3)^2+(y-2)^2=57$$

or
$$x^2+y^2-6x-4y-44=0.$$

Ex. 7. *Find the coordinates of the limiting points of the coaxial system to which the circles*

$$x^2+y^2+4x+2y+5=0$$

and $$x^2+y^2+2x+4y+7=0 \text{ belong.}$$ *(Utkal 1946)*

The radical axis of the given circles is $x-y-1=0$.

Hence the equation to a system of circles coaxial with the given circles is

$$x^2+y^2+2x+4y+7+\lambda(x-y-1)=0$$

or $$x^2+y^2+x(2+\lambda)+y(4-\lambda)+7-\lambda=0. \quad \ldots \quad \ldots (1)$$

The radius

$$=\sqrt{\left\{\left(\frac{2+\lambda}{2}\right)^2+\left(\frac{4-\lambda}{2}\right)^2+\lambda-7\right\}}$$

$$=\tfrac{1}{2}\sqrt{(2\lambda^2-8)}.$$

Hence the limiting points are given by the value of λ such that $$2\lambda^2-8=0.$$

$$\therefore \quad \lambda=\pm 2.$$

Substituting $\lambda=+2$ in (1), we get

$$x^2+y^2+4x+2y+5=0$$

or $$(x+2)^2+(y+1)^2=0$$

which gives the limiting point to be $(-2, -1)$.

Substituting $\lambda=-2$ in (1) again, we get

$$x^2+y^2+6y+9=0$$

or $$x^2+(y+3)^2=0$$

which gives $(0, -3)$ as the limiting point.

Ex. 8. *Find the equation to the common tangents to the circles*

$$x^2+y^2=c^2 \text{ and } (x-a)^2+y^2=b^2.$$

The centres of the circles are $(0, 0)$ and $(a, 0)$ and their radii are c and b.

The point of intersection of direct common tangents (P) is the point of external division of the line joining the centres in the ratio $c : b$. Hence the coordinates of P are

$$\frac{ca-b.0}{c-b}, \frac{c.0-b.0}{c-b}$$

or
$$\frac{ca}{c-b}, 0.$$

Similarly the point of intersetion of transverse common tangents, Q, the point of internal division of the line joining the centres in the ratio $c : b$.

Hence the coordinates of ϕ are

$$\frac{ca+b.0}{c+b}, \frac{c.0+b.0}{c+b}$$

or
$$\frac{ca}{c+b}, 0.$$

Any line through P is

$$y=m\left(x-\frac{ca}{c-b}\right).$$

The perpendicular from the point (0, 0) on this line should be equal to c.

$$\therefore \quad \frac{m.\dfrac{ca}{c-b}}{\sqrt{(1+m^2)}}=c$$

or
$$\frac{m^2.a^2}{(c-b)^2}=1+m^2$$

or
$$m^2\left[\frac{a^2}{(c-b)^2}-1\right]=1$$

or
$$m^2\frac{(a-c+h)(a+c-h)}{(c-b)^2}=1$$

or
$$m=\pm\frac{c-b}{(a-c+b)(a+c-b)}$$

$$=\pm\frac{b-c}{\sqrt{[a^2-(b-c)^2]}}.$$

Hence the equation to direct common tangent is

$$y=\pm\frac{(b-c)}{\sqrt{[a^2-(b-c)^2]}}\left(x-\frac{ca}{c-b}\right).$$

Similarly the equation to the transverse common tangent is

$$y=m\left(x-\frac{ca}{c+b}\right)$$

where

$$\frac{m.\dfrac{ca}{c+b}}{\sqrt{(1+m^2)}}=c$$

or

$$(1+m^2)=m^2\ \frac{a^2}{(b+c)^2}$$

or

$$m^2\left\{\frac{a^2}{(b+c)^2}-1\right\}=1$$

or

$$m=\pm\frac{b+c}{\sqrt{[a^2-(b+c)^2]}}.$$

Ex. 9. *Prove that the circle of similitude of the two circles* $x^2+y^2-2gx+c=0$ *and* $x^2+y^2-2g_1x+c=0$ *is the circle*

$$x^2+y^2-\frac{2\ (gg_1+c)}{g+g_1}\ x+c=0$$

The centres of the circles are $(g, 0)$, $(g_1, 0)$ and their radii are $\sqrt{(g^2-c)}$ and $\sqrt{(g_1^2-c)}$.

$\therefore$ the centres of similitude are the points

$$\left\{\frac{g_1\sqrt{(g^2-c)}-g\sqrt{(g_1^2-c)}}{\sqrt{(g^2-c)}-\sqrt{(g_1^2-c)}},\ 0\right\}$$

and

$$\left\{\frac{g_1\sqrt{(g^2-c)}+g\sqrt{(g_1^2-c)}}{\sqrt{(g^2-c)}+\sqrt{(g_1^2-c)}},\ 0\right\}.$$

Hence the equation of the circle of similitude is

$$\left\{x-\frac{g_1\sqrt{(g^2-c)}-g\sqrt{(g_1^2-c)}}{\sqrt{(g^2-c)}-\sqrt{(g_1^2-c)}}\right\}\left\{x-\frac{g_1\sqrt{(g^2-c)}+g\sqrt{(g_1^2-c)}}{\sqrt{(g^2-c)}+\sqrt{(g_1^2-c)}}\right\}+y^2=0$$

or

$$x^2+y^2-2x\left(\frac{gg_1+c}{g+g_1}\right)+c=0.$$

Ex. 10. *If the circle* $x^2+y^2-2\alpha x-2\beta y+c=0$ *is the circle of a coaxial system, having the origin as a limiting point, show that* *the other limiting point is* $\left(\frac{\alpha c}{\alpha^2+\beta^2}, \frac{\beta c}{\alpha^2+\beta^2}\right)$ *and prove that the equations of this system and the conjugate system are respectively*

$$\lambda(x^2+y^2)-2\alpha x-2\beta y+c=0,$$
$$(\alpha+\mu\beta)(x^2+y^2)-c(x+\mu y)=0$$

where λ *and* μ *are variable parameters.*

If (0, 0) is one of the limiting points of the system

$$x^2+y^2-2\alpha x-2\beta y+c=0,$$
$$c=\alpha^2+\beta^2 \text{ and } \alpha=0, \beta=0.$$

Hence one of the circles is $x^2+y^2=0$.

Hence the equation to the coaxial system is

$$x^2+y^2-2\alpha x-2\beta y+c+k(x^2+y^2)=0$$

or $$(1+k)(x^2+y^2)-2\alpha x-2\beta y+c=0$$

or $$\lambda(x^2+y^2)-2\alpha x-2\beta y+c=0 \qquad \ldots(1)$$

where $$\lambda=1+k.$$

The centre of this circle is $(\alpha/\lambda, \beta/\lambda)$ and its radius is

$$\sqrt{\left(\frac{\alpha^2}{\lambda^2}+\frac{\beta^2}{\lambda^2}-\frac{c}{\lambda}\right)}.$$

If the radius is 0, $\lambda=\dfrac{\alpha^2+\beta^2}{c}$.

Hence the limiting points become $(\alpha/\lambda, \beta/\lambda)$.

$$\left(\frac{\alpha c}{\alpha^2+\beta^2}, \frac{\beta c}{\alpha^2+\beta^2}\right).$$

Let the equation to the circle which cuts (1) orthogonally be

$$x^2+y^2+2gx+2fy+t=0 \qquad \ldots(2)$$

when the condition of orthogonality gives

$$-2\frac{\alpha}{\lambda}.g-\frac{2\beta}{\lambda}.f=\frac{c}{\lambda}+t$$

or $$-2\alpha g-2\beta f=c+\lambda t$$

or $$-2\alpha g-2\beta f-c=\lambda.t$$

and this must be true for all values of λ.

Hence if $t=0$, $2f=-\frac{2\alpha g+c}{\beta}$.

Hence the equation to the orthogonal system becomes

$$x^2+y^2+2gx-\left(\frac{2\alpha g+c}{\beta}\right)y=0$$

or

$$x^2+y^2+2g\left(x-\frac{2\alpha g+c}{2\beta g}y\right)=0.$$

Writing $-\frac{2\alpha g+c}{2\beta g}=\mu$ or $2g=-\frac{c}{\alpha+\beta\mu}$, we get

$$x^2+y^2-\frac{c}{\alpha+\beta\mu}(x+\mu y)=0$$

or

$$(\alpha+\beta\mu)(x^2+y^2)-c(x+\mu y)=0.$$

Ex. 10. *Find the coordinates of the centre and radii of the circle $r^2+ar\cos\theta+b^2=0$.*

Let (R, ϕ) be the polar coordinates of the centre and k its radius. Then the given equation should be identical with

$$r^2-2rR\cos(\theta-\phi)+R^2-k^2=0.$$

$$\therefore\quad -2R\cos\phi=a \text{ and } 2R\sin\phi=0.$$

$$R^2-k^2=b^2.$$

$$\therefore\quad \tan\phi=0,\ R^2=a^2/4.$$

$$\therefore\quad k^2=(a^2/4)-b^2.$$

$\therefore$ the centre of the circle is $(a/2, 0)$ and the radius is

$$\frac{\sqrt{(a^2-4b^2)}}{2}.$$

Ex. 11. *If the line $1/r=a\cos\theta+b\sin\theta$ touch the circle $r=2c\cos\theta$, show that $c^2b^2+2ca=1$.*

The centre of the circle is $(c, 0)$ and its radius is c.

The equation of the straight line in cartesian coordinates becomes $1=ax+by$.

The perp. from $(c, 0)$ on this line should be equal to the radius c.

$$\therefore \quad \frac{ac-1}{\sqrt{(a^2+b^2)}}=c$$

or $$(ac-1)^2=c^2(a^2+b^2)$$

or $$-2ac+1=b^2c^2$$

or $$b^2c^2+2ac=1.$$

Ex. 12. *A circle passes through the point (r_1, θ_1) and touches the initial line at a distance c from the pole. Show that its polar equation is*

$$\frac{r^2-2cr\cos\theta+c^2}{r\sin\theta}=\frac{r_1{}^2-2cr_1\cos\theta_1+c^2}{r_1\sin\theta_1}. \qquad \textit{(Luck. '46)}$$

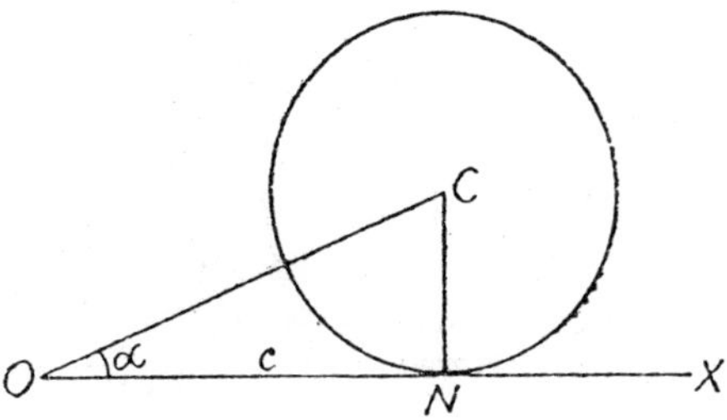

$ON=c$. let $\angle CON=\alpha$.

Hence the distance $OC=c\sec\alpha$ and radius of the circle is $CN=c\tan\alpha$.

Hence the polar coordinates of the centre are

$$(c\sec\alpha, \alpha).$$

The equation to the circle is therefore

$$r^2-2rc\sec\alpha\cos(\theta-\alpha)+c^2\sec^2\alpha=c^2\tan^2\alpha$$

or $$r^2-2rc(\cos\theta+\sin\theta\tan\alpha)+c^2=0. \qquad \ldots(1)$$

But it passes through (r_1, θ_1).

$$\therefore \quad r_1{}^2-2r_1c(\cos\theta_1+\sin\theta_1\tan\alpha)+c^2=0. \qquad \ldots(2)$$

Eliminating $\tan\alpha$ from (1) and (2), we get

$$\frac{r^2-2rc\cos\theta+c^2}{r\sin\theta}=\frac{r_1{}^2-2r_1\cos\theta+c^2}{r_1\sin\theta_1}$$

which is the equation to the required circle.

Examples X

1. Find the radical axis and length of the common chord of the circles

$$x^2+y^2-2x-8y+13=0$$

and $$x^2+y^2-8x-4y+11=0.$$

2. If from any point P tangents PA and PB are drawn to two given circles with centres O_1 and O_2 respectively and if PL is the perpendicular from P on their radical axis, prove that $PA^2-PB^2=2PL.O_1O_2$.

3. If two circles $x^2+y^2+\alpha x+\beta y=0$ and $x^2+y^2+\alpha' x+\beta' y=0$ touch each other, show that $\alpha'\beta=\alpha\beta'$.

4. Prove that the two circles

$$x^2+y^2+2ax+c=0 \text{ and } x^2+y^2+2by+c=0$$

touch if $1/a^2+1/b^2=1/c$. *(Pb. 1945)*

5. Find the radical centre of the circles

$$x^2+y^2+4x+7=0,\ 2x^2+2y^2+3x+5y+9=0$$

and $$x^2+y^2+y=0.$$ *(Agra '44, Banaras '49)*

6. Find the general equation of a circle cutting $x^2+y^2=c^2$ orthogonally and passing through the point (a, b). Hence show that it must pass through the point

$$\left(\frac{c^2a}{a^2+b^2}, \frac{c^2b}{a^2+b^2}\right).$$

7. If two conics

$$a_1x^2+2h_1xy+b_1y^2+2g_1x+2f_1y+c_1=0$$

and $$a_2x^2+2h_2xy+b_2y^2+2g_2x+2f_2y+c_2=0$$

intersect in four concyclic points, prove that

$$\frac{a_1b_1}{h_1}=\frac{a_2-b_2}{h_2}.$$

8. Show that as λ varies, the circles

$$x^2+y^2+2ax+2by+2\lambda(ax-by)=0$$

form a coaxal system. Find the equation of the

radical axis and of the coaxal system that is orthogonal to the first system. *(Allahabad '50)*

9. Find the equation of the circle whose diameter is the common chord of the circles

$$x^2+y^2+2x+3y+1=0 \text{ and } x^2+y^2+4x+3y+2=0.$$

(Utkal 1947)

10. Find the general equation of all circles any pair of which has the same radical axis as the circles

$$x^2+y^2=4 \text{ and } x^2+y^2+2x+4y=6.$$ *(Banaras 1946)*

11. Find the coordinates of the limiting points of the coaxal system determined by the circles

$$x^2+y^2+2x-6y=0 \text{ and } 2x^2+2y^2-10y+5=0.$$

(Mysore 1945)

12. If A, B, C be the centres of three coaxal circles, and t_1, t_2, t_3 be the tangents to them from any point, prove the relation $BC.t_1^2+CA.t_2^2+AB.t_3^2=0$. *(Dehli 1947)*

13. Show that the system of circles

$$x^2+y^2+2\lambda x+2\mu y+\nu=0$$

have a common orthogonal circle if λ, μ, ν satisfy the equation of the form $A\lambda+B\mu+C\nu+D=0$.

(Travancore 1947)

14. If the origin be at one of the limiting point of a system of coaxal circles of which $x^2+y^2+2gx+2fy+c=0$ is a member, the equation of the system of circles cutting them all orthogonally is $(x^2+y^2)(g+\mu f)+c(x+\mu y)=0$.

(Agra 1937)

15. If a circle cuts three given circles $S_1=0$, $S_2=0$ and $S_3=0$ orthogonally, prove that it will cut any circle of the system $\lambda_1S_1+\lambda_2S_2+\lambda_3S_3=0$ orthogonally.

16. If the circumcircle of the triangle formed by the lines

$$\beta x+\gamma y+\alpha=0, \; \gamma x+\alpha y+\beta=0, \; \alpha x+\beta y+\gamma=0$$

passes through the origin, show that

$$(\beta^2+\gamma^2)(\gamma^2+\alpha^2)(\alpha^2+\beta^2)=\alpha\beta\gamma(\alpha+\beta)(\beta+\gamma)(\gamma+\alpha).$$

17. Prove that the circle of similitude of the circles

$$S\equiv(x-h)^2+(y-k)^2-r^2=0$$

and $$S'=(x-h')^2+(y-k')^2-r'^2=0$$

is $$\frac{S}{r'^2}=\frac{S'}{r^2}.$$

18. Prove that the equation to two given circles can always be put in the form

$$x^2+y^2+ax+b=0,\ x^2+y^2+a'x+b=0$$

and that one of the circles will be within the other if aa' and b are both positive. *(Agra 1942, '53, Dehli '47)*

19. Find the common external tangents to two circles

$x^2+y^2=16$ and $x^2+y^2+6x-8y=0$. *(Dehli 1945)*

20. Find the common tangents of the circles

$x^2+y^2=1$ and $(x-1)^2+(y-3)^2=4$. *(Punjab '46)*

21. Find the equation of the straight lines which touch both the circles $x^2+y^2=4$ and $(x-4)^2+y^2=1$. Hence or otherwise obtain the coordinates of the centres of similitude. *(Annamalai 1947)*

22. If $x\cos\theta+y\sin\theta=2$ is the equation of a common tangent to $x^2+y^2=4$ and $x^2+y^2-6\sqrt{3}x-6y+20=0$, find the value of θ. Prove also that the angle between the pair of direct common tangents to two circles is

$$\tan^{-1}(4\sqrt{\tfrac{2}{7}}).$$ *(Roorkee 1947)*

23. Prove that the equation of the circle cutting the three given circles $x^2+y^2+2g_rx+2f_ry+c_r=0$ $(r=1, 2, 3)$ orthogonally is

$$\begin{vmatrix} x^2+y^2 & -x & -y & 1 \\ c_1 & g_1 & f_1 & 1 \\ c_2 & g_2 & f_2 & 1 \\ c_3 & g_3 & f_3 & 1 \end{vmatrix}=0.$$

By writing the last equation as

$$\begin{vmatrix} c_1+g_1x+f_1y & g_1+x & f_1+y \\ c_2+g_2x+f_2y & g_2+x & f_2+y \\ c_3+g_3x+f_3y & g_3+x & f_3+y \end{vmatrix}=0,$$

prove that the circle in question is the locus of the point whose polars with respect to the three given circles are concurrent. *(Banaras 1950)*

24. If A, B, C be the centres and a, b, c be the radii of three coaxal circles, prove that

$$a^2BC+b^2CA+c^2AB+BC.CA.AB=0. \quad (Pb.\ 1953)$$

25. Show that the circles

$$r=2a\cos(\theta-\alpha) \text{ and } r=2b\cos(\theta-\beta)$$

intersect at an angle $\alpha-\beta$.

26. Find the equation for determining the values of r at the points of intersection of the circle $r=2a\cos\theta$ and the straight line $r\cos(\theta-\beta)=p$. Deduce the value of p when the straight line becomes a tangent. *(Allahabad 1949)*

27. Show that for all values of a, the straight line

$$r\cos(\theta-\alpha)=a+r_1\cos\alpha$$

is a tangent to the circle $r^2-2rr_1\cos\theta+r_1^2-a^2=0$.

28. Show that the polar equation of the circle drawn on the line joining the points (a, α) and (b, β) as diameter is $r^2-r\{a\cos(\theta-\alpha)+b\cos(\theta-\beta)\}+ab\cos(\alpha-\beta)=0$.

29. Show that the equation $r^2-\lambda r\cos(\theta-\alpha)+\lambda a=0$ where λ is a variable represents a system of coaxal circles, and find the limiting points.

30. O is a fixed point. P is any point on a given circle. OP is joined and on it is taken a point Q such that $OP.OQ$=constant $(=k)$. Prove that the locus of Q is a circle which degenerates into a straight line if the point O is taken on the original circle.

CHAPTER VI

THE PARABOLA

6·1. Conic Sections. Conic sections are the curves that can be obtained by the intersections of a cone with a plane in various ways. They are the parabola, ellipse and the hyperbola.

The locus of a point which moves so that its distance from a fixed point bears a constant ratio with its distance from a fixed straight line is called a conic.

The fixed ratio is called the eccentricity of the conic and is denoted by the letter e.

If the eccentricity e is less than unity the conic is an ellipse.

If it is greater than unity the conic is a hyperbola.

If it is equal to unity, the conic is a parabola, so that in this case the distances of a point on the parabola from the fixed straight line are equal.

The fixed point is called the focus of the conic.

The fixed straight line is called the directrix.

The straight line passing through the focus and perpendicular to the directrix is called the axis of the conic.

The point of intersection of the axis and the conic is called the vertex of the conic,

6·2. Standard equation of a parabola.

Let S be the fixed point and ZM the fixed line. Let ZS

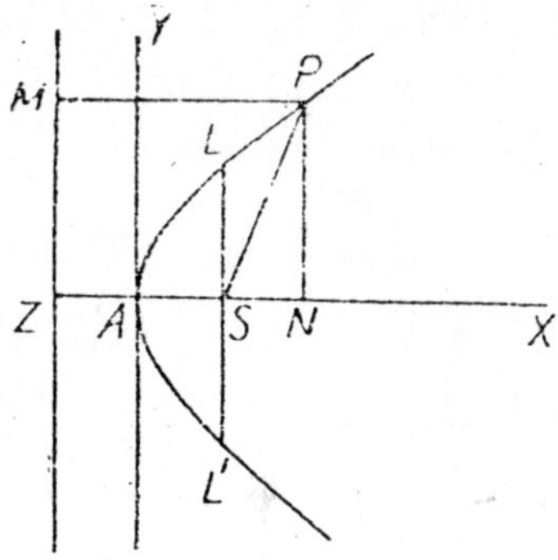

be the x-axis. Take A as the middle point of ZS. Then because $ZA=AS$, the point A satisfies the condition of being on the locus. Hence A is the vertex of the parabola.

Let a line through A perpendicular to AX be the axis of Y. Then A is the origin of coordinates.

Let P be a point on the locus and let its coordinates be (x, y). As P is on the parabola,

$$PS=PM. \qquad \ldots \quad \ldots(1)$$

Let AS be equal to a; hence the coordinates of S are $(a, 0)$, so that

$$PS=\sqrt{\{(x-a)^2+y^2\}}.$$

Also $$PM=NZ=NA+AZ=x+a.$$

Hence from (1),

$$\sqrt{\{(x-a)^2+y^2\}}=x+a$$

or $$y^2=(x+a)^2-(x-a)^2$$
$$=4ax$$

which is a relation between x and y, the coordinates of a point on the locus, and is therefore the equation to the parabola.

The coordinates of the focus S are $(a, 0)$. The equation to the directrix is $x=-a$. PN is the ordinate of the point P. The double ordinate LSL' passing through the focus S is called the latus-rectum of the parabola.

The abscissa of S is a. Hence

$$LS^2=4a.a$$

from the equation of the parabola.

$$\therefore \quad LS=2a \text{ and } LSL'=4a.$$

6·3. Tracing the parabola $y^2=4ax$.

If $x=0$, y is also 0, so that the curve passes through the origin.

For one value of x, there are two equal and opposite values of y.

Hence the curve is symmetrical about the X-axis.

As X increasee to infinity y also increases to infinity.

If x is negative, both the values of y are imaginary, which means that there is no part of the curve on the negative side of the x-axis.

Hence the shape of the curve is as shown in the figure.

Similarly the parabolas

$$y^2=-4ax \quad \ldots \quad \ldots(1)$$

$$x^2=4ay \quad \ldots \quad \ldots(2)$$

$$x^2=-4ay \quad \ldots \quad \ldots(3)$$

can be traced. Their figures are given below respectively :—

(1)

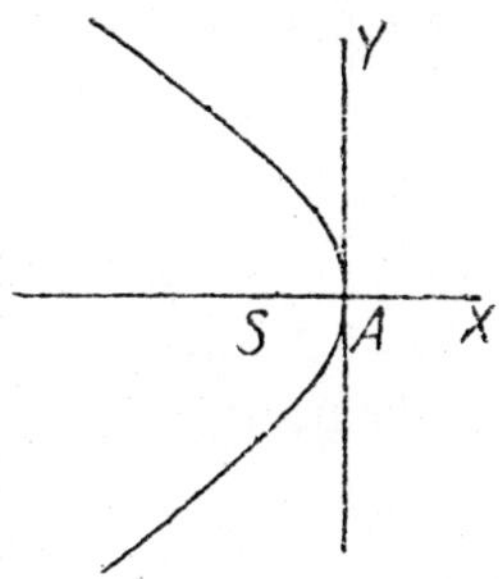

(2)

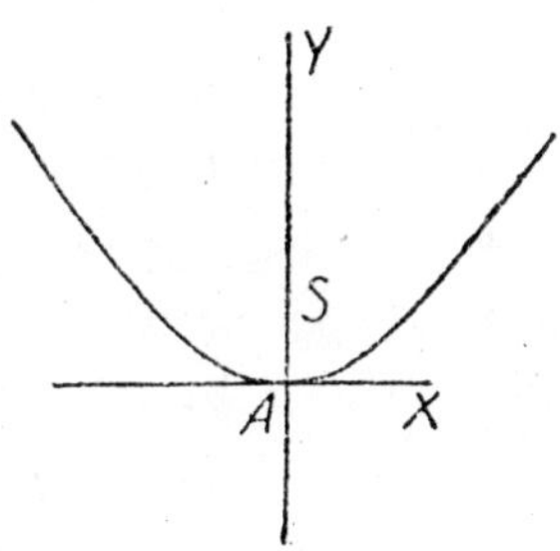

(3)

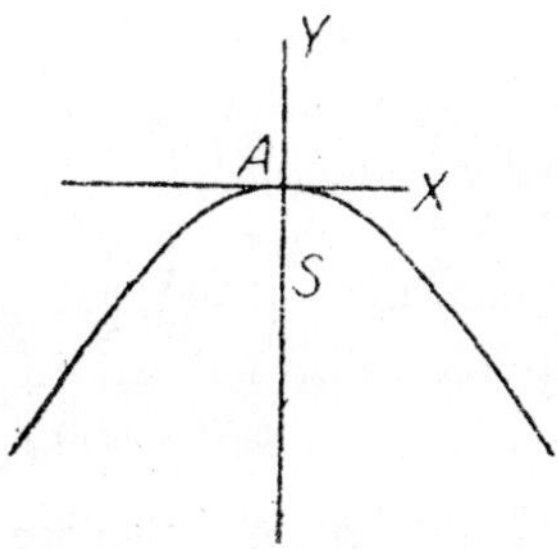

6·4. Position of a point relative to the parabola $y^2=4ax$.

Let P' be the point (x', y'). Draw $P'N$ perpendicular to the axis and suppose it cuts the parabola at the point P. The point P' lies outside the parabola if

$$P'N > PN.$$

But $PN^2=4ax'$, x' being the abscissa of N. Also

$$P'N=y'.$$

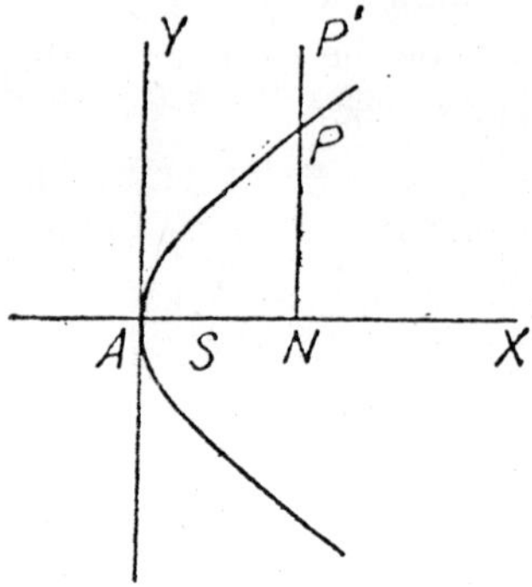

Hence the point lies outside the parabola if $y'^2 > 4ax'$. Similarly the point (x', y') lies inside the parabola if

$$y'^2 < 4ax'.$$

We know already that if the point lies on the parabola,

$$y'^2=4ax'$$

6·5. General equation to a parabola.

(i) **Rectangular axes.** Let the coordinates of the focus S be (h, k) and let $lx+my+n=0$ be the equation to the directrix. Let P be any point (x, y) on the parabola.

According to the given condition

$$SP=\sqrt{(x-h)^2+(y-k)^2}$$

=perpendicular from (x, y) on the line

$$lx+my+n=0$$

$$=\frac{lx+my+n}{\sqrt{(l^2+m)^2}}$$

or $$(l^2+m^2)\{(x-h)^2+(y-k)^2\}=(lx+my+n)^2$$

or $$l^2y^2+m^2x^2-2lmxy-2x\{h(l^2+m^2)+nl\}-2y\{k(l^2+m^2)+mn\}$$
$$+(l^2+m^2)(h^2+k^2)-n^2=0\ldots\text{(A)}$$

which is the required equation of the parabola.

(ii) **Axes inclined at an angle ω.** Again let the coordinates of the focus S be (h, k) and $lx+my+n=0$ the equation to the directrix. If the coordinates of any point on the parabola are (x, y).

$$SP=\sqrt{(x-h)^2+(y-k)^2+2\,(x-h)\,(y-k)\cos\omega}$$

= perpendicular from (x, y) on the line

$$lx+my+n=0$$

$$=\frac{(lx+my+n)\sin\omega}{\sqrt{(l^2+m^2-2lm\cos\omega)}}$$

or $$(l^2+m^2-2lm\cos\omega)\,\{(x-h)^2+(y-k)^2+2\,(x-h)\,(y-k)\cos\omega\}=(lx+my+n)^2\sin^2\omega$$

or $$\{x\,(m-l\cos\omega)\,y-(l-m\cos\omega)\}^2$$
$$-2x\,\{(h+k\cos\omega)\,(l^2+m^2-2lm\cos\omega)+nl\sin^2\omega\}$$
$$-2y\,\{(k+h\cos\omega)\,(l^2+m^2-2lm\cos\omega)+mx\sin^2\omega\}$$
$$+(h^2+k^2+2hk\cos\omega)\,(l^2+m^2-2lm\cos\omega)-n^2\sin^2\omega=0$$

which is the required equation to the parabola.

From equation (A) it is clear that the general equation to a parabola can be written in the form

$$ax^2+by^2+2\sqrt{(ab)}\,xy+2gx+2fy+c=0.$$

Hence it is obvious that the general equation of second degree $ax^2+by^2+2hxy+2gx+2fy+c=0$ will always represent a parabola if $h^2=ab$ that is the terms of the second degree form a perfect square.

If $h^2=ab$ and also $s=abc+2gfh-af^2-bg^2-ch^2=0$ the general equation represents only a pair of straight lines.

Ex. 1. *Trace the parabola $x=ay^2+by+c$.*

The above equation can be written as

$$x=a\left\{\left(y+\frac{b}{2a}\right)^2+\left(\frac{c}{a}-\frac{b^2}{4a^2}\right)\right\}$$

or $$\left(y+\frac{b}{2a}\right)^2=\frac{1}{a}\left(x+\frac{b^2-4ac}{4a}\right).$$

Transferring the origin to the point $\left(-\frac{b^2-4ac}{4a}, -\frac{b}{2a}\right)$, the equation becomes $y^2=\frac{x}{a}$, which is the standard form.

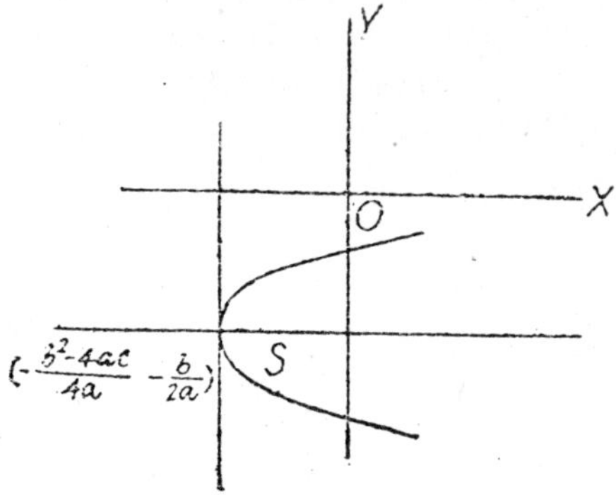

The latus rectum of the parabola is $\frac{1}{a}$. The vertex is the point $\left(-\frac{b^2-4ac}{4a}, -\frac{b}{2a}\right)$. The axis is the line through the vertex parallel to the x-axis. Hence its equation is $y=-\frac{b}{2a}$. The focus is a distance $\frac{1}{4a}$ ahead of the vertex as shown in the figure. Its coordinates are

$$-\frac{b^2-4ac}{4a}+\frac{1}{4a}, \quad -\frac{b}{2a}.$$

If b^2-4ac is a negative quantity, the vertex and focus will lie in the fourth quadrant. The directrix will pass through the point $-\frac{b^2-4ac}{4a}-\frac{1}{4a}, -\frac{b}{2a}$ and is a line parallel to the y-axis. Hence its equation is $x=-\frac{b^2-4ac}{4a}-\frac{1}{4a}$.

5·6. Intersection of a line with the parabola.

Let the line be $y=mx+c$ ... ...(1)

The coordinates of the points of intersection satisfy the above equation and the equation of the parabola

$$y^2=4ax \qquad \ldots \qquad \ldots(2)$$

Eliminating y from (1) and (2),

$$(mx+c)^2=4ax$$

or

$$m^2x^2+2x\,(mc-2a)+c^2=0 \qquad \ldots \qquad \ldots(3)$$

which is a quadratic equation in x and gives two roots real, coincident or imaginary according as

$$\{2\,(mc-2a)\}^2 >, = \text{ or } < 4m^2c^2$$

or

$$a^2 >, = \text{ or } < amc$$

or

$$a >, = \text{ or } < mc$$

or

$$c <, = \text{ or } > a/m.$$

The straight line $y=mx+c$, therefore touches the parabola if $c=a/m$. Hence $y=mx+a/m$ is the equation of a tangent to the parabola for all values of m.

6·7. Length of the chord intercepted on a line.

Let the line be $y=mx+c$. Then the abscissae of the points of intersection of this line with the parabola are given by the equation (3) of the last article. If the roots of this equation are x_1 and x_2,

$$x_1+x_2=\frac{2\,(2a-mc)}{m^2} \qquad \ldots \qquad \ldots(1)$$

and

$$x_1x_2=\frac{c^2}{m^2} \qquad \ldots \qquad \ldots(2)$$

Let the coordinates of the points of intersection be (x_1, y_1) and (x_2, y_2).

Length of the chord joining (x_1, y_1) and (x_2, y_2)

$$=\sqrt{[(x_1-x_2)^2+(y_1-y_2)^2]} \qquad \ldots \qquad \ldots(3)$$

But (x_1, y_1) and (x_2, y_2) are also the points on the line $y=mx+c$. They satisfy this equation.

$$\therefore \quad y_1=mx_1+c \qquad \ldots \qquad \ldots(4)$$

and

$$y_2=mx_2+c \qquad \ldots \qquad \ldots(5)$$

Subtracting (5) from (4),

$$y_1-y_2=m\,(x_1-x_2).$$

Substituting for y_1-y_2 in (3), the length of the chord

$$=\sqrt{[(x_1-x_2)^2+m^2\,(x_1-x_2)^2]}$$

$$=(x_1-x_2)\sqrt{(1+m^2)} \qquad \ldots \quad \ldots(6)$$

But from (1) and (2),

$$(x_1-x_2)^2=(x_1+x_2)^2-4x_1x_2$$

$$=\frac{4\,(2a-mc)^2}{m^4}-\frac{4c^2}{m^2}$$

$$=\frac{4}{m^4}\{(2a-mc)^2-c^2m^2\}$$

$$=\frac{16a}{m^4}\,(a-mc) \qquad \ldots \quad \ldots(7)$$

From (6), by substitution from (7), length of the chord

$$=\frac{4}{m^2}\sqrt{[a\,(a-mc)\,(1+m^2)]}.$$

6·8. Tangent at a point on the parabola.

Let $P\,(x_1, y_1)$ and $Q\,(x_2, y_2)$ be two points on the parabola.

The equation to the line joining the points P and Q is

$$y-y_1=\frac{y_2-y_1}{x_2-x_1}\,(x-x_1). \qquad \ldots \quad \ldots(1)$$

As the points P and Q lie on the parabola $y^2=4ax$,

$$y_1^2=4ax_1, \qquad \ldots \quad \ldots(2)$$

$$y_2^2=4ax_2. \qquad \ldots \quad \ldots(3)$$

Subtracting (3) from (2)

$$y_1^2-y_2^2=4a\,(x_1-x_2)$$

or

$$\frac{y_1-y_2}{x_1-x_2}=\frac{4a}{y_1+y_2} \qquad \ldots \quad \ldots(4)$$

Substituting the value of $\dfrac{y_1-y_2}{x_1-x_2}$ from (4) in (1), the equation of the chord PQ is given by

$$y-y_1=\frac{4a}{y_1+y_2}(x-x_1) \qquad \ldots \quad \ldots (5)$$

The chord PQ becomes a tangent to the parabola at the point P; if the point Q coincides with P so that $x_2=x_1$ and $y_2=y_1$.

Hence the equation of the tangent at P becomes from (5) above,

$$y-y_1=\frac{2a}{y_1}(x-x_1)$$

or

$$yy_1=2a\ (x-x_1)+y_1^2$$

$$=2a\ (x-x_1)+4ax_1 \text{ from (2)}$$

$$=2a\ (x+x_1).$$

6·9. Point of contact.

We have seen that $y=mx+(a/m)$ is a tangent to the parabola.

We have to find the point where it touches the parabola.

Let the point of contact be (x_1, y_1). Then the equation of the tangent is

$$yy_1=2a\ (x+x_1). \qquad \ldots \quad \ldots (1)$$

But the given equation is

$$y=mx+\frac{a}{m} \qquad \ldots \quad \ldots (2)$$

Hence (1) and (2) represent the same line and comparing coefficients,

$$\frac{y_1}{1}=\frac{2a}{m}=\frac{2ax_1}{\frac{a}{m}}$$

which gives $x_1=\dfrac{a}{m^2}$ and $y_1=\dfrac{2a}{m}$.

The point of contact is $\left(\dfrac{a}{m^2}, \dfrac{2a}{m}\right)$.

6·10. Normal at a point on the parabola.

The normal is a line through the point (x_1, y_1) perpendicular to the tangent $yy_1 = 2a(x+x_1)$ ... ...(1)

The slope of the line perpendicular to (1) is $-\frac{y_1}{2a}$.

Hence the equation to the normal is

$$y - y_1 = -\frac{y_1}{2a}(x - x_1).$$

6·11. Normal in terms of its slope.

Suppose the slope of the normal is m; then

$$-\frac{y_1}{2a} = m$$

or $$y_1 = -2am.$$

But the point (x_1, y_1) is on the parabola

$$4a^2m^2 = 4ax_1$$

or $$x_1 = am^2.$$

Substituting $x_1 = am^2$ and $y_1 = -2am$ in the equation to the normal we get $y = mx - 2am - am^3$ as the normal to the parabola at the point $(am^2, -2am)$.

6·12. Parametric Coordinates.

We have seen that $\left(\frac{a}{m^2}, \frac{2a}{m}\right)$ and $(am^2, -2am)$ are points on the parabola as they satisfy its equation $y^2 = 4ax$.

In general we signify any point on a parabola by the coordinates $(at^2, 2at)$ where t is a variable parameter for different points on the parabola. These coordinates also satisfy the equation of the parabola.

Equation to a chord joining two points $(at_1^2, 2at_1)$, $(at_2^2, 2at_2)$ on the parabola is

$$y - 2at_1 = \frac{2a(t_2 - t_1)}{a(t_2^2 - t_1^2)}(x - at_1^2)$$

or
$$y-2at_1=\frac{2}{t_1+t_2}(x-at_1^2)$$
or
$$(t_1+t_2)\,y=2x+2at_1t_2.$$

If this chord becomes a tangent at the point $(at_1^2, 2at_1)$, $t_2=t_1$ and the equation to the tangent becomes
$$t_1\,y=x+at_1^2.$$

The equation to the normal at the point $(at_1^2, 2at_1)$ is
$$y-2at_1=-t_1\,(x-at_1^2)$$
or
$$y=-t_1x+2at_1+at_1^3.$$

The tangents at the points t_1 and t_2 of the parabola are
$$t_1\,y=x+at_1^2,$$
$$t_2\,y=x+at_2^2.$$
Solving these equations $x=at_1t_2$ and $y=a\,(t_1+t_2)$ which are the coordinates of the point of intersection of the tangents at t_1 and t_2.

6·13. Some properties of the parabola.

(1) The tangent at any point of a parabola bisects the angle between the focal distance of the point and the perpendicular on the directrix from the point.

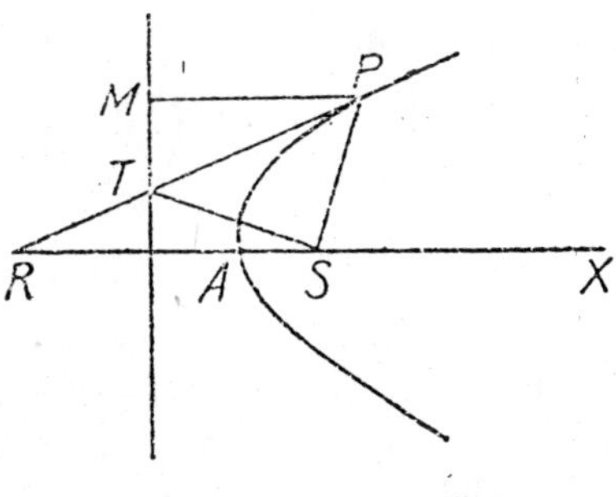

Let S be the focus and P a point on the parabola. PT is a tangent at P. It is to prove that PT bisects the angle SPM, PM being perpendicular from P on the directrix. Let the coordinates of P be $(at^2, 2at)$.

The equation of the tangent at P is
$$ty=x+at^2 \qquad \ldots \quad \ldots \quad \ldots(1)$$
This meets the x-axis where $y=0$.

$$\therefore \quad x = -at^2.$$

Hence the cocrdinates of R are $(-at^2, 0)$.

The coordinates of S are $(a, 0)$.

$$\therefore \quad RS = a(1+t^2).$$

Also
$$PS = \sqrt{[\{a(1-t^2)\}^2 + 4a^2t^2]}$$
$$= a\sqrt{\{(1+t^2)^2\}}$$
$$= a(1+t^2).$$
$$\therefore \quad RS = SP.$$
$$\therefore \quad \angle SRP = \angle RPS.$$

But
$$\angle SRP = \angle RPM \text{ (alternate)}.$$
$$\therefore \quad \angle RPM = \angle RPS.$$

(2) The portion of a tangent to a parabola intercepted between the directrix and the curve subtends a right angle at the focus.

The tangent at P

$$ty = x + at^2$$

meets the directrix $x = -a$ at T whose coordinates are

$$\left(-a, \frac{a(t^2-1)}{t}\right).$$

Hence the direction tangent of the line ST is

$$\frac{a\left(\frac{t^2-1}{t}\right)}{-a-a} \quad \text{or} \quad \frac{1-t^2}{2t}.$$

Also the direction tangent of SP is

$$\frac{-2at}{a(1-t^2)} \quad \text{or} \quad \frac{-2t}{1-t^2}.$$

But
$$\frac{1-2t^2}{2t} \times \frac{-2t}{1-t^2} = -1.$$

Hence SP and ST are perpendicular to each other.

(3) The tangents at the extremities of a focal chord of a parabola intersect at right angles on the directrix.

Let the extremities of the focal chord be $(at_1^2, 2at_1)$, $(at_2^2, 2at_2)$. The point of intersection of the tangents at these points is

$$at_1t_2,\ a\,(t_1+t_2).$$

The equation to the line joining the points $(at_1^2, 2at_1)$, $(at_2^2, 2at_2)$ is

$$y-2at_1=\frac{2at_1-2at_2}{at_1^2-at_2^2}\,(x-at_1^2)$$

or
$$y-2at_1=\frac{2}{t_1+t_2}\,(x-at_1^2).$$

But this passes through focus $(a, 0)$.

$$\therefore \quad -2at_1=\frac{2a}{t_1+t_2}\,(1-t_1^2)$$

or
$$-2t_1\,(t_1+t_2)=2\,(1-t_1^2)$$

or
$$t_1t_2=-1.$$

Hence the abscissa of the point of intersection is at_1t_2 or $-a$, that is the point lies on the directrix.

(4) If SY be perpendicular to the tangent at a point P of a parabola, then Y lies on the tangent at the vertex and $\quad SY^2=AS.SP$.

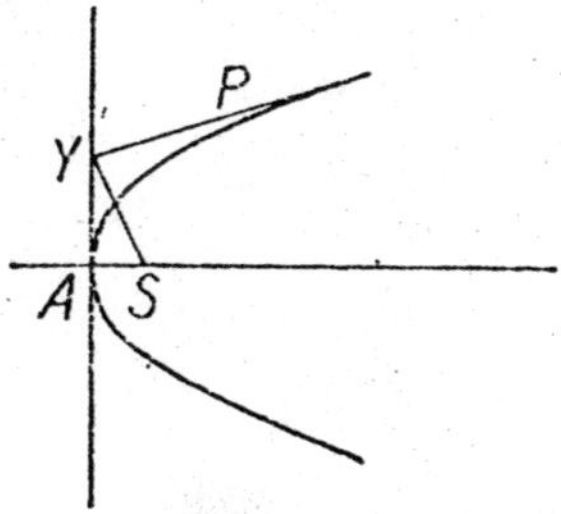

Let the point P be $(at^2, 2at)$.

The tangent at P is

$$ty=x+at^2.$$

The point where it meets the tangent at the vertex is $(0, at)$.

The slope of SY is $\dfrac{at}{-a}=-t$.

The slope of the tangent is $=\dfrac{1}{t}$.

Hence SY is perpendicular to the tangent PY.

Also $$SY^2 = a^2 + a^2t^2$$
$$= a^2(1+t^2)$$
$$= a.a(1+t^2)$$
$$= AS.SP.$$

(5) The subtangent of a point on a parabola is bisected at the vertex and the sub-normal is of constant length.

NT the projection of the tangent at P on the axis is called the subtangent and NG the projection of the normal at P on the axis is called the sub-normal.

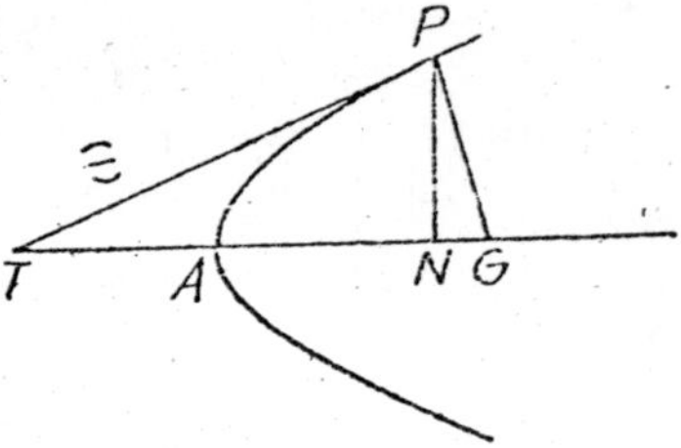

The tangent at P is
$$tz = x + at^2.$$

The coordinates of T the point where it meets the axis are $(-at^2, 0)$.
$$\therefore \quad AT = at^2.$$
Also $$AN = at^2.$$
Hence A is the middle point of TN.

The equation of the normal at P is
$$y = -tx + 2at + at^3.$$

This meets the axis at G whose coordinates are $(2a + at^2, 0)$, so that
$$AG = 2a + at^2.$$
$$\therefore \quad NG = AG - AN$$
$$= 2a + at^2 - at^2$$
$$= 2a$$
$$= \text{semi-latus rectum.}$$

Ex. 2. *If l and l' are the lengths of segments of any focal chord of the parabola $y^2 = 4ax$, prove that*
$$\frac{1}{l} + \frac{1}{l'} = \frac{1}{a}.$$ *(C. U. 1951)*

Let the extremities of the focal chord be $(at_1^2, 2at_1)$ and $(at_2^2, 2at_2)$.

As proved before in property (3) the line joining these points passes through the focus if

$$t_1t_2=-1. \qquad \ldots(1)$$

$$l=\sqrt{[\{a\,(t_1^2-1)\}^2+4a^2t_1^2]}$$
$$=a\,(1+t_1^2).$$

Also $\qquad l'=a\,(1+t_2^2).$

$$\therefore \quad \frac{1}{l}+\frac{1}{l'}=\frac{1}{a\,(1+t_1^2)}+\frac{1}{a\,(1+t_2^2)}$$
$$=\frac{1+t_1^2+1+t_2^2}{a\,(1+t_1^2)\,(1+t_2^2)}$$
$$=\frac{1+t_1^2+t_2^2+t_1^2t_2^2}{a\,(1+t_1^2)\,(1+t_2^2)} \text{ by (1)}$$
$$=\frac{(1+t_1^2)\,(1+t_2^2)}{a\,(1+t_1^2)\,(1+t_2^2)}$$
$$=\frac{1}{a}.$$

Ex. 3. *A chord is drawn through the vertex of the parabola $y^2=4ax$ making an angle θ with its axis. Prove that its length is equal to $4a\cos\theta/\sin^2\theta$.*

Hence or otherwise prove that if r_1, r_2 be the lengths of two perpendicular chords of the parabola drawn through its vertex, then

$$(r_1r_2)^{4/3}=16a^2\,(r_1^{2/3}+r_2^{2/3}).$$

The slope of the chord is $\tan\theta$.

$\therefore$ its equation is $y=x\tan\theta$

It cuts the parabola at the points given by

$$y^2=4ax=x^2\tan^2\theta.$$
$$\therefore \quad x=4a\cot^2\theta$$

and $\qquad y=4a\cot\theta.$

$\therefore$ length of the chord $=\sqrt{(16a^2 \cot^4 \theta+16a^2 \cot^2 \theta)}$

$$=4a \cot \theta \sqrt{(\cot^2 \theta+1)}$$

$$=4a \cot \theta \operatorname{cosec} \theta$$

$$=\frac{4a \cos \theta}{\sin^2 \theta}.$$

For a chord passing through the vertex and perpendicular to the above chord the slope will be $-\cot \theta$ and its equation will be $y=-x \cot \theta=x \tan (\pi/2+\theta)$, that is θ will change into $\pi/2+\theta$.

Hence if
$$r_1=\frac{4a \cos \theta}{\sin^2 \theta} \qquad \ldots(1)$$

$$r_2=\frac{4a \cos (\pi/2+\theta)}{\sin^2 (\pi/2+\theta)}$$

$$=+\frac{4a \sin \theta}{\cos^2 \theta}. \qquad \ldots(2)$$

From (1) and (2), $(r_1r_2)^{4/3}=\left(+\dfrac{16a^2}{\sin \theta \cos \theta}\right)^{4/3}$

$$=\left(\frac{16a^2}{\sin \theta \cos \theta}\right)^{4/3}.$$

Also $r_1{}^{2/3}+r_2{}^{2/3}=\dfrac{(4a \cos \theta)^{2/3}}{\sin^{4/3} \theta}+\dfrac{(4a \sin \theta)^{2/3}}{\cos^{4/3} \theta}$

$$=\frac{(16a^2)^{1/3} \cos^2 \theta+(16a^2)^{1/3} \sin^2 \theta}{(\sin \theta \cos \theta)^{4/3}}.$$

$$\therefore \quad 16a^2 (r_1{}^{2/3}+r_2{}^{2/3})=\left(\frac{16a^2}{\sin \theta \cos \theta}\right)^{4/3}.$$

Ex. 4. *Show that the locus of the middle point of a variable chord of the parabola $y^2=4ax$, such that the focal distances of its extremities are in the ratio 2 : 1 is*

$$9\,(y^2-2ax)^2=4a^2\,(2x-a)\,(4x+a). \qquad \textit{(Luck. 1941)}$$

Let the extremities of the variable chord be $P\,(at_1{}^2, 2at_1)$ and $Q\,(at_2{}^2, 2at_2)$.

Let (x, y) be the middle point of this chord, then

$$\frac{a(t_1^2+t_2^2)}{2}=x. \qquad \ldots(1)$$

$$\frac{2a(t_1+t_2)}{2}=y. \qquad \ldots(2)$$

Focal distance of P is $a(1+t_1^2)$.
Focal distance of Q is $a(1+t_2^2)$.
Hence by the given condition

$$1+t_1^2=2(1+t_2^2)$$

or $$t_1^2=1+2t_2^2. \qquad \ldots(3)$$

From (1) and (3),

$$1+3t_2^2=\frac{2x}{a}$$

or $$3t_2^2=\frac{2x-a}{a}$$

or $$t_2^2=\frac{2x-a}{3a} \qquad \ldots(4)$$

and $$t_1^2=\frac{4x+a}{3a} \qquad \ldots(5)$$

From (2), $$t_1+t_2=\frac{y}{a}$$

or $$t_1^2+t_2^2+2t_1t_2=\frac{y^2}{a^2}$$

or $$\frac{2x-a}{3a}+\frac{4x+a}{3a}+2\sqrt{\left\{\frac{(2x-a)(4x+a)}{9a^2}\right\}}=\frac{y^2}{a^2}$$

or $$\frac{y^2}{a^2}-\frac{2x}{a}=2\sqrt{\left\{\frac{(2x-a)(4x+a)}{9a^2}\right\}}$$

or $$\frac{(y^2-2ax)^2}{a^4}=\frac{4(2x-a)(4x+a)}{9a^2}$$

or $$9(y^2-2ax)^2=4a^2(2x-a)(4x+a)$$

Ex. 5. *Prove that the circle circumscribing the triangle formed* *by any three tangents to a parabola passes through the focus.*

(Punjab 1954)

Let the tangents be drawn at the points t_1, t_2 and t_3 of the parabola. Their points of intersection are $\{at_1t_2, a(t_1+t_2)\}$, $\{at_2t_3, a(t_2+t_3)\}$ $\{at_1t_3, a(t_1+t_3)\}$.

Now we have to find a circle passing through the above three points.

Let the circle be

$$x^2+y^2+2gx+2fy+c=0 \quad \ldots(1)$$

The above points lie on it

$$a^2t_1^2t_2^2+a^2(t_1+t_2)^2+2gat_1t_2+2fa(t_1+t_2)+c=0 \quad \ldots(2)$$

$$a^2t_2^2t_3^2+a^2(t_2+t_3)^2+2gat_2t_3+2fa(t_2+t_3)+c=0 \quad \ldots(3)$$

and $$a^2t_3^2t_1^2+a^2(t_1+t_3)^2+2gat_1t_3+2fa(t_1+t_3)+c=0 \quad \ldots(4)$$

Subtracting (3) from (2), we get

$$at_2^2(t_1+t_3)+a(t_1+2t_2+t_3)+2gt_3+2f=0 \quad \ldots(5)$$

Subtracting (3) from (4), we get

$$at_3^3(t_1+t_2)+a(t_1+2t_3+t_3)+2gt_3+2f=0 \quad \ldots(6)$$

Subtracting (6) from (5), we get

$$2g(t_2-t_3)+a(t_2-t_3)\{1+t_1t_2+t_2t_3+t_3t_1\}=0.$$

$$\therefore \quad 2g=-a\{1+t_1t_2+t_2t_3+t_3t_1\}.$$

Subtracting for g in (5) or (6),

$$2f=-a\{t_1+t_2+t_3-t_1t_2t_3\}.$$

Subtracting for g and f in any one of (2), (3) or (4), we get

$$c=a^2(t_1t_2+t_2t_3+t_3t_1)$$

Hence the required equation of the circle is

$$x^2+y^2-ax(1+t_1t_2+t_2t_3+t_3t_1)-ay(t_1+t_2+t_3-t_1t_2t_3)$$
$$+a^2(t_1t_2+t_2t_3+t_3t_1)=0$$

which passes through the focus $(a, 0)$.

Ex. 6. *The locus of the middle points of normal chords of the parabola $y^2=4ax$, is*

$$\frac{y^2}{2a}+\frac{4a^3}{y^2}=x-2a.$$

(Lucknow 1956)

Let the normal be

$$y = -tx + 2at + at^3 \quad \ldots(1)$$

The intersection of this normal with the parabola $y^2 = 4ax$ is given by the quadratic in y obtained from (1) by substituting $x = \frac{y^2}{4a}$.

$$\frac{ty^2}{4a} + y - 2at - at^3 = 0$$

or

$$ty^2 + 4ay - 8a^2t - 4a^2t^3 = 0 \quad \ldots(2)$$

If the two roots are y_1 and y_2

$$y_1 + y_2 = \frac{-4a}{t} = 2y \quad \ldots(3)$$

if (x, y) are the coordinates of the middle point of the normal chord. Also because (x_1, y_1), (x_2, y_2) lie on the normal (1)

$$y_1 = -tx_1 + 2at + at^3 \quad \ldots(4)$$

and

$$y_2 = -tx_2 + 2at + at^3 \quad \ldots(5)$$

From (4) and (5),

$$y_1 + y_2 = -t(x_1 + x_2) + 4at + 2at^3 \quad \ldots(6)$$

From (3) and (6),

$$x_1 + x_2 = 2x = \frac{4a}{t_2} + 4at + 2at^2 \quad \ldots(7)$$

Substituting for t from (3) in (7), we get

$$\frac{y^2}{a} + 4a + \frac{8a^3}{y^2} = 2x,$$

or

$$\frac{y^2}{2a} + \frac{4a^3}{y^2} = x - 2a.$$

Ex. 7. *Prove that the area of the triangle formed by the normals to the parabola at the points $(at_1^2, 2at_1)$, $(at_2^2, 2at_2)$ and $(at_3^2, 2at_3)$ is*

$$\frac{a^2}{2}(t_2 - t_3)(t_3 - t_1)(t_1 - t_2)(t_1 + t_2 + t_3)^2.$$

(Agra 1957)

The equation to the normal at t_1 is

$$y=-t_1x+2at_1+at_1^3 \qquad \ldots(1)$$

and at t_2 the equation is

$$y=-t_2x+2at_2+at_2^3. \qquad \ldots(2)$$

The point of intersection of (1) and (2) is

$$2a+a(t_1^2+t_1t_2+t_2^2),\ -at_1t_2(t_1+t_2)$$

Similarly the other points of intersection are

$$2a+a(t_2^2+t_3^2+t_2t_3),\ -at_2t_3(t_2+t_3)$$

and

$$2a+a(t_3^2+t_1^2+t_3t_1),\ -at_3t_1(t_3+t_1).$$

Hence the area of the triangle whose vertices are the above points is

$$=\frac{1}{2}\begin{vmatrix} 2a+a(t_1^2+t_2^2+t_1t_2) & -at_1t_2(t_1+t_2) & 1 \\ 2a+a(t_2^2+t_3^2+t_2t_3) & -at_2t_3(t_2+t_3) & 1 \\ 2a+a(t_3^2+t_1^2+t_3t_1) & -at_3t_1(t_3+t_1) & 1 \end{vmatrix}$$

$$=\frac{1}{2}\begin{vmatrix} a(t_1^2+t_1t_2-t_3^2-t_2t_3) & -at_2(t_1^2+t_1t_2-t_2t_3-t_3^2) & 0 \\ a(t_2^2+t_2t_3-t_1^2-t_3t_1) & -at_3(t_2^2+t_2t_3-t_1t_3-t_1^2) & 0 \\ 2a+a(t_3^2+t_1^2+t_3t_1) & -at_3t_1(t_3+t_1) & 1 \end{vmatrix}$$

$$=\frac{1}{2}\begin{vmatrix} a(t_1-t_3)(t_1+t_3+t_2) & -at_2(t_1-t_3)(t_1+t_3+t_2) \\ a(t_2-t_1)(t_2+t_1+t_3) & -at_3(t_2-t_1)(t_2+t_1+t_3) \end{vmatrix}$$

$$=\frac{1}{2}a^2(t_1+t_2+t_3)^2(t_1-t_3)(t_2-t_1)\begin{vmatrix} 1 & -t_2 \\ 1 & -t_3 \end{vmatrix}$$

$$=\frac{1}{2}a^2(t_1+t_2+t_3)^2(t_1-t_3)(t_2-t_1)(t_2-t_3).$$

$$=\frac{1}{2}a^2(t_2-t_3)(t_3-t_1)(t_1-t_2)(t_1+t_2+t_3)^2.$$

Examples XI

1. Find the vertex, focus, axis and directrix of the parbolas,
   (i) $y^2=2x+4y$, (ii) $x^2+2y=8x-5$, (3) $x^2-2ax+4ay=0$.

2. Find the equation to the parabola whose focus is the origin and whose directrix is the line $2x+y-1=0$. *(C. U. 1954)*

3. Prove that the equation $y^2+2ax+2by+c=0$ represents a parabola, whose axis is parallel to the axis of x. Find its vertex. *(C. U. 1953)*

4. For the parabola $4(y-1)^2=-7(x-3)$ find (i) the latus rectum, and (ii) the coordinates of the focus and vertex. *(C. U. 1956)*

5. PQ is a double ordinate of a parabola. Find the locus of its points of trisection.

6. Prove that the locus of the middle points of all chords of the parabola $y^2=4ax$ which are drawn through the vertex is the parabola $y^2=2ax$. *(C. U. 1946)*

7. Find the length of the side of an equilateral triangle inscribed in the parabola $y^2=4ax$ so that one angular point is at the vertex. *(Utkal 1946)*

8. A straight line touches both $x^2+y^2=2a^2$ and $y^2=8ax$; show that its equation is $y=\pm(x+2a)$. *(C. U. 1955)*

9. Find the coordinates of that particular point on the parabola, the normal at which—terminated by the axis—is equal in length to the latus rectum. *(C. U. 1956)*

10. If a chord which is normal to the parabola at one end subtend a right angle at the vertex, prove that it is inclined at an angle $\tan^{-1}\sqrt{2}$ to the axis. *(C. U. 1953)*

11. A double ordinate of the curve $y^2=4px$ is of length $8p$; prove that the lines from the vertex to its two ends are at right angles. *(C. U. 1952)*

12 If a normal to a parabola make an angle ϕ with the axis, show that it will cut the curve again at an angle $\tan^{-1}(\frac{1}{2}\tan\phi)$. *(C. U. 1952)*

13. Prove that the line $2x+4y=9$ is a normal to the parabola $y^2=8x$. Also find the coordinates of the foot of this normal. *(C.U. 1950)*

14. Show that the centre of the circle which touches the line $x+y=0$ and which passes through the point (a, a) lies on the parabola

$$x^2-2xy+y^2-4ax-4ay+4a^2=0.$$

15. A parabola is drawn to pass through A and B, the ends of a diameter of a given circle of radius a, and to have as directrix a tangent to a concentric circle of radius b ; the axes being AB and a perpendicular diameter, prove that the locus of the focus of the parabola is

$$\frac{x^2}{b^2}+\frac{y^2}{b^2-a^2}=1.$$

16. Show that the equation of the tangent to the parabola $y^2=4ax$ which makes an angle θ with its axis is

$$y=x \tan \theta+a \cot \theta.$$ *(Lucknow 1943)*

17. Find the locus of the foot of the perpendicular drawn from a fixed point to any tangent to a parabola. *(Patna 1947)*

18. PQ is a chord of a parabola normal at P, AQ is drawn from the vertex A, and through P a line is drawn parallel to AQ meeting the axes in R. Show that AR is double the focal distance of P. *(Aligarh 1945)*

19. If the normal at the point $(at^2, 2at)$ on a parabola $y^2=4ax$ meets it again at $(at_1^2, 2at_1)$, prove that

$$t_1=-\left(t+\frac{2}{t}\right).$$ *(Sagar 1949)*

20. Prove that the chord of the parabola $y^2=4ax$ whose equation is $y-x\sqrt{2}+4a\sqrt{2}=0$ is a normal to the curve and that its length is $6\sqrt{3}a$. *(Punjab 1951)*

21. Find the locus of the point of intersection of two normals to the parabola $y^2=4ax$ which are at right angles to one another. *(Agra 1944)*

22. Prove that the locus of the intersection of the normals at the ends of a focal chord of a parabola is another parabola whose latus rectum is one-fourth that of the given parabola. Find the position of its vertex.

(Mysore 1945)

23. Prove that the two parabolas $y^2=ax$, $x^2=by$ cut one another at an angle $\tan^{-1}\dfrac{3a^{1/3}b^{1/3}}{2\,(a^{2/3}+b^{2/3})}$. *(Agra 1949)*

24. Find the condition that the line $lx+my+n=0$ may be a normal to the parabola $y^2=4ax$. *(Lucknow 1948)*

25. Equilateral triangles are inscribed in the parabola $y^2=4ax$. Prove that the locus of their centroids is $9y^2=4a\,(x-8a)$.

26. Prove that the area of a triangle inscribed in a parabola is twice the area of the triangle formed by the tangents at its vertices. *(Agra 1955)*

27. Show that if two tangents to a parabola intercept a constant length on any fixed tangent, the locus of their intersection is another equal parabola. *(Agra 1956)*

28. Show that the normals at the ends of a focal chord of $y^2=4ax$ intersect on $y^2=a\,(x-3a)$. *(Rajputana 1951)*

29. If the normals at two points of a parabola be inclined to the axis at angles θ and ϕ, such that

$$\tan\theta\tan\phi=2,$$

show that they intersect on the parabola.

(Agra Supp. 1955)

30. If the tangents at two points of the parabola $y^2=4ax$ intersect at the point (x_1, y_1), prove that the correspond-

ing normals intersect at the point

$$\left(2a - x_1 + \frac{y_1^3}{a},\ -\frac{x_1 y_1}{a}\right).$$ *(Travancore 1941)*

31. Tangents are drawn to a parabola at the points whose abscissae are in the ratio $\mu : 1$. Prove that they intersect on the curve

$$y^2 = (\mu^{1/4} + \mu^{-1/4})\,ax.$$

32. If the tangent to the parabola $y^2 = 4ax$ meets the axis in T and the tangent at the vertex A in Y, and the rectangle $TAYG$ is completed, show that the locus of G is $y^2 + ax = 0$.

33. Prove that the length of the intercept on the normal at the point $P\,(at^2, 2at)$ of a parabola $y^2 = 4ax$ made by the circle described on the line joining the focus and P as diameter is $a\sqrt{(1+t^2)}$. *(Roorkee 1949)*

34. The normals at the points Q, R on $y^2 = 4ax$ meet the parabola at the point P. Prove that the locus of the circum-centre of the triangle PQR is the parabola,

$$2y^2 - ax + a^2 = 0.$$ *(Utkal 1948)*

35. If P, Q, R be three points on a parabola, whose ordinates are in geometrical progression, prove that the tangents at P and R meet on the ordinate of Q.

6·14. Tangents through a point.

Let the coordinates of the point through which the tangent passes be (x_1, y_1).

Any tangent to the parabola $y^2 = 4ax$ is $y = mx + a/m$.

As it passes through (x_1, y_1),

$$y_1 = mx_1 + a/m$$

or

$$m^2 x_1 - my_1 + a = 0 \qquad \ldots(1)$$

This equation is a quadratic in m and gives two values of m. Hence in general two tangents can be drawn from a point to the parabola.

The two tangents are real and distinct, coincident or imaginary according as the roots of the above equation are real and distinct, coincident or imaginary which gives

$$y_1^2 - 4ax_1 > = \text{ or } < 0.$$

i.e. as the point (x_1, y_1) lies outside, on or inside the parabola.

6·15. Equation of the pair of tangents from an external point to the parabola $y^2 = 4ax$.

Let the external point be (x_1, y_1) and let (h, k) be any other point on the tangent drawn from (x_1, y_1).

The equation of the straight line joining the points (x_1, y_1) and (h, k) is given by

$$y - y_1 = \frac{k - y_1}{h - x_1}(x - x_1)$$

or
$$y = \frac{k - y_1}{h - x_1}x + \frac{hy_1 - kx_1}{h - x_1} \quad \ldots(1)$$

As it is tangent to the parabola it must be of the form

$$y = mx + \frac{a}{m} \quad \ldots(2)$$

Comparing (1) and (2),

$$m = \frac{k - y_1}{h - x_1}, \quad \frac{a}{m} = \frac{hy_1 - kx_1}{h - x_1} \quad \ldots(3)$$

Eliminating m in equations (3), we get

$$a = \frac{(k - y_1)(hy_1 - kx_1)}{(h - x_1)^2},$$

or
$$a(h - x_1)^2 = (k - y_1)(hy_1 - kx_1)$$

The above is a relation in (h, k) on the basis of the given condition.

Hence the locus of (h, k) is

$$(y - y_1)(xy_1 - yx_1) = a(x - x_1)^2 \quad \ldots(4)$$

As it is a second degree equation in (x, y) it represents a pair of tangents drawn from (x_1, y_1) to the parabola.

Equation (4) can be rewritten as

$$(y^2-4ax)(y_1^2-4ax_1)=\{yy_1-2a(x+x_1)\}^2$$

or

$$SS_1=T^2$$

where

$$S\equiv y^2-4ax,\ S_1\equiv y_1^2-4ax_1$$

and

$$T\equiv yy_1-2a(x+x_1).$$

6·16. Chord of Contact :—To find the equation of the chord of contact of tangents drawn from a point (x_1, y_1) to the parabola.

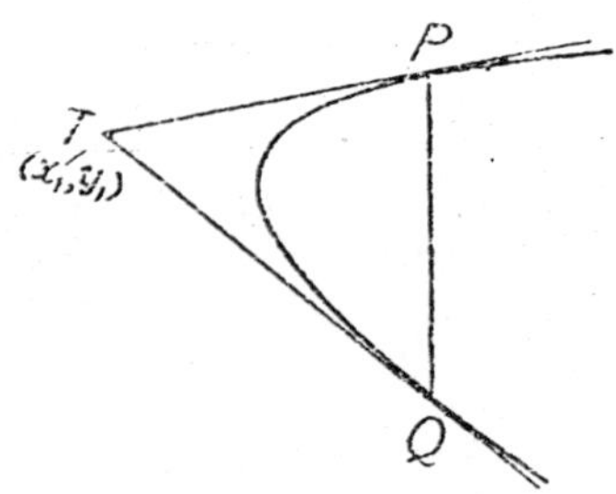

Let $P(p_1, q_1)$ and $Q(p_2, q_2)$ be the points of contact of two tangents drawn from $T(x_1, y_1)$ to the parabola.

The equation to the tangent at P is,

$$yq_1=2a(x+p_1)$$

It passes through $T(x_1\ y_1)$.

$$\therefore\quad y_1q_1=2a(x_1+p_1) \qquad \ldots(1)$$

Similarly the tangent at Q passes through (x_1, y_1).

$$\therefore\quad y_1q_2=2a(x_1+p_2) \qquad \ldots(2)$$

The equation to the chord of contact PQ is given by

$$y-q_1=\frac{q_2-q_1}{p_2-p_1}(x-p_1) \qquad \ldots(3)$$

from which p_1, q_1, p_2, q_2 are to be eliminated with the help of equations (1) and (2).

Subtracting (1) from (2), we get

$$y_1(q_2-q_1)=2a(p_2-p_1)$$

or

$$\frac{q_2-q_1}{p_2-p_1}=\frac{2a}{y_1}.$$

Substituting this value of $\frac{q_2-q_1}{p_2-p_1}$ in (3), we get

$$y-q_1=\frac{2a}{y_1}(x-p_1)$$

or
$$yy_1=y_1q_1+2a(x-p_1)$$
$$=2ax+y_1q_1-2ap_1$$
$$=2ax+2ax_1 \qquad \text{from (1)}$$
$$=2a(x+x_1),$$

which is the equation of the chord of contact.

Note. If the point (x_1, y_1) lies within the parabola, the two tangents and their points of contact are imaginary. Let the chord of contact be real.

If (x_1, y_1) lies on the parabola, the chord of contact is the tangent at the point.

6·17. Polar : The locus of the points of intersection of tangents drawn to a conic at the ends of the chords passing through a fixed point is called the polar of the point. The polar is always a straight line. The fixed point is called the pole of the polar.

To find the equation of the polar of a point (x_1, y_1) with respect to the parabola $y^2=4ax$.

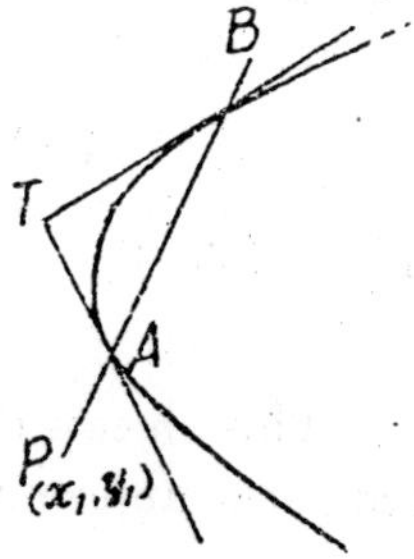

Let a chord PAB be drawn through the point P (x_1, y_1). Let the tangents at A and B meet each other at $T(h, k)$. Then AB is the chord of contact of tangents drawn from $T(h, k)$. Hence its equation is

$$yk=2a(x+h) \qquad \dots \quad \dots(1)$$

As it passes through the point (x_1, y_1),

$$y_1k=2a(x_1+h).$$

This is a relation between h and k on the basis of the given condition.

Hence the locus of the point (h, k) is given by

$$yy_1=2a\,(x+h)$$

which is the polar of the point (x_1, y_1).

Corollary. The polar of the focus $(a, 0)$ is $(x+a)=0$, the directrix of the parabola.

6·18. Pole of a straight line.

To find the pole of the straight line $lx+my=n$ with respect to the parabola.

Let the pole be (x_1, y_1). Then the polar is

$$yy_1=2a\,(x+x_1) \qquad \ldots \quad \ldots(1)$$

But the given polar is $lx+my=n \qquad \ldots \quad \ldots(2)$

Comparing (1) and (2),

$$\frac{-2a}{l}=\frac{y_1}{m}=\frac{2ax_1}{n};$$

$$\therefore \quad y_1=\frac{-2am}{l} \text{ and } x_1=\frac{-n}{l}$$

which give the coordinates of the pole.

6·19. Propositions on Polars.

(1) The chord of a parabola having a given point as its middle point is parallel to the polar of that point.

Let (h, k) be the middle point of the chord. The equation to the chord is

$$y-k=m\,(x-h).$$

This meets the parabola $y^2=4ax$ in points whose ordinates are given by

$$y-k=m\left(\frac{y^2}{4a}-h\right)$$

or

$$m\frac{y^2}{4a}-y+k-mh=0.$$

If y_1 and y_2 are the ordinates of the points of intersection, then

$$y_1+y_2=\frac{4a}{m}. \qquad \ldots \quad \ldots(1)$$

But

$$\frac{y_1+y_2}{2}=k. \qquad \ldots \quad \ldots(2)$$

From (1) and (2),

$$k=\frac{2a}{m} \quad \text{or} \quad m=\frac{2a}{k}.$$

Hence the slope of the chord is $2a/k$ which is also the slope of the polar $ky=2a(x+h)$ of (h, k).

(2) If the polar of P with respect to a parabola passes through Q, then the polar of Q passes through P.

Let the points P and Q be (x_1, y_1) and (x_2, y_2) respectively.

The polar of P is $yy_1=2a(x+x_1)$.

If this passes through Q (x_2, y_2),

$$y_1.y_2=2a(x_1+x_2).$$

But this is also the condition that the point (x_1, y_1) lies on $yy_2=2a(x+x_2)$, the polar of the point (x_2, y_2).

6·20. Chord with a given middle point.

To find the equation of a chord whose middle point is (x_1, y_1).

The required chord is parallel to the polar of (x_1, y_1) which is

$$yy_1=2a(x+x_1).$$

It passes through the point (x_1, y_1).

Hence its equation is

$$y-y_1=\frac{2a}{y_1}(x-x_1).$$

6·21. Diameters : The locus of the middle points of a system of parallel chords of a parabola is called a diameter and the chords are called its double ordinates.

To find the locus of the middle point of a system of parallel chords of the parabola $y^2=4ax$.

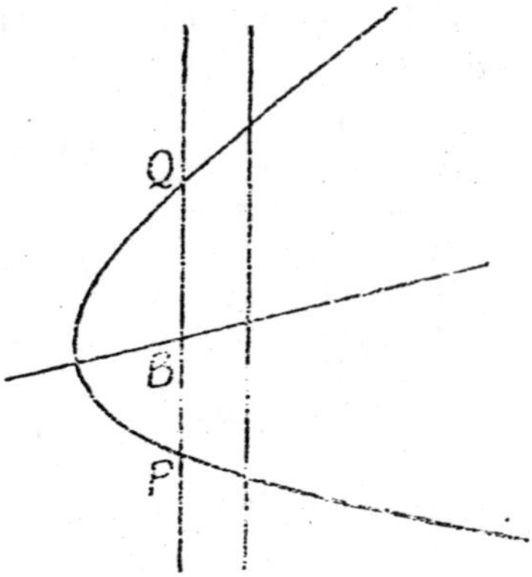

Let the equation to one of the chords of the system be $y=mx+c$ where m is constant and C the variable for different chords. Let the coordinates of the middle point B be (h, k). Let the coordinates of P and Q the point of intersection with the parabola be (x_1, y_1) and (x_2, y_2) respectively. The ordinates of the points of intersection are given by the equation

$$y=m.\frac{y^2}{4a}+c \quad \text{or} \quad my^2-4ay+4ac=0 \qquad \ldots(1)$$

obtained from $y=mx+c$ and $y^2=4ax$ by eliminating x.

If y_1 and y_2 are the roots of (1), then

$$y_1+y_2=\frac{4a}{m}. \qquad \ldots(2)$$

Also (h, k) is the middle point of the line joining (x_1, y_1) and (x_2, y_2).

$$\therefore \quad \frac{y_1+y_2}{2}=k=\frac{2a}{m} \text{ from (2).} \qquad \ldots(3)$$

Hence for all chords the ordinate of the middle point satisfies equation (3) which does not contain the variable c.

$\therefore$ the locus of the middle point (h, k) is the line

$$y=\frac{2a}{m}$$

which is parallel to the axis of the parabola $y=\frac{2a}{m}$ meets the parabola at the point $\left(\dfrac{a}{m^2}, \dfrac{2a}{m}\right)$.

The tangent at this point is

$$y=mx+\frac{a}{m},$$

which is parallel to the given system of chords.

Hence the tangent at the point where the diameter meets the curve is parallel to the ordinates.

6·22. The tangents at the extremities of a chord meet on the diameter which bisects the chord.

Let the chord be $y=mx+c$ ...(1)

Let the tangents at the extremities of the chord meet at (x_1, y_1).

Then the chord is the chord of contact of (x_1, y_1) and its equation is

$$yy_1=2a(x+x_1) \qquad \text{...(2)}$$

Equations (1) and (2) represent the same line. Hence on comparison,

$$m=\frac{2a}{y_1}.$$

or
$$y_1=\frac{2a}{m},$$

so that the point (x_1, y_1) lies on the diameter of the chord

$$y=mx+c.$$

6·23. Solved Examples.

Ex. 1. *Show that if tangents be drawn to the parabola $y^2=4ax$ from a point on the line $x+4a=0$, their chords of contact will subtend a right angle at the vertex.* (*Agra 1947*)

Let the coordinates of the point be (x_1, y_1). It lies on the line $x+4a=0$.

$$\therefore \quad x_1+4a=0 \qquad \text{...(1)}$$

The equation to the chord of contact of the point (x_1, y_1) is

$$yy_1 = 2a(x + x_1) \quad \ldots(2)$$

The straight lines joining the vertex to the points of intersection of the line (2) and the parabola $y^2 = 4ax$ is given by

$$y^2 = 4ax\left(\frac{yy_1 - 2ax}{2ax_1}\right)$$

or

$$4ax^2 + x_1 y^2 = 2xyy_1$$

If the above lines are at right angles

$$x_1 + 4a = 0.$$

But this is already satisfied by (1) Hence the proposition.

Ex. 2. *Show that the polar of any point on the circle*

$$x^2 + y^2 - 2ax - 3a^2 = 0.$$

with respect to the circle $x^2 + y^2 + 2ax - 3a^2 = 0$ *will touch the parabola*

$$y^2 + 4ax = 0. \quad \textit{(Agra 1947)}$$

Let (x_1, y_1) be a point on the circle

$$x^2 + y^2 - 2ax - 3a^2 = 0.$$

Then

$$x_1^2 + y_1^2 - 2ax_1 - 3a^2 = 0 \quad \ldots(1)$$

The polar of (x_1, y_1) with respect to the circle

$$x^2 + y^2 + 2ax - 3a^2 = 0$$

is

$$xx_1 + yy_1 + a(x + x_1) - 3a^2 = 0$$

or

$$yy_1 = -x(x_1 + a) - ax_1 + 3a^2 \quad \ldots(2)$$

If this line touches the parabola, it must be of the form

$$y = mx - \frac{a}{m}. \quad \ldots(3)$$

Comparing (2) and (3), we get

$$\frac{1}{y_1} = \frac{m}{-(x_1 + a)} = \frac{1}{m(x_1 - 3a)}. \quad \ldots(4)$$

Eliminating m in (4), we get

$$\frac{-(x_1+a)}{y_1}=\frac{y_1}{(x_1-3a)},$$

or
$$y_1^2=-(x_1-3a)(x_1+a)$$
$$=-(x_1^2-2ax_1-3a^2)$$

or
$$x_1^2+y_1^2-2ax_1-3a^2=0$$

which is the same as (1).

Ex. 3. *Find the locus of the poles of normal chords of the parabola $y^2=4ax$.*

Let a normal to the parabola be $y=mx-2am-am^3$...(1)

Let the coordinates of the pole be (x_1, y_1).

Then the polar is

$$yy_1=2a(x+x_1) \qquad \text{...(2)}$$

so that (1) and (2) represent the same straight line.

Hence comparing coefficients,

$$\frac{1}{y_1}=\frac{m}{2a}=\frac{-2am-am^3}{2ax_1} \qquad \text{...(3)}$$

We have to find the locus of (x_1, y_1). Hence we have to find a relation between (x_1, y_1) which does not contain the variable m.

From equation (3), we get

$$m=\frac{2a}{y_1} \qquad \text{...(4)}$$

and
$$-2m-m^3=\frac{2x_1}{y_1} \qquad \text{...(5)}$$

Eliminating m in the above equation, we get

$$-\frac{4a}{y_1}-\left(\frac{2a}{y_1}\right)^3=\frac{2x_1}{y_1}$$

or
$$-\frac{4a}{y_1}-\frac{8a^3}{y_1^3}=\frac{2x_1}{y_1}$$

or $$-2ay_1^2-4a^3=x_1y_1^2$$

or $$y_1^2(x_1+2a)+4a^3=0.$$

Hence the locus of (x_1, y_1) is

$$y^2(x+2a)+4a^3=0.$$

Examples XII

1. Prove that the locus of the poles of the focal chords of the parabola is the directrix.

2. Find the locus of the poles of chords of the parabola $y^2=4ax$ which subtend a constant angle α at the vertex. *(Utkal 1948)*

3. Show that the locus of the poles of chords of the parabola $y^2=4ax$ which are at a constant distance c from the vertex is $y^2+4a^2(1-x^2/c^2)=0$.

4. Prove that the length of the chord of contact of the tangents drawn from the point (x_1, y_1) to the parabola $y^2=4ax$ is $\sqrt{[(y_1^2+4a^2)(y_1^2-4ax_1)]}/a$. *(Punjab 1947)*

5. Prove that the area of the triangle formed by the tangents from the point (x_1, y_1) to the parabola $y^2=4ax$ and their chord of contact is $\dfrac{(y_1^2-4ax_1)^{3/2}}{2a}$.

6. Prove that the locus of the poles of tangents to the parabola $y^2=4ax$ with respect to the circle $x^2+y^2=2ax$ is the circle $x^2+y^2=ax$. *(C. U. 1948)*

7. Tangents are drawn from a variable point P to the parabola $y^2=4ax$ such that they form a triangle of constant area c^2 with the tangent at the vertex. Show that the locus of P is $x^2(y^2-4ax)=4c^4$. *(Travancore 1949)*

8. Show that the locus of the point of intersection of tangents to $y^2=4ax$ which intercept a constant length d on the directrix is $(y^2+4ax)(x+a)^2=d^2x^2$. *(Travancore 1947)*

9. If tangents be drawn to the parabola $y^2=4ax$ from any point on the parabola $y^2=4bx$, the normals at the

points of contact will meet on the curve

$$y^2 (a-4b)^3+4ab\,(x-2a)^3=0. \qquad \textit{(Saugor 1948)}$$

10. Show that the locus of the middle points of chords of a parabola which passes through a fixed point is a parabola whose latus rectum is half that of the given parabola. *(Agra 1954)*

11. Show that the locus of the middle points of chords of the parabola $y^2=4ax$ which are of constant length $2d$ is

$$(4ax-y^2)\,(y^2+4a^2)=4a^2d^2.$$

12. Find the locus of the middle points of a variable chord of the parabola $y^2=4ax$, which in all positions touches the parabola $y^2+4bx=0$. *(Andhra 1947)*

13. The perpendicular from a point P upon its polar with respect to a parabola is a constant length. Find the locus of P.

14. Show that the locus of the middle points of a variable chord of the parabola $y^2=4ax$, such that the focal distances of its extremities are in the ratio 2 : 1 is

$$9\,(y^2-2ax)^2=4a^2\,(2x-a)\,(4x+a). \qquad \textit{(Poona 1952)}$$

15. The polar of P with respect to a parabola touches a circle whose centre is the focus and whose diameter is equal to the latus rectum. Find the locus of P.

6·24. Oblique axes. To find the equation of a parabola, the axes being a diameter and the tangent at the extremity of that diameter.

With reference to AX and AY as axes let the equation of the parabola be $y^2=4ax$ and let P be any point (x_1, y_1) on it. Let PX' be the diameter at P and let PY' be the tangent at P,

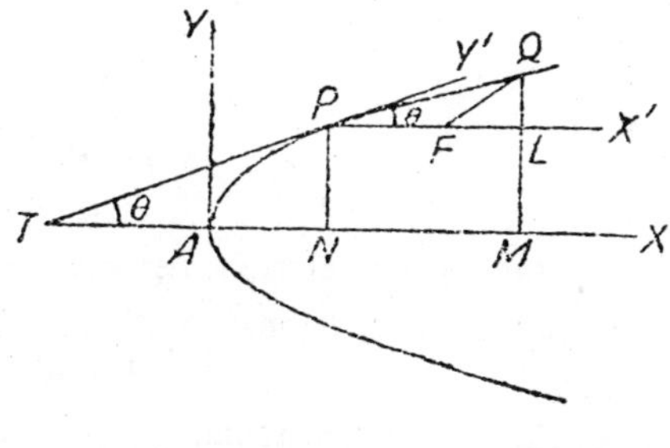

Let PX' and PY' be the new axes and with reference to these axes let the coordinates of any point Q on the parabola be (x, y) so that $PF=x$ and $FQ=y$. The angle between the axes PX' and PY' is now θ, the angle of inclination of the tangent at P to the old axis AX.

Draw QM perpendicular to AX cutting PX' in L and draw PN perpendicular to AX.

Coordinates of Q with reference to old axes are AM and MQ.

But $$AM=AN+PF+FL$$
$$=x_1+x+y\cos\theta$$

and $$MQ=ML+LQ$$
$$=y_1+y\sin\theta.$$

But as Q lies on the parabola $y^2=4ax$, AM and MQ must satisfy this equation.

$$\therefore\quad (y_1+y\sin\theta)^2=4a\,(x_1-x+y\cos\theta)$$

or $$y_1^2+2yy_1\sin\theta+y^2\sin^2\theta=4ax_1+4a\,(x+y\cos\theta)$$

or $$2yy_1\sin\theta+y^2\sin^2\theta=4a\,(x+y\cos\theta)\text{ as }y_1^2=4ax_1\quad\ldots(1)$$

But $$\tan\theta=2a/y_1\quad\ldots(2)$$

being the direction tangent of the tangent at $P\ (x_1, y_1)$ whose equation is $yy_1=2a\,(x+x_1)$.

$$\therefore\quad 2yy_1\sin\theta=2y.2a\cos\theta\qquad\text{from (2).}$$

$\therefore$ equation (1), becomes

$$y^2\sin^2\theta=4ax$$

or $$y^2=4px\quad\ldots(8)$$

writing $p=\dfrac{a}{\sin^2\theta}$.

But (3) is a relation between x and y, the coordinates of any point on the curve with reference to the diameter and tangent at P as the axes of coordinates. Hence (3) is the equation to the parabola.

Note. We know that if we change the origin of coordinates, the direction of axes and the angle between them, we have to substitute expressions of the form $lx+my+n$ and $l_1x+m_1y+n_1$ for x and y respectively. Hence the equation $y^2=4ax$ will change into

$$(l_1x+m_1y+n_1)^2=4a(lx+my+n). \qquad ...(1)$$

This shows that whatever the axes of reference, in the equation of a parabola, the second degree terms always form a perfect square.

Conversely, an equation of the form (1), will always represent a parabola, for if we take $l_1x+m_1y+n_1=0$, as the new y-axis and $lx+my+n=0$ as the new x-axis, then equation (1) can be interpreted to mean that the square of the perpendicular from any point on the parabola to $l_1x+m_1y+n_1=0$ varies as the perpendicular from the same point on the line $lx+my+n=0$. But these perpendiculars are respectively the ordinate and abcissae of the point. Hence the equation with reference to new axes will be of the form $y^2=4px$.

It is obvious that the line $y=mx+c$ will still be a tangent to the parabola $y^2=4px$ if $c=\dfrac{p}{m}$.

Also the tangent at (x_1, y_1) to the parabola is $yy_1=2p(x+x_1)$. The same will be the equation of polar of (x_1, y_1) and the locus of middle points of chords parallel to $y=mx$ is $y=\dfrac{2p}{m}$.

6·25. Conormal points. If the normals at the point $P, Q, R, \ldots$ of a conic pass through a common point, then the points $P, Q, R\ldots$ are conormal points.

(1) To prove that three normals can be drawn from a given point to a parabola.

Let the equation to the parabola be $y^2=4ax$. Then the equation to any normal to the parabola is

$$y=mx-2am-am^3 \qquad \ldots(1)$$

If it passes through a point (h, k),

$$k=mh-2am-am^3$$

or $$am^3+m(2a-h)+k=0 \qquad \ldots(2)$$

which is a cubic in m and therefore gives three values of m.

Hence three normals can be drawn from a point to a parabola.

The three values of m are either all real or one real and two imaginary as imaginary roots exist in pairs.

(2) The sum of the ordinates of the conormal points is zero.

Let the conormal points be P, Q, R and let their coordinates be $(am_1^2, -2am_1)$, $(am_2^2, -2am_2)$ and $(am_3^2, -2am_3)$ respectively.

From equation (2) the sum of roots is zero.

$$\therefore \quad m_1+m_2+m_3=0$$

or $$-2a(m_1+m_2+m_3)=0$$

or $$y_1+y_2+y_3=0, \qquad \ldots(8)$$

if y_1, y_2 and y_3 are the ordinates of the points P, Q and R.

(3) The locus of the point of intersection of the normals at the ends of a system of parallel chords of a parabola is a straight line which is a normal to the parabola.

If P and Q are the extremities of one chord of a system of parallel chords, then we have shown before that $y_1+y_2=\dfrac{2a}{m}$=constant ; when y_1 and y_2 are the coordinates of P and Q, so that from equation (3), y_3=const. *i.e.* R must be a fixed point.

Hence the point of intersection of the normals at P and Q must lie on the normal at R.

(4) *To find the equation to the circle through conormal points.*

If the normals at the points P, Q, R of the parabola meet in the point (h, k), then

$$am^3+m(2a-h)+k=0. \qquad \ldots(1)$$

If y is the ordinate of any point $(am^2, -2am)$ on the parabola, then $\qquad y=-2am.$

so that equation (1) becomes

$$y^3+4a(2a-h)y-8a^2k=0 \qquad \ldots(2)$$

Let the equation to the circle passing through the points P, Q and R be

$$x^2+y^2+2gx+2fy+c=0.$$

The ordinates of the points of intersection of the above circle with the parabola $y^2=4ax$ are given by the equation

$$\frac{y^4}{16a^2}+y^2+2g\frac{y^2}{4a}+2fy+c=0$$

or

$$y^4+8ay^2(g+2a)+32a^2fy+16a^2c=0 \qquad \ldots(3)$$

Three of the roots y_1, y_2, y_3, y_4 of (3) must be the same as the roots of (2). Let y_1, y_2, y_3 be the common roots. Then we know that $\qquad y_1+y_2+y_3=0.$

Also from (3), $\qquad y_1+y_2+y_3=0.$

$$\therefore \quad y_4=0.$$

Hence the circle through the points P, Q, R whose ordinates are y_1, y_2 and y_3 respectively passes through another point on the parabola whose ordinate is zero. This point must be the vertex of the parabola. Hence the circle through the conormal points passes through the origin so that in the equation to the circle, $c=0$.

Equation (3) therefore becomes

$$y^3+8ay(g+2a)+32a^2f=0, \qquad \ldots(4)$$

so that equations (2) and (4) must be identical.

Comparing coefficients,

$$2a-h=2g+4a \qquad \ldots(5)$$

and $$4f=-k. \qquad \ldots(6)$$

$$\therefore \quad 2g=-(h+2a)$$

and $$2f=-\frac{k}{2}.$$

Substituting these values of f and g in the equation to the circle, we get $x^2+y^2-x(h+2a)-\frac{k}{2}y=0$ as the equation to the circle passing through the conormal points P, Q and R.

6·26. Parabola touching two oblique axes :

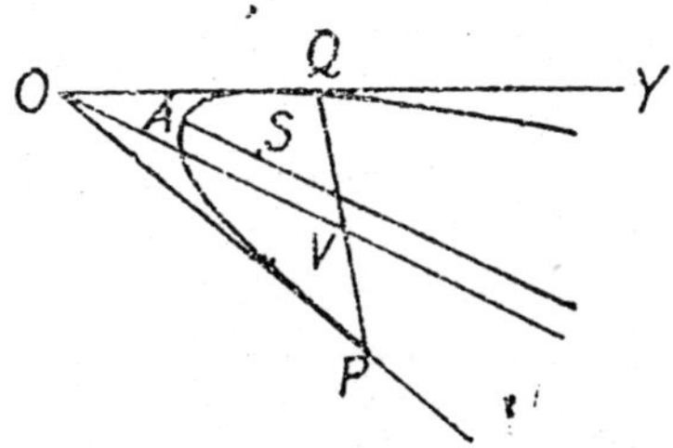

We have shown that the second degree terms in the equation to a parabola form a perfect square. Hence the general equation to a parabola can be written as

$$(Ax+By)^2+2gx+2fy+c=0. \qquad \ldots(1)$$

This meets the x-axis in points whose abscissae are given by

$$A^2x^2+2gx+c=0. \qquad \ldots(2)$$

If the parabola touches the axis of x at a distance h from the origin, then the roots of the above equation are each equal to h so that

$$2h=-\frac{2g}{A^2} \text{ and } h^2=\frac{c}{A^2};$$

$$\therefore \quad g=-A^2h \text{ and } c=A^2h^2. \qquad \ldots(3)$$

Similarly if the parabola touches the axis of y at a distance k from the origin,

$$f=-B^2k \text{ and } c=B^2k^2 \qquad \ldots(4)$$

Equating the values of c from (3) and (4), we get

$$B^2k^2=A^2h^2;$$

$$\therefore \quad B = \pm A\frac{h}{k}. \qquad \ldots(5)$$

Considering the negative sign,

$$B = -A\frac{h}{k},\ g = -A^2h,\ f = -A^2\frac{h^2}{k} \text{ and } c = A^2h^2.$$

Substituting these values in (1), the equation to the parabola becomes

$$\left(Ax - A\frac{h}{k}y\right)^2 - 2A^2hx - \frac{2A^2h^2}{k}y + A^2h^2 = 0$$

or
$$\left(x - \frac{h}{k}y\right)^2 - 2hx - \frac{2h^2}{k}y + h^2 = 0$$

i. e.
$$\left(\frac{x}{h} - \frac{y}{k}\right)^2 - \frac{2x}{h} - \frac{2y}{k} + 1 = 0 \qquad \ldots(6)$$

The above equation can be written as

$$\left(\frac{x}{h} + \frac{y}{k}\right)^2 - \frac{4xy}{hk} - 2\left(\frac{x}{h} + \frac{y}{k}\right) + 1 = 0$$

or
$$\left(\frac{x}{h} + \frac{y}{k} - 1\right)^2 = \frac{4xy}{hk}$$

or
$$\left(\frac{x}{h} + \frac{y}{k} - 1\right) = \pm 2\sqrt{\left(\frac{xy}{hk}\right)}$$

or
$$\sqrt{\left(\frac{x}{h}\right)} + \sqrt{\left(\frac{y}{k}\right)} = 1. \qquad \ldots(6)$$

If we take the positive sign for B, the equation would reduce to

$$\left(\frac{x}{h} + \frac{y}{k} - 1\right)^2 = 0.$$

which represents a pair of coincident straight lines.

Note. The radical signs in (6) can be either positive or negative. In the figure the abscissa of any point on the portion PAQ of the curve is $< a$ and the ordinate $< b$. Hence for this portion of the curve both signs are positive. For points beyond P the abscissa is $> a$ and the ordinates less than b, so that $\frac{x}{a} > \frac{y}{b}$ and the signs must be $+$ and $-$.

The equation for this portion becomes

$$\sqrt{\left(\frac{x}{h}\right)}-\sqrt{\left(\frac{y}{k}\right)}=1.$$

Similarly for points beyond Q, the equation becomes

$$-\sqrt{\left(\frac{x}{h}\right)}+\sqrt{\left(\frac{y}{k}\right)}=1.$$

There is no part of the curve corresponding to two negative signs of the radicals.

6·27. Focus of the parabola

$$\sqrt{\left(\frac{x}{h}\right)}+\sqrt{\left(\frac{y}{k}\right)}=1.$$

Let S be the focus of the parabola. It can be shown that angles OSP and OSQ are equal and $OS^2=SP.SQ$, so that triangles OSP and OSQ are similar and therefore angles SOP and SQO are equal.

If a circle is drawn through the points O, S and Q, the line OP will touch it at O.

The focus S therefore lies on the circle through the point $(0, k)$ and touches the axis of X at the origin. The equation of this circle is

$$x^2+2xy\cos\omega+y^2=ky$$

angle POQ being ω.

The focus S also lies on the circle through P and touching the axis of y at the origin.

This circle is

$$x^2+2xy\cos\omega+y^2=hx.$$

The point of intersection of these circles is the focus and hence its coordinates are

$$\left(\frac{hk^2}{h^2+2hk\cos\omega+k^2}, \frac{h^2k}{h^2+2hk\cos\omega+k^2}\right).$$

6·28. Axis of the parabola

$$\sqrt{\left(\frac{x}{h}\right)}+\sqrt{\left(\frac{y}{k}\right)}=1.$$

We know that the tangents at the ends of a chord of a parabola meet on the diameter of the chord ; the line through Q and the middle point V of PQ will be parallel to the axis AS of the parabola.

The coordinates of V are therefore

$$\left(\frac{h}{2}, \frac{k}{2}\right).$$

The equation of OR is $y=\frac{k}{h}x$.

The axis is a line parallel to this and is therefore

$$hy-kx=c.$$

But it passes through the focus

$$\left(\frac{hk^2}{h^2+2hk\cos\omega+k^2}, \frac{h^2k}{h^2+2hk\cos\omega+k^2}\right)$$

$$\therefore \quad c=\frac{h^3k-hk^3}{h^2+2hk\cos\omega+k^2}.$$

Hence the equation to the axis is

$$hy-kx=\frac{hk\,(h^2-k^2)}{h^2+2hk\cos\omega+k^2}.$$

6·29. Vertex of the parabola

$$\sqrt{\left(\frac{x}{h}\right)}+\sqrt{\left(\frac{y}{k}\right)}=1.$$

The vertex is the point of intersection of the axis

$$hy-kx=\frac{hk\,(h^2-k^2)}{h^2+2hk\cos\omega+k^2} \qquad \ldots \quad \ldots (1)$$

and the parabola

$$\sqrt{\left(\frac{x}{h}\right)}+\sqrt{\left(\frac{y}{k}\right)}=1.$$

The equation to the parabola can be written as

$$\left(\frac{x}{h}-\frac{y}{k}+1\right)^2=\frac{4x}{h} \qquad \ldots \quad \ldots (2)$$

Solving (1) and (2), the vertex is

$$\left(\frac{hk^2\,(k+h\cos\omega)^2}{(h^2+2hk\cos\omega+k^2)^2},\ \frac{h^2k\,(h+k\cos\omega)^2}{(h^2+2hk\cos\omega+k^2)^2}\right).$$

6·30. Directrix and latus rectum of the parabola

$$\sqrt{\left(\frac{x}{h}\right)}+\sqrt{\left(\frac{y}{k}\right)}=1.$$

We know that the point of intersection of perpendicular tangents lies on the directrix. Let $(f, 0)$ be a point on OX.

A straight line through the point $(f, 0)$ perpendicular to OX is

$$y=m\,(x-f) \text{ where } 1+m\cos\omega=0.$$

The equation to this line is therefore

$$x+y\cos\omega=f \qquad \ldots \quad \ldots(1)$$

This straight line touches the parabola, if

$$\frac{f}{h}+\frac{f}{k\cos\omega}=1$$

i.e. if

$$f=\frac{hk\cos\omega}{h+k\cos\omega}.$$

The point $\left(\dfrac{hk\cos\omega}{h+k\cos\omega}, 0\right)$ therefore lies on the directrix.

Similary, we can find a point $(0, g)$ on OY such that the line through this point perpendicular to OY touches the parabola. The coordinates of that point are

$$\left(0, \frac{kh\cos\theta}{k+h\cos\omega}\right).$$

The directrix is the straight line joining the points $(f, 0)$ and $(0, g)$. Hence its equation is

$$x\,(h+k\cos\omega)+y\,(k+h\cos\omega)=hk\cos\omega \ldots \quad \ldots(2)$$

The latus rectum is twice the perpendicular from the focus on the directrix. Hence it is

$$\frac{4h^2k^2 \sin^2 \omega}{(h^2+2hk \cos \omega+k^2)^{3/2}}.$$

6·31. Tangent at the vertex of the parabola

$$\sqrt{\left(\frac{x}{h}\right)}+\sqrt{\left(\frac{y}{k}\right)}=1.$$

This is a line parallel to the directrix and passing through the vertex. Hence its equation is

$$(h+k \cos \omega)\left[x-\frac{hk^2\,(k+h \cos \omega)^2}{(h^2+2hk \cos \omega+k^2)^2}\right]$$

$$+(k+h \cos \omega)\left[y-\frac{h^2k\,(h+k \cos \omega)^2}{(h^2+2hk \cos \omega+k^2)^2}\right]=0$$

or
$$\frac{x}{k+h \cos \omega}+\frac{y}{h+k \cos \omega}=\frac{hk}{h^2+2hk \cos \omega+k^2}.$$

6·32. Solved Examples.

Ex. 1. *Show that the locus of points such that two of the three normals drawn from them to the parabola $y^2=4ax$ coincide is*

$$27ay^2=4\,(x-2a)^3 \qquad \textit{(Agra 1954)}$$

If the normal at the point $(am^2, -2am)$ passes through the point (h, k),

$$am^3+m\,(2a-h)+k=0 \qquad \ldots \quad \ldots(1)$$

If m_1, m_1, m_2 are the roots of this equation, two being coincident.

$$\text{sum of roots}=2m_1+m_2=0 \qquad \ldots \quad \ldots(2)$$

product of roots in pairs

$$=m_1^2+m_1m_2+m_1m_2=\frac{2a-h}{a}, \qquad \ldots \quad \ldots(3)$$

$$\text{product of the roots}=m_1^2m_2=-\frac{k}{a} \qquad \ldots \quad \ldots(4)$$

From (2), $\qquad m_2=-2m_1$

Substituting in (3),

$$m_1^2 - 2m_1 . 2m_1 = \frac{2a-h}{a}.$$

or

$$-3m_1^2 = \frac{2a-h}{a} \qquad \ldots \quad \ldots (5)$$

From (4),

$$-2m_1^3 = -\frac{k}{a} \qquad \ldots \quad \ldots (6)$$

Eliminating m_1 between (5) and (6), we get

$$\frac{k^2}{4a^2} = \frac{(h-2a)^3}{27a^3}$$

or

$$27ak^2 = 4\,(h-2a)^3.$$

Hence the locus of (h, k) is $27ay^2 = 4\,(x-2a)^3$.

Ex. 2. *The normals at the points Q, R on $y^2 - 4ax = 0$ meet the parabola at the point P. Prove that the locus of the circum-centre of the triangle PQR is the parabola $2y^2 - ax + a^2 = 0$.*

(Utkal 1946)

The equation to the circle through PQR is

$$x^2 + y^2 - x\,(h+2a) - \frac{k}{2}\,y = 0.$$

If α and β are the coordinates of the centre of this circle

$$\alpha = \frac{h+2a}{2} \qquad \ldots \quad \ldots (1)$$

and

$$\beta = +\frac{k}{4} \qquad \ldots \quad \ldots (2)$$

Also because two normals at Q and R meet at P

$$k^2 = 4ah \qquad \ldots \quad \ldots (3)$$

From (1), (2) and (3)

$$16\beta^2 = 8a\,(\alpha - a)$$

or

$$2\beta^2 = a\,(\alpha - a)$$

Hence the locus of (α, β) is

$$2y^2 - ax + a^2 = 0.$$

Ex. 3. *If the three normals from a point to the parabola* $y^2=4ax$ *cut the axis in points whose distances from the vertex are in A. P. Show that the point lies on the curve* $27ay^2=2(x-2a)^3$.

(*Agra 51*)

As before if m_1, m_2, m_3 are the roots of the equation

$$am^3+m(2a-h)+k=0,$$

$$m_1+m_2+m_3=0 \qquad \ldots \qquad \ldots(1)$$

$$m_1m_2+m_2m_3+m_3m_1=\frac{2a-h}{a}, \qquad \ldots \qquad \ldots(2)$$

$$m_1m_2m_3=-\frac{k}{a} \qquad \ldots \qquad \ldots(3)$$

The equation to the normal is

$$y=mx-2am-am^3.$$

It cuts the x-axis at points whose abscissae are $2a+am_1^2$, $2a+am_2^2$ and $2a+am_3^2$. These distances are in A. P.

$$\therefore \quad 4a+2am_2^2=4a+a(m_1^2+m_3^2)$$

or $$m_1^2+m_3^2=2m_2^2$$

or $$(m_1+m_3)^2-2m_1m_3=2m_2^2$$

or $$m_2^2+\frac{2k}{am_2}=2m_2^2 \text{ [from (1) and (3)]}$$

or $$m_2^3=\frac{2k}{a} \qquad \ldots \qquad \ldots(4)$$

From (2), $m_2(m_1+m_3)+m_1m_3=\dfrac{2a-h}{a}$

or $$-m_2^2-\frac{k}{am_2}=\frac{2a-h}{a}$$

or $$k+am_2^3=m_2(h-2a)$$

or $$3k=m_2(h-2a) \quad \text{[from (4)]}$$

or $$27k^3=m_2^3(h-2a)^3$$

$$=\frac{2k}{a}(h-2a)^3 \qquad \text{[from (4)]}$$

or $$27ak^2=2(h-2a)^3$$

Hence the locus of (h, k) is $27ay^2=2(x-2a)^3$.

Ex. 4. *Three normals at P, Q, R on the parabola $y^2=4ax$ meet in a point on the line $y=k$. Prove that the sides of the triangle PQR touch the parabola $x^2=2ky$.*

The line joining the points $(am_1^2, -2am_1)$ and $(am_2^2, -2am_2)$ is given by

$$y+2am_1=-\frac{2}{m_1+m_2}(x-am_1^2).$$

This touches the parabola $x^2-2ky=0$ if the equation

$$x^2=-2k\left[-\frac{2(x-am_1^2)}{m_1+m_2}-2am_1\right]$$

has equal roots,

i. e. if $$\frac{k}{a}=(m_1+m_2)\,m_1m_2$$

or if $$-\frac{k}{a}=m_1m_2m_3 \text{ as } m_1+m_2=-m_3$$

which is always true.

Ex. 5. *Find the equation to the tangent at the point P (x_1, y_1) of the parabola*

$$\sqrt{\left(\frac{x}{h}\right)}+\sqrt{\left(\frac{y}{k}\right)}=1$$

Take another point Q (x_2, y_2) on the curve. The line joining P and Q is

$$y-y_1=\frac{y_2-y_1}{x_2-x_1}(x-x_1) \qquad \ldots \quad \ldots (1)$$

But P and Q lie on the parabola

$$\sqrt{\left(\frac{x}{h}\right)}+\sqrt{\left(\frac{y}{k}\right)}=1.$$

$$\therefore \quad \sqrt{\left(\frac{x_1}{h}\right)}+\sqrt{\left(\frac{y_1}{k}\right)}=1 \qquad \ldots \quad \ldots (2)$$

and $$\sqrt{\left(\frac{x_2}{h}\right)}+\sqrt{\left(\frac{y_2}{k}\right)}=1 \qquad \ldots \quad \ldots (3)$$

Subtracting (2) from (3), we get

$$\frac{\sqrt{x_2}-\sqrt{x_1}}{\sqrt{h}}+\frac{\sqrt{y_2}-\sqrt{y_1}}{\sqrt{k}}=0$$

or
$$\frac{\sqrt{y_2}-\sqrt{y_1}}{\sqrt{x_2}-\sqrt{x_1}}=-\frac{\sqrt{k}}{\sqrt{h}}$$

Substituting this in (1), we get

$$y-y_1=-\frac{\sqrt{k}}{\sqrt{h}}\frac{\sqrt{y_2}+\sqrt{y_1}}{\sqrt{x_2}+\sqrt{x_1}}(x-x_1).$$

This line becomes the tangent at P if Q approaches P *i.e.* $x_2=x_1$ and $y_2=y_1$.

Hence the equation of the tangent at P is given by

$$y-y_1=-\frac{\sqrt{k}}{\sqrt{h}}\frac{\sqrt{y_1}}{\sqrt{x_1}}(x-x_1)$$

or
$$\frac{x}{\sqrt{(hx_1)}}+\frac{y}{\sqrt{(ky_1)}}=\sqrt{\left(\frac{y_1}{k}\right)}+\sqrt{\left(\frac{x_1}{h}\right)}$$
$$=1 \text{ from (2).}$$

The equation of the tangent at (x_1, y_1) is

$$\frac{x}{\sqrt{(hx_1)}}+\frac{y}{\sqrt{(ky_1)}}=1.$$

Ex. 6. *Find the condition that the line* $\frac{x}{a}+\frac{y}{b}=1$ *may touch the parabola*

$$\sqrt{\left(\frac{x}{h}\right)}+\sqrt{\left(\frac{y}{k}\right)}=1.$$

Let (x_1, y_1) be the point of contact. The equation to the tangent at (x_1, y_1) is

$$\frac{x}{\sqrt{(hx_1)}}+\frac{y}{\sqrt{(ky_1)}}=1 \qquad \ldots \quad \ldots \quad \ldots(1)$$

But the tangent is given to be

$$\frac{x}{a}+\frac{y}{b}=1. \qquad \ldots \quad \ldots \quad \ldots(2)$$

Comparing (1) and (2), we get

$$\frac{\sqrt{(hx_1)}}{a}=\frac{\sqrt{(ky_1)}}{b}=1.$$

$$\therefore \quad \sqrt{x_1}=\frac{a}{\sqrt{h}} \quad \text{or} \quad \sqrt{\left(\frac{x_1}{h}\right)}=\frac{a}{h}.$$

Also
$$\sqrt{\left(\frac{y_1}{k}\right)}=\frac{b}{k}.$$

But (x_1, y_1) is a point on the parabola

$$\sqrt{\left(\frac{x_1}{h}\right)}+\sqrt{\left(\frac{y_1}{k}\right)}=1.$$

$$\therefore \quad \frac{a}{h}+\frac{b}{k}=1.$$

Ex. 7. *A parabola is so drawn that it touches two rectangular axes and its focus is at a distance c from the origin. Show that locus of its vertex is the curve*

$$x^{2/3}+y^{2/3}=c^{2/3}.$$

If h and k are the distances of the points of contact with the axes from the origin, the coordinates of focus are

$$\left(\frac{hk^2}{h^2+k^2}, \frac{h^2k}{h^2+k^2}\right).$$

The distance of the focus from the origin, c is given by

$$c^2=\frac{h^2k^2\,(h^2+k^2)}{(h^2+k^2)}$$

$$=\frac{h^2k^2}{(h^2+k^2)}. \qquad \ldots \quad \ldots \quad \ldots (1)$$

If (x, y) are the coordinates of the vertex, then

$$x=\frac{hk^4}{(h^2+k^2)^2} \qquad \ldots \quad \ldots \quad \ldots (2)$$

and
$$y=\frac{h^4k}{(h^2+k^2)^2}. \qquad \ldots \quad \ldots \quad \ldots (3)$$

Now
$$x^{2/3}+y^{2/3}=\frac{h^{2/3}k^{8/3}+h^{8/3}k^{2/3}}{(h^2+k^2)^{4/3}}$$

$$=\frac{h^{2/3}k^{2/3}(h^2+k^2)}{(h^2+k^2)^{4/3}}$$

$$=\frac{h^{2/3}k^{2/3}}{(h^2+k^2)^{1/3}}$$

$$=c^{2/3} \text{ from (1).}$$

Ex. 8. *A parabola of latus rectum 'l' slides between two rectangular axes, show that the locus of the focus is*

$$16x^2y^2=l^2(x^2+y^2).$$

If h and k are the distances of the points of contact of the parabola with the axes, form the origin, the coordinates of the focus (x, y) are given by

$$x=\frac{hk^2}{h^2+k^2}, \qquad \ldots \quad \ldots \quad \ldots (1)$$

$$y=\frac{h^2k}{h^2+k^2}, \qquad \ldots \quad \ldots \quad \ldots (2)$$

ω being 90°.

The latus rectum is l.

$$\therefore \quad l=\frac{4h^2k^2}{h^2+k^2}. \qquad \ldots \quad \ldots \quad \ldots (3)$$

From (1) and (2)

$$x^2+y^2=\frac{h^2k^2(h^2+k^2)}{(h^2+k^2)^2}$$

$$=\frac{h^2k^2}{h^2+k^2}.$$

Also $$x^2y^2=\frac{h^6k^6}{(h^2+k^2)^4}.$$

$$\therefore \quad l^2=\frac{16h^4k^4}{(h^2+k^2)^2}$$

$$=16\frac{x^2y^2}{x^2+y^2}$$

or $$16x^2y^2=l^2(x^2+y^2).$$

Examples XIII.

1. Show that three normals can be drawn to a parabola from any point, and that one of these at least must be real. What is the condition that only two normals may be drawn from a point (α, β) to the parabola $y^2=4ax$? Find the locus of such points (α, β). *(Agra 1946)*

2. P, Q are the points t_1, t_2 on the parabola $y^2=4ax$; the normals at P and Q meet on the parabola; show that $t_1t_2=2$ and that the middle point of PQ lies on the parabola $x^2=2a\ (x+2a)$. *(Agra 1938)*

3. From the point where any normal to the parabola $y^2=4ax$ meets the axis is drawn a line perpendicular to this normal. Prove that this line always touches an equal parabola. *(C. U. 1945)*

4. If the normals at three points P, Q, R of a parabola meet at a point O and S be the focus, prove that
$$SP.SQ.SR=a.OS^2.$$

5. The normals at P, Q, R on the parabola $y^2=4ax$ meet at a point on the line $x=\alpha$. Prove that the sides of the triangle PQR touch the parabola
$$y^2=16a\ (x+2a-\alpha).$$ *(Utkal 1947)*

6. Find the orthocentre of the triangle formed by joining the three points on a parabola, the normals at which pass through a given point (h, k).

   If the point lies on the parabola itself, show that the locus of the orthocentre is the parabola
$$y^2=a\ (x+6a).$$

7. If the normals at P, Q, R to a parabola are concurrent, then the centroid of the $\triangle PQR$ lies on the axis.

8. From a point (h, k) three normals are drawn to the

parabola $y^2=4ax$. If two of the normals make complementary angles with the axis, find the locus of (h, k).

9. Prove that the circle circumscribing the triangle formed by any three tangents to a parabola passes through the focus.

10. A parabola whose latus rectum is l, slides in contact with each of two fixed perpendicular straight lines. Determine the locus of its vertex referred to these lines as axes.

11. A parabola touches two given straight lines if its axis passes through the point (h, k), referred to these lines as axes, prove that the focus lies on the curve

$$x^2-y^2-hx+ky=0.$$

12. If the directrices of the parabolas

$$\sqrt{\frac{x}{h}}+\sqrt{\frac{y}{k}}=1$$

where h and k are variable, all pass through a fixed point (p, q), prove that all parabolas touch the line

$$\frac{x}{p+q \sec \omega}+\frac{y}{q+p \sec \omega}=1,$$

ω being the angle between the coordinate axes.

13. If the axes are rectangular, the locus of the focus of the parabola $\sqrt{\frac{x}{h}}+\sqrt{\frac{y}{k}}=1$, h and k being variable such that $hk=c^2$ is the curve $(x^2+y^2)^2=c^2xy$.

14. Find the focus and directrix of the parabola

$$\sqrt{(ax)}+\sqrt{(by)}=1.$$

If the axis of the parabola $\sqrt{(ax)}+\sqrt{(by)}=1$ pass through a fixed point, prove that the locus of the focus will be a rectangular hyperbola.

[*Allahabad M. Sc. (Pre.) 1949*]

15. If a and b be the lengths of two tangents to a parabola

which intersect at right angles and m be one-quarter of the latus rectum, show that

$$\frac{a^{2/3}}{b^{4/3}}+\frac{b^{2/3}}{a^{4/3}}=\frac{1}{m^{2/3}}.$$

16. The normals at P, Q, R are concurrent and PQ meets the diameter through R on the directrix. Prove that it touches the parabola $y^2+16a(x+a)=0$.

17. If tangents be drawn to $y^2-4ax=0$ from any point on $y^2-4bx=0$, the normals at the points of contact will meet on the curve $y^2(a-4b)^3+4ab(x-2a)^3=0$.

18. The triangle PQR is inscribed in the parabola $y^2-4ax=0$ and PQ, PR pass through the point $(0, 4a)$, $(0, -4a)$ respectively. Prove that PQ touches the circle $x^2+y^2-4ax=0$.

CHAPTER VII

THE ELLIPSE

7·1. Definition.

The locus of a point which moves so that its distance from a fixed point bears to its distance from a fixed straight line a constant ratio, which is less than unity, is an *ellipse.*

The fixed point is called the focus, the fixed line, the directrix and the fixed ratio, the eccentricity of the ellipse. Eccentricity is denoted by letter e, so that in case of ellipse

$$e < 1.$$

To find the standard equation of an ellipse.

Let S be the fixed point, the focus and ZK is the fixed line, the directrix. Draw SZ perpendicular from S on the fixed line ZK. Since $e < 1$ the straight line ZS can be divided

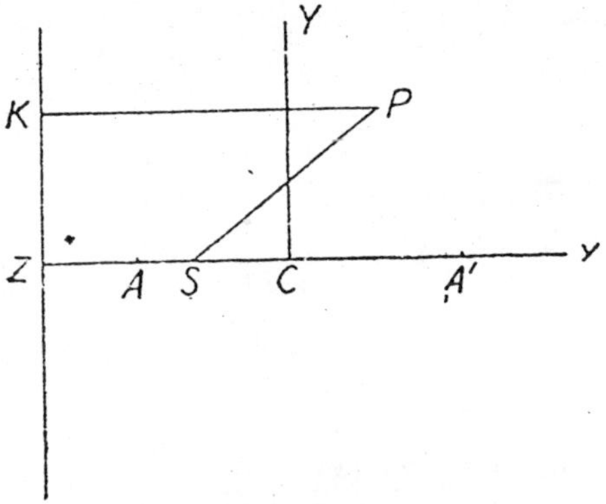

internally and externally and at the points A and A' respectively in the ratio $e : 1$. Hence A is a point on the locus so that

$$\frac{AS}{AZ}=e$$

or $$AS=e.AZ \qquad \ldots(1)$$

and $$\frac{A'S}{A'Z}=e$$

or $$A'S=e.A'Z \qquad \ldots(2)$$

Let $AA'=2a$ and let C be the middle point of AA'.

From (1) and (2) by addition,

$$AS+A'S=e\,(AZ+A'Z)$$

or $$2a=e\,(CZ-CA+CZ+CA')$$
$$=2e.CZ \qquad \text{as } CA=CA'$$

$$\therefore \quad CZ=\frac{a}{e}.$$

Also from (1) and (2) by subtraction,

$$A'S-AS=e\,(CZ+CA'-CZ+CA)$$

or $$(CS+CA'-CA+CS)=2ae$$

$$\therefore \quad 2CS=2ae,$$

or $$CS=ae$$

Take C as the origin and CA' and a perpendicular line CY as the axes of X and Y respectively. Then the coordinates of S are $(-ae, 0)$ and the equation to the directrix ZK

is $$x=-\frac{a}{e}.$$

Let the coordinates of P, any point on the ellipse be (x, y).

Now $$SP=e.PK.$$

But $$SP=\sqrt{[(x+ae)^2+y^2]}$$

and $$PK=x+\frac{a}{e}.$$

Substituting in (3), we get

$$(x+ae)^2+y^2=e^2\left[x+\frac{a}{e}\right]^2,$$

or $$x^2+2xae+a^2e^2+y^2=e^2x^2+2xae+a^2$$

or $$x^2(1-e^2)+y^2=a^2(1-e^2)$$

or $$\frac{x^2}{a^2}+\frac{y^2}{a^2(1-e^2)}=1.$$

Putting $b^2=a^2(1-e^2)$, the above equation becomes

$$\frac{x^2}{a^2}+\frac{y^2}{b^2}=1,$$

which is the equation of the ellipse.

7·2. Tracing of the ellipse, $\frac{x^2}{a^2}+\frac{y^2}{b^2}=1$.

Solving the equation for y, we get

$$y=\pm\frac{b}{a}\sqrt{(a^2-x^2)}.$$

For every value of x, there are two equal and opposite values of y.

Hence the curve is symmetrical about the X-axis.

Similarly $$x=\pm\frac{a}{b}\sqrt{(b^2-y^2)}$$

which shows that the curve is also symmetrical about the axis of y. Hence the shape of the curve can be completely determined by tracing it only in the first quadrant.

If $x=0$, $y=b$, so that $(0, b)$ is a point on the ellipse. Similarly $(a, 0)$ is a point on the ellipse. As x increases from 0 to a, y goes on decreasing from b to 0.

If $x > a$, y is imaginary which shows that there is no part of the curve, so the right of point $(a, 0)$. Thus the ellipse is a closed curve as shown below,

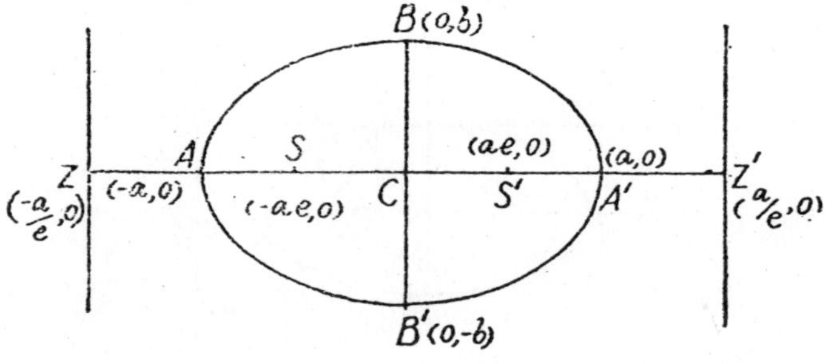

As the curve is symmetrical about the axis of Y, there is another focus S' to the right of C and another directrix to the right of A' passing through the point Z' whose coordivates are $\left(\frac{a}{e}, 0\right)$.

AA' is called the major axis, its length is $2a$. BB' is called the minor axis its length is $2b$. The coordinates of the foci are $(ae, 0)$ and $(-ae, 0)$. The directrices are

$$x=\frac{a}{e} \text{ and } x=-\frac{a}{e}.$$

LSL' is the latus rectum. Coordinates of S' are $(ae, 0)$. Let the ordinate of the point L be l. Then the point (ae, l) lies on the ellipse.

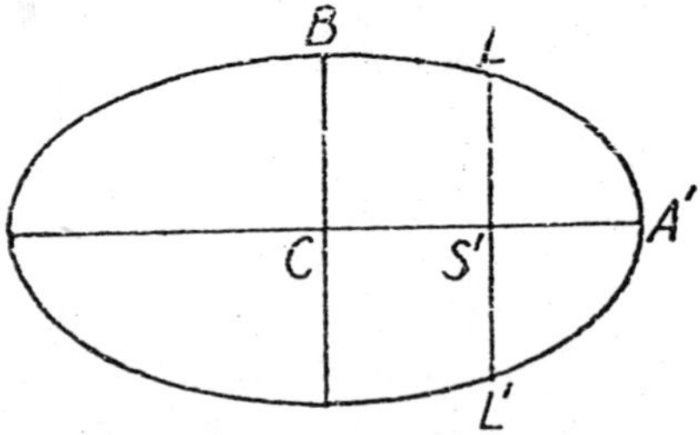

$$\therefore \quad \frac{a^2e^2}{a^2}+\frac{l^2}{b^2}=1,$$

or $$l^2=b^2(1-e^2)$$
but $$b^2=a^2(1-e^2);$$

$$\therefore \quad l^2=\frac{b^4}{a^2}.$$

or semi latus rectum $$l=\frac{b^2}{a}.$$

7·3. Focal distances of a point.

Let P be a point (x_1, y_1) on the ellipse in the given figure,

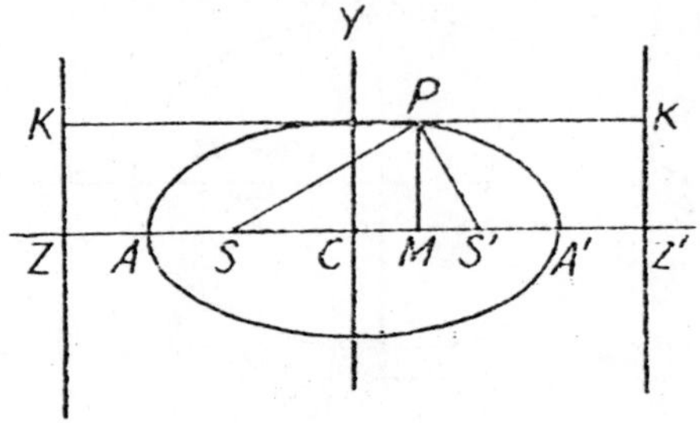

$$\begin{aligned} SP &= e.PK \\ &= e.MZ \\ &= e\,(CM+CZ) \\ &= e\left(x_1+\frac{a}{e}\right) \\ &= a+ex_1 \\ S'P &= e.(PK') \\ &= e\,(MZ') \\ &= e\,(CZ'-CM) \\ &= e\left(\frac{a}{e}-x_1\right) \\ &= a-ex_1. \end{aligned}$$

$$\therefore \quad SP+S'P=2a$$

Hence the sum of the focal distances of any point on the ellipse is equal to major axis.

This property gives a method of drawing an ellipse mechanically.

Take a thread equal to length $2a$. Fasten the ends of the thread to two fixed points S and S' by means of pins. Keep a pencil point in contact with the thread and move it on a piece of paper, such that the two portions of the thread remain tight. The pencil point will describe an ellipse whose foci are S and S' and whose major axis is equal to the length of the thread.

7·4. Position of a point relative to the ellipse. It is obvious that the point (x_1, y_1) will be outside or inside the ellipse if $\frac{x_1^2}{a^2}+\frac{y_1^2}{b^2} >$ or < 1 respectively.

7·5. General equation of an ellipse.

Let the focus be the point (h, k) and let the equation

of the directrix be $px+qy+r=0$, e being the eccentricity of the ellipse.

Let (x, y) be any point on the ellipse. Then the distance of the point (x, y) from the focus must be equal to e times its distance from the line $px+qy+r=0$, so that

$$\sqrt{[(x-h)^2+(y-k)^2]}=e.\frac{px+qy+r}{\sqrt{(p^2+q^2)}}$$

or $$(x-h)^2+(y-k)^2=e^2\ \frac{(px+qy+r)^2}{(p^2+q^2)}$$

or $$x^2\ (p^2+q^2-p^2e^2)+y^2\ (p^2+q^2-q^2e^2)$$
$$-2x\ [h\ (p^2+q^2)+pre^2]-2y\ [k\ (p^2+q^2)+qre^2]$$
$$-2pqe^2xy+(h^2+k^2)\ (p^2+q^2)-e^2r^2=0$$

which is of the form

$$ax^2+2hxy+by^2+2gx+2fy+c=0.$$

7·6. Ex. 1. *Obtain the equation to an ellipse whose focus is the point $(1, -1)$, whose directrix is the line $x-y+3=0$ and whose eccentricity is $\frac{1}{2}$.* *(C. U. 1945)*

Let any point on the ellipse be (x, y), then

$$[(x-1)^2+(y+1)^2]=\tfrac{1}{2}\ \frac{x-y+3}{\sqrt{2}}$$

or $$8\ (x-1)^2+8\ (y+1)^2=(x-y+3)^2$$

or $$8x^2+8y^2-16x+16y=x^2+y^2+9-2xy+6x-6y$$

or $$7x^2+7y^2+2xy-22x+22y-9=0.$$

Ex. 2. *Find the latus rectum, eccentricity and the coordinates of the foci of the ellipse $9x^2+5y^2-30y=0$.*

$$9x^2+5y^2-30y=0$$

or $$9x^2+5\ (y^2-6y)=0$$

or $$9x^2+5\ (y-3)^2=45$$

or $$\frac{x^2}{5}+\frac{(y-3)^2}{9}=1.$$

Transferring the origin to the point (0, 3) the equation becomes $\frac{x^2}{5}+\frac{y^2}{9}=1$.

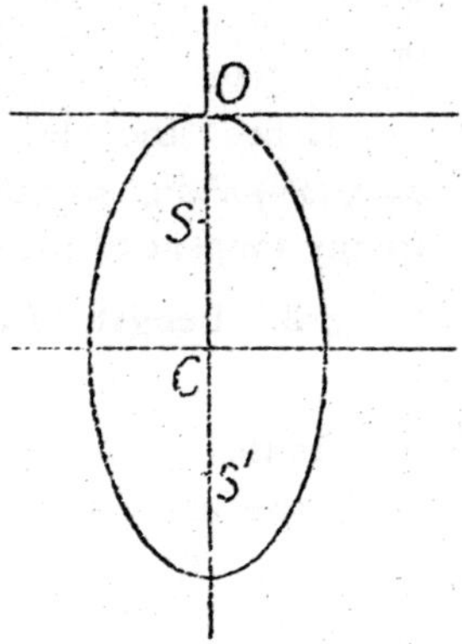

The major axis lies on the y-axis and is of length 3.

The minor axis is of length $\sqrt{5}$.

As $b^2=a^2(1-e^2)$,

$5=9(1-e^2)$.

$\therefore \quad e^2=\frac{4}{9}$.

$\therefore \quad e=\frac{2}{3}$.

$\therefore \quad es=ae=3\times\frac{2}{3}=2$.

Coordinates of S are (0, 1).

Coordinates of S' are (0, 5).

Latus rectum$=\frac{2b^2}{a}=\frac{2\times 5}{3}=\frac{10}{3}$.

7·7. Intersection of an ellipse and a straight line.

Let the ellipse be $\frac{x^2}{a^2}+\frac{y^2}{b^2}=1$ ...(1)

and let the line be $y=mx+c$. ...(2)

Substituting the value of y from (2) in (1), we get

$$\frac{(mx+c)^2}{b^2}+\frac{x^2}{a^2}=1$$

or $x^2(b^2+a^2m^2)+2mcxa^2+a^2(c^2-b^2)=0$ ...(3)

which is a quadratic in x and gives two values of x. These are the abscissae of the points of intersection.

The roots are real and different, real and coincident or imaginary according as

$$4m^2c^2a^4 > = \text{ or } < 4(b^2+a^2m^2)(c^2-b^2)a^2$$

or $m^2c^2a^4 > = \text{ or } < a^2b^2c^2+a^4m^2c^2-a^2b^4-a^4b^2m^2$

or $a^2b^2(b^2+a^2m^2) > = \text{ or } < a^2b^2c^2$

or $$(b^2+a^2m^2) >= \text{ or } < c^2$$

or $$c^2 <= \text{ or } > b^2+a^2m^2.$$

Hence the line $y=mx+c$ touches the ellipse if $c=\sqrt{(b^2+a^2m^2)}$, so that $y=mx+\sqrt{(b^2+a^2m^2)}$ is the equation to any tangent to the ellipse.

7·8. Length of the chord intercepted on the line y=mx+c.

As the points (x_1, y_1), (x_2, y_2), the points of intersection lie on the line $y=mx+c$,

$$y_1=mx_1+c,$$

$$y_2=mx_2+c.$$

$$\therefore \quad y_1-y_2=m\,(x_1-x_2). \qquad \ldots(1)$$

Also from equation (3) in the last article

$$x_1+x_2=-\frac{2mca^2}{b^2+a^2m^2}$$

and $$x_1x_2=\frac{a^2\,(c^2-b^2)}{b^2+a^2m^2}.$$

$$\therefore \quad (x_1-x_2)^2=(x_1+x_2)^2-4x_1x_2$$

$$=\frac{4m^2c^2a^4}{(b^2+a^2m^2)^2}-\frac{4a^2\,(c^2-b^2)}{(b^2+a^2m^2)^2}$$

$$=\frac{4\,\{m^2c^2a^4-a^2\,(c^2-b^2)\,(b^2+a^2m^2\,\}}{(b^2+a^2m^2)^2}$$

$$=\frac{4\,\{a^2b^2\,(b^2-c^2+a^2m^2)\}}{(b^2+a^2m^2)^2}.$$

$$\therefore \quad x_1-x_2=\frac{2ab\,(b^2-c^2+a^2m^2)^{1/2}}{(b^2+a^2m^2)} \qquad \ldots(2)$$

Length of the chord

$$=\sqrt{[(x_1-x_2)^2+(y_1-y_2)^2]}$$

$$=\sqrt{[(x_1-x_2)^2+m^2\,(x_1-x_2)^2]} \qquad \text{from (1)}$$

$$=[\sqrt{(1+m^2)}].(x_1-x_2)$$

$$=\sqrt{(1+m^2)}.\,\frac{2ab\sqrt{(b^2-c^2+a^2m^2)}}{(b^2+a^2m^2)} \qquad \text{from (2)}$$

$\therefore$ length of the chord

$$=\frac{2ab}{(b^2+a^2m^2)}\sqrt{[(1+m^2)(b^2-c^2+a^2m^2)]}.$$

7·9. Tangent at a given point of the ellipse

$$\frac{x^2}{a^2}+\frac{y^2}{b^2}=1.$$

Let the given point be (x_1, y_1).

Let (x_2, y_2) be another point on the ellipse.

Then the equation to the line joining these points is given by

$$y-y_1=\frac{y_2-y_1}{x_2-x_1}(x-x_1) \quad \ldots(1)$$

But (x_1, y_1) and (x_2, y_2) lie on the ellipse,

$$\therefore \quad \frac{x_1^2}{a^2}+\frac{y_1^2}{b^2}=1 \quad \ldots(2)$$

and

$$\frac{x_2^2}{a^2}+\frac{y_2^2}{b^2}=1 \quad \ldots(3)$$

Subtracting (2) from (3), we get

$$\frac{(x_2^2-x_1^2)}{a^2}+\frac{(y_2^2-y_1^2)}{b^2}=0$$

or

$$\frac{y_2-y_1}{x_2-x_1}=-\frac{b^2}{a^2}\,\frac{x_2+x_1}{y_2+y_1}.$$

Substituting for $\frac{y_2-y_1}{x_2-x_1}$ in (1), we get

$$y-y_1=-\frac{b^2}{a^2}\cdot\frac{x_2+x_1}{y_2+y_1}(x-x_1).$$

As the point (x_2, y_2) approaches the point (x_1, y_1), the above straight line becomes a tangent at the point (x_1, y_1).

Hence putting $x_2=x_1$ and $y_2=y_1$ in the above equation,

$$y-y_1=-\frac{b^2}{a^2}\,\frac{x_1}{y_1}(x-x_1)$$

or

$$\frac{yy_1}{b^2}-\frac{y_1^2}{b^2}=-\frac{xx_1}{a^2}+\frac{x_1^2}{a^2}$$

or $$\frac{xx_1}{a^2}+\frac{yy_1}{b^2}=\frac{x_1^2}{a^2}+\frac{y_1^2}{b^2}=1 \qquad \text{from (2).}$$

The equation to the tangent at the point (x_1, y_1) is therefore $$\frac{xx_1}{a^2}+\frac{yy_1}{b^2}=1.$$

7·10. Point of contact of a line with the ellipse $\frac{x^2}{a^2}+\frac{y^2}{b^2}=1$.

Let the line be $lx+py=n$. ...(1)

Let the point of contact be (x_1, y_1).

Then the equation to the tangent is

$$\frac{xx_1}{a^2}+\frac{yy_1}{b^2}=1. \qquad \text{...(2)}$$

Equations (1) and (2) represent the same line. Comparing coefficients in both, $\frac{l}{x_1/a^2}=\frac{p}{y_1/b^2}=\frac{n}{1}$.

$$\therefore \quad x_1=\frac{a^2l}{n} \quad \text{and} \quad y_1=\frac{pb^2}{n} \qquad \text{...(3)}$$

To find the condition that the line $lx+my=n$ touches the ellipse, we eliminate x_1 and y_1 from (3) and the equation

$$\frac{x_1^2}{a^2}+\frac{y_1^2}{b^2}=1$$

so that $$\frac{a^4l^2}{a^2x^2}+\frac{b^4p^2}{b^2n^2}=1$$

or $$a^2l^2+b^2p^2=n^2.$$

Otherwise :—

We know that the condition that the line $y=mx+c$ touches the ellipse $x^2/a^2+y^2/b^2=1$ is $c=\sqrt{(a^2m^2+b^2)}$.

Putting the line $lx+my=n$ in the form $y=mx+c$, we get

$$y=-\frac{l}{p}x+\frac{n}{p}$$

so that c corresponds to n/p and m to $-l/p$.

The condition becomes

$$\frac{n}{p}=\sqrt{\left(\frac{a^2l^2}{p^2}+b^2\right)}$$

or $$n^2=a^2l^2+b^2p^2.$$

7·11. Normal at a point of the ellipse $\frac{x^2}{a^2}+\frac{y^2}{b^2}=1$.

Let the point be (x_1, y_1).

The equation to the tangent at this point is

$$\frac{xx_1}{a^2}+\frac{yy_1}{b^2}=1 \qquad \ldots(1)$$

The normal is a line perpendicular to (1). Hence its equation is $$\frac{xy_1}{b^2}-\frac{yx_1}{a^2}=c \qquad \ldots(2)$$

But it passes through (x_1, y_1).

$$\therefore \quad \frac{x_1y_1}{b^2}-\frac{x_1y_1}{a^2}=c.$$

Substituting this value of c in (2), we get

$$\frac{xy_1}{b^2}-\frac{yx_1}{a^2}=x_1y_1\left(-\frac{1}{a^2}+\frac{1}{b^2}\right)$$

or $$\frac{xy_1}{b^2}-\frac{x_1y_1}{b^2}=\frac{x_1 y}{a^2}-\frac{x_1y_1}{a^2}$$

or $$\frac{y_1}{b^2}(x-x_1)=\frac{x_1}{a^2}(y-y_1)$$

or $$\frac{x-x_1}{x_1/a^2}=\frac{y-y_1}{y_1/b^2} \qquad \ldots(3)$$

which is the equation to the normal at the point (x_1, y_1).

7·12. Auxiliary Circle.

A circle described on the major axis of an ellipse as diameter is called the auxiliary circle of the ellipse.

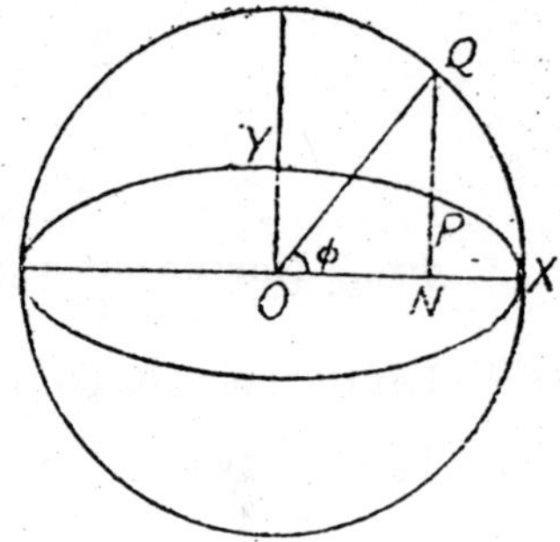

Let $P(x_1, y_1)$ be a point on the ellipse $\frac{x^2}{a^2}+\frac{y^2}{b^2}=1$ and let Q be the corresponding point on the auxiliary circle.

As (x_1, y_1) lies on the ellipse

$$\frac{x_1^2}{a^2}+\frac{y_1^2}{b^2}=1 \qquad \ldots(1)$$

Let y_2 be the ordinate of Q.

Then the point Q with coordinates (x_1, y_2) lies on the circle $x^2+y^2=a^2$ so that

$$x_1^2+y_2^2=a^2 \qquad \ldots(2)$$

Subtracting (2) from (1), we get

$$\frac{y_1^2}{b^2}-\frac{y_2^2}{a^2}=0.$$

or

$$\frac{y_2}{y_1}=\frac{a}{b}.$$

This gives the ratio of the ordinates of any point on the ellipse and the corresponding point on the auxilary circle.

7·13. Eccentric angle.

The angle ϕ which the radius of point Q on the auxiliary circle makes with the major axis is called the eccentric angle of the corresponding point P on the ellipse. In the figure $\angle QON$ is the eccentric angle of the point P.

From ΔQON, $$\frac{QN}{OQ}=\sin\phi.$$

$$\therefore \quad QN=a\sin\phi$$

and

$$ON=a\cos\phi$$

Now from last article

$$\frac{QN}{PN}=\frac{a}{b}.$$

$$\therefore \quad PN=\frac{b}{a}\,QN=\frac{b}{a}\times a\sin\phi$$

$$=b\sin\phi.$$

Hence the coordinates of the point P are $(a\cos\phi, b\sin\phi)$. These are the parametric coordinates of the point P.

7·14. Tangent and normal at the point ϕ.

The equation to the straight line joining the points $(a\cos\phi_1, b\sin\phi_1)$ and $(a\cos\phi_2, b\sin\phi_2)$ is given by

$$y-b\sin\phi_1=\frac{b}{a}\cdot\frac{\sin\phi_2-\sin\phi_1}{\cos\phi_2-\cos\phi_1}(x-a\cos\phi_1)$$

$$=\frac{b}{a}\cdot\frac{2\cos\frac{\phi_1+\phi_2}{2}\sin\frac{\phi_2-\phi_1}{2}}{2\sin\frac{\phi_1+\phi_2}{2}\sin\frac{\phi_1-\phi_2}{2}}(x-a\cos\phi_1)$$

$$=-\frac{b}{a}\cdot\frac{\cos\frac{\phi_1+\phi_2}{2}}{\sin\frac{\phi_1+\phi_2}{2}}(x-a\cos\phi_1)$$

or $$\frac{x}{a}\cos\frac{\phi_1+\phi_2}{2}+\frac{y}{b}\sin\frac{\phi_1+\phi_2}{2}=\cos\phi_1\cos\frac{\phi_1+\phi_2}{2}+\sin\phi_1\sin\frac{\phi_1+\phi_2}{2}$$

$$=\cos\left(\frac{\phi_1-\phi_2}{2}\right) \qquad \ldots(1)$$

If ϕ_2 approaches ϕ_1, this becomes the equation of the tangent at the point ϕ_1. Hence the equation of the tangent at the point ϕ_1 is given by

$$\frac{x}{a}\cos\phi_1+\frac{y}{b}\sin\phi_1=1 \qquad \ldots(2)$$

Normal at the point $(a\cos\phi_1, b\sin\phi_1)$ passes through this point and is perpendicular to (2). Hence its equation is

$$(y-b\sin\phi_1)=\frac{a}{b}\cdot\frac{\sin\phi_1}{\cos\phi_1}(x-a\cos\phi_1)$$

or $$b(y-b\sin\phi_1)\operatorname{cosec}\phi_1=a(x-a\cos\phi_1)\sin\phi_1$$

or $$ax\sec\phi_1-by\operatorname{cosec}\phi_1=a^2-b^2.$$

7·15. Solved Examples.

Ex. 1. *Find the condition that the line $lx+my=n$ may be a normal to the ellipse.*

The equation to the normal is given by

$$ax\sec\phi-by\operatorname{cosec}\phi=a^2-b^2.$$

Comparing this with $lx+my=n$, we get

$$\frac{a\sec\phi}{l}=\frac{b\operatorname{cosec}\phi}{-m}=\frac{a^2-b^2}{n}.$$

$$\therefore\quad \cos\phi=\frac{na}{l(a^2-b^2)}\text{ and }\sin\phi=\frac{nb}{-m(a^2-b^2)}.$$

Eliminating ϕ, we get

$$\frac{n^2a^2}{l^2(a^2-b^2)^2}+\frac{n^2b^2}{m^2(a^2-b^2)^2}=1,$$

or $$n^2(a^2m^2+b^2l^2)=l^2m^2(a^2-b^2)^2.$$

Ex. 2. *A circle of radius r is concentric with the ellipse $\frac{x^2}{a^2}+\frac{y^2}{b^2}=1$. Prove that the common tangent is inclined to the major axis at an angle $\tan^{-1}\sqrt{\left(\frac{r^2-b^2}{a^2-r^2}\right)}$.* *(Agra 1958)*

Let the common tangent be $y=mx+c$.

If this is a tangent to the ellipse,

$$c=\sqrt{(a^2m^2+b^2)}. \qquad \ldots(1)$$

If t is a tangent to the circle,

$$x^2+y^2=r^2.$$

$$c=r\sqrt{(1+m^2)} \qquad \ldots(2)$$

$$\therefore\quad r\sqrt{(1+m^2)}=\sqrt{(am^2+b^2)}$$

or $$r^2(1+m^2)=a^2m^2+b^2$$

or $$m^2 (a^2-r^2)=r^2-b^2$$

or $$m^2=\frac{r^2-b^2}{a^2-r^2}$$

or $$m=\sqrt{\left(\frac{r^2-b^2}{a^2-r^2}\right)}.$$

If ϕ is the angle made by the tangent with the major axis,

$$\tan\phi=\sqrt{\left(\frac{r^2-b^2}{a^2-r^2}\right)}$$

$$\therefore \quad \phi=\tan^{-1}\sqrt{\left(\frac{r^2-b^2}{a^2-r^2}\right)}$$

Ex. 3. *If the normal at the end of a latus rectum pass through one extremity of the minor axis, show that the eccentricity of the curve is given by the equation $e^4+e^2-1=0$.* (*Agra 1949*)

The coordinates of one end of the latus rectum are

$$(ae, b^2/a).$$

The equation to the normal at this point is

$$\frac{x-ae}{(ae)/a^2}=\frac{y-b^2/a}{b^2/(ab^2)}$$

or $$\frac{a\,(x-ae)}{e}=a\left(y-\frac{b^2}{a}\right).$$

It passes through the extremity of the minor axis $(0, b)$.

$$\therefore \quad -a^2=a\,(b-b^2/a)$$

or $$-a=b-a\,(1-e^2)$$

or $$-ae^2=b$$

or $$-ae^2=a\sqrt{(1-e^2)}$$

or $$e^4=1-e^2 \quad \text{or} \quad e^4+e^2-1=0.$$

Ex. 4. *If any two chords be drawn through two points on the major axis of an ellipse equidistant from the centre, show that*

$$\tan\frac{\alpha}{2}\tan\frac{\beta}{2}\tan\frac{\gamma}{2}\tan\frac{\delta}{2}=1,$$

where $\alpha, \beta, \gamma, \delta$ are the eccentric angles of the extremities of the chords.

(Agra 1946)

The equation to the line joining the points

$$(a \cos \alpha, b \sin \alpha) \text{ and } (a \cos \beta, b \sin \beta)$$

is
$$\frac{x}{a} \cos \frac{\alpha+\beta}{2} + \frac{y}{b} \sin \frac{\alpha+\beta}{2} = \cos \frac{\alpha-\beta}{2}. \qquad \ldots(1)$$

If it passes through a point $(d, 0)$ on the major axis,

$$\frac{d}{a} \cos \frac{\alpha+\beta}{2} = \cos \frac{\alpha-\beta}{2}. \qquad \ldots(2)$$

Similarly if the equation to the line joining the points $(a \cos \gamma, b \sin \gamma)$, $(a \cos \delta, b \sin \delta)$ passes through an equidistant point $(-d, 0)$ on the major axis,

$$-\frac{d}{a} \cos \frac{\gamma+\delta}{2} = \cos \frac{\gamma-\delta}{2} \qquad \ldots(3)$$

Eliminating d from (2) and (3), we get

$$-\frac{\cos \frac{\alpha+\beta}{2}}{\cos \frac{\gamma+\delta}{2}} = \frac{\cos \frac{\alpha-\beta}{2}}{\cos \frac{\gamma-\delta}{2}}$$

or
$$\cos \frac{\alpha+\beta}{2} \cos \frac{\gamma-\delta}{2} + \cos \frac{\alpha-\beta}{2} \cos \frac{\gamma+\delta}{2} = 0$$

or
$$\left(\cos \frac{\alpha}{2} \cos \frac{\beta}{2} - \sin \frac{\alpha}{2} \sin \frac{\beta}{2}\right)$$
$$\times\left(\cos \frac{\gamma}{2} \cos \frac{\delta}{2} + \sin \frac{\gamma}{2} \sin \frac{\delta}{2}\right)$$
$$+\left(\cos \frac{\alpha}{2} \cos \frac{\beta}{2} + \sin \frac{\alpha}{2} \sin \frac{\beta}{2}\right)$$
$$\times\left(\cos \frac{\gamma}{2} \cos \frac{\delta}{2} - \sin \frac{\gamma}{2} \sin \frac{\delta}{2}\right) = 0$$

or
$$\left(1 - \tan \frac{\alpha}{2} \tan \frac{\beta}{2}\right)\left(1 + \tan \frac{\gamma}{2} \tan \frac{\delta}{2}\right)$$
$$+\left(1 + \tan \frac{\beta}{2} \tan \frac{\alpha}{2}\right)\left(1 - \tan \frac{\gamma}{2} \tan \frac{\delta}{2}\right) = 0$$

or $$1-\tan\frac{\alpha}{2}\tan\frac{\beta}{2}+\tan\frac{\gamma}{2}\tan\frac{\delta}{2}$$
$$-\tan\frac{\alpha}{2}\tan\frac{\beta}{2}\tan\frac{\gamma}{2}\tan\frac{\delta}{2}$$
$$+1+\tan\frac{\beta}{2}\tan\frac{\alpha}{2}-\tan\frac{\gamma}{2}\tan\frac{\delta}{2}$$
$$-\tan\frac{\alpha}{2}\tan\frac{\beta}{2}\tan\frac{\gamma}{2}\tan\frac{\delta}{2}=0$$

or $$\tan\frac{\alpha}{2}\tan\frac{\beta}{2}\tan\frac{\gamma}{2}\tan\frac{\delta}{2}=1.$$

Ex. 5. *Prove that the locus of the middle point of the portion of a tangent to the ellipse $x^2/a^2+y^2/b^2=1$ included between the axes is the curve $a^2/x^2+b^2/y^2=4$.* *(Patna 1949)*

Let the tangent to the ellipse be
$$\frac{x\cos\phi}{a}+\frac{y\sin\phi}{b}=1.$$

It cuts the X and Y axes at points $\left(\frac{a}{\cos\phi}, 0\right)$ and $\left(0, \frac{b}{\sin\phi}\right)$ respectively.

Let (h, k) be the coordinates of the middle point of the chord intercepted between the axes.

Then $$\frac{a}{2\cos\phi}=h \qquad \ldots(1)$$

and $$\frac{b}{2\sin\phi}=k. \qquad \ldots(2)$$

Eliminating ϕ between (1) and (2),
$$\frac{a^2}{4h^2}+\frac{b^2}{4k^2}=1.$$

Hence the locus of (h, k) is the curve $(a^2/x^2)+(b^2/y^2)=4$.

Ex. 6. *If a triangle is inscribed in an ellipse and two of its* *sides are parallel to given straight lines, the envelope of the third side is another ellipse.* *(Agra 1953)*

Let the eccentric angles of P, Q, R lie at θ_1, θ_2, θ_3. Then if PQ and PR are parallel to given straight lines,

$$\tan \frac{\phi_1+\phi_2}{2}=\text{constant}$$

or $$\phi_1+\phi_2=2\alpha \quad \ldots(1)$$

and similarly $$\phi_1+\phi_3=2\beta. \quad \ldots(2)$$

$$\therefore \quad \phi_2-\phi_3=2(\alpha-\beta). \quad \ldots(3)$$

The equation to the third side QR is

$$\frac{x}{a}\cos\left(\frac{\phi_2+\phi_3}{2}\right)+\frac{y}{b}\sin\left(\frac{\phi_2+\phi_3}{2}\right)=\cos\frac{\phi_2-\phi_3}{2}$$

$$=\cos(\alpha-\beta) \qquad \text{from (3).}$$

If $\phi_2+\phi_3=2\phi$, then the equation to QR becomes

$$\frac{x}{a}\cos\phi+\frac{y}{b}\sin\phi=\cos(\alpha-\beta)$$

or $$\frac{x}{a}\left[\cos^2\frac{\phi}{2}-\sin^2\frac{\phi}{2}\right]+\frac{y}{b}.2\sin\frac{\phi}{2}.\cos\frac{\phi}{2}$$
$$-\cos(\alpha-\beta)\left(\cos^2\frac{\phi}{2}+\sin^2\frac{\phi}{2}\right)=0$$

or $$\frac{x}{a}\left(1-\tan^2\frac{\phi}{2}\right)+\frac{y}{b}\,2\tan\frac{\phi}{2}$$
$$-\cos(\alpha-\beta)\left(1+\tan^2\frac{\phi}{2}\right)=0$$

or $$\frac{x}{a}(1-t^2)+\frac{y}{b}.2t-\cos(\alpha-\beta)(1+t^2)=0$$

if $\tan\phi/2=t$

or $$t^2\left[-\frac{x}{a}-\cos(\alpha-\beta)\right]+\frac{2y}{b}t+\left[\frac{x}{a}-\cos(\alpha+\beta)\right]=0.$$

This is a quadratic in parameter t.

Hence the envelope of the above line is given by

$$\frac{y^2}{b^2}=-\left[\frac{x}{a}+\cos(\alpha-\beta)\right]\left[\frac{x}{a}-\cos(\alpha-\beta)\right]$$

$$\frac{y^2}{b^2}=-\frac{x^2}{a^2}+\cos^2(\alpha-\beta)$$

or $$\frac{x^2}{a^2}+\frac{y^2}{b^2}=\cos^2(\alpha-\beta)$$

which is another ellipse.

7·16. Subtangent and subnormal of a point on the ellipse, $\frac{x^2}{a^2}+\frac{y^2}{b^2}=1$.

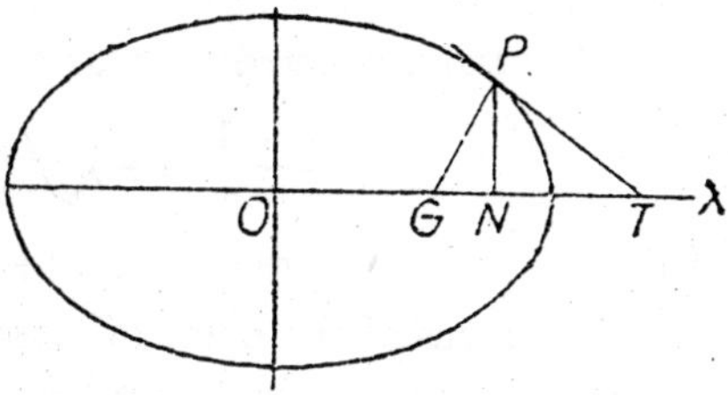

Let the point P on the ellipse be (x_1, y_1). The equation to the tangent at P is

$$\frac{xx_1}{a^2}+\frac{yy_1}{b^2}=1 \ldots (1)$$

It cuts the X-axis at the point $T\left(\frac{a^2}{x_1}, 0\right)$.

Length of the subtangent$=NT$

$$=OT-ON$$

$$=\frac{a^2}{x_1}-x_1=\frac{a^2-x_1^2}{x_1}$$

The equation to the normal at P is

$$\frac{x-x_1}{\frac{x_1}{a^2}}=\frac{y-y_1}{\frac{y_1}{b^2}}$$

It cuts the X-axis at G whose coordinates are

$$\left\{\frac{(a^2-b^2)\,x_1}{a^2}, 0\right\}.$$

Length of the subnormal$=GN$

$$=ON-OG$$

$$\therefore \quad GN = x_1 - \frac{(a^2 - b^2)\, x_1}{a^2}$$

$$= \frac{b^2}{a^2}\, x_1.$$

7·17. Locus of the point of intersection of two perpendicular tangents.

Any tangent to the ellipse $\frac{x^2}{a^2} + \frac{y^2}{b^2} = 1$ is given by

$$y > mx + \sqrt{(a^2m^2 + b^2)}$$

or

$$y - mx = \sqrt{(a^2m^2 + b^2)}. \qquad \ldots(1)$$

Any tangent perpendicular to (1) is

$$y = -\frac{1}{m}\, x + \sqrt{\left(\frac{a^2}{m^2} + b^2\right)}$$

or

$$my - x = \sqrt{(a^2 + b^2m^2)} \qquad \ldots(2)$$

The locus of the point of intersection of (1) and (2) will be obtained by eliminating m between (1) and (2).

Squaring and adding (1) and (2), we get

$$(1 + m^2)\, y^2 + (1 + m^2)\, x^2 = a^2\, (1 + m^2) + b^2\, (1 + m^2)$$

or

$$x^2 + y^2 = a^2 + b^2$$

which is called the director circle.

7·18. Some properties of an ellipse.

(1) The tangent and normal at any point of an ellipse bisect the internal and external angles between the focal distances of the point.

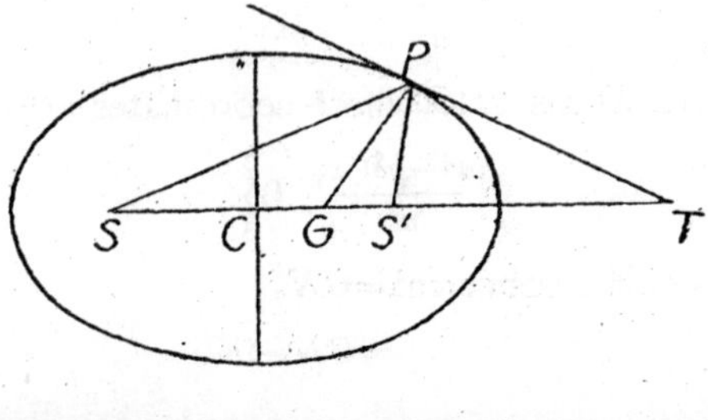

Let P be any point (x_1, y_1) on the ellipse

$$\frac{x^2}{a^2}+\frac{y^2}{b^2}=1.$$

The normal at (x_1, y_1) is given by the equation

$$\frac{x-x_1}{\frac{x_1}{a^2}}=\frac{y-y_1}{\frac{y_1}{b^2}}.$$

This meets the major axis $y=0$ in a point G, where

$$CG=-\frac{b^2x_1}{a^2}+x_1.$$

$$=x_1\left(1-\frac{b^2}{a^2}\right)=e^2x_1$$

Now $$S'G=CS'-CG$$

$$=ae-e^2x_1$$

$$=e\,(a-ex_1)$$

$$=e\,.\,PS'.$$

Similarly $$GS=e\,.\,PS.$$

$$\therefore\quad \frac{PS}{PS'}=\frac{GS}{GS'}.$$

$\therefore$ the normal bisects the angle SPS'.

Since the tangent is perpendicular to the normal, PT bisects the external angle between PS and PS'.

(2) *The product of the perpendiculars from the foci upon the tangent at any point of the ellipse is equal to the square of the semi-minor axis and the feet of the perpendiculars lie on the auxiliary circle.*

Let the equation to any tangent to the ellipse be

$$y=mx+\sqrt{(a^2m^2+b^2)} \qquad \ldots(1)$$

Length of the perpendicular from the focus $(ae, 0)$ on the above line is

$$\frac{ame+\sqrt{(a^2m^2+b^2)}}{\sqrt{(1+m^2)}}.$$

Also the length of the perpendicular from the other focus $(-ae, 0)$ on this line is

$$\frac{-ame+\sqrt{(a^2m^2+b^2)}}{\sqrt{(1+m^2)}}.$$

The product of the perpendiculars

$$=\frac{a^2m^2+b^2-a^2m^2e^2}{(1+m^2)}$$

$$=\frac{b^2+a^2m^2(1-e^2)}{1+m^2}=\frac{b^2+b^2m^2}{1+m^2}$$

$$=b^2.$$

The straight line passing through the focus $(ae, 0)$, perpendicular to line (1) is given by

$$y-0=-\frac{1}{m}(x-ae)$$

or $$x+my=ae=\sqrt{(a^2-b^2)} \quad \ldots(2)$$

We have to find the locus of the point of intersection of (1) and (2). We have to eliminate m from (1) and (2).

Squaring and adding them, we get

$$(y-mx)^2+(x+my)^2=a^2(1+m^2)+b^2-b^2$$

or $$x^2(1+m^2)+y^2(1+m^2)=a^2(1+m^2)$$

$$\therefore \quad x^2+y^2=a^2$$

which is the auxilary circle.

(3) *If the tangent at any point P meet the major and minor axes in T and T′, then CN.CT=a^2 and CM.CT′=b^2 where N and M are the feet of the perpendiculars from P on the major and minor axes respectively.*

The tangent at the point (x_1, y_1) to the ellipse

$$\frac{x^2}{a^2}+\frac{y^2}{b^2}=1$$

is $$\frac{xx_1}{a^2}+\frac{yy_1}{b^2}=1 \quad \ldots(1)$$

This meets the major axis at T where $CT=a^2/x_1$.

Also $$CN=x_1.$$

$$\therefore \quad CN.CT=a^2.$$

Similarly the point of intersection of (1) with the minor axis T' is given by

$$CT'=b^2/y_1 \quad \text{and} \quad CM=y_1.$$

$$\therefore \quad CM.CT'=b^2.$$

Examples XIV

1. Find the equation of an ellipse, whose focus is (0, 3), corresponding directrix is $x+y+1=0$ and eccentricity $\frac{1}{3}$.
2. Find the equation of the ellipse whose latus rectum is 5 and whose eccentricity is $\frac{2}{3}$, referred to the principal axes as axes of coordinates. *(Dacca 1945)*
3. Find the eccentricity, the foci and the latus rectum of the ellipse $2x^2+5y^2=20$.
4. Trace the following curve :—
$$\frac{(x-1)^2}{16}+\frac{(y-5)^2}{9}=1.$$
5. Find the centre, foci, eccentricity and the lengths of the axes of the ellipse $x^2+3y^2-4x+6y+6=0$.
6. Prove that if a bar of given length moves with its extremities on two fixed straight lines at right angles, any point on it describes an ellipse. *(Utkal 1946)*
7. Find the eccentricity of an ellipse if its latus rectum is equal to one half of its major axis.
8. Find the locus of the middle points of chords of the ellipse $x^2/a^2+y^2/b^2=1$ which are drawn through the positive end of the minor axis.
9. If any tangent to the ellipse $x^2/a^2+y^2/b^2=1$ intercepts lengths h and k on the axes, prove that
$$a^2/h^2+b^2/k^2=1.$$ *(Calcutta 1951)*

10. If an angle between the lines joining the foci of an ellipse to an extremity of the minor axis is 90°, find the eccentricity. Find also the equation of the ellipse if the major axis is $2\sqrt{2}$ units in length.

(Roorkee 1947)

11. If the normal at a point P on the ellipse of semi-axes a, b, and centre C cuts the major and minor axes at G and g, show that $a^2.CG^2+b^2.cg^2=(a^2-b^2)^2$. Also prove that $CG=e^2.CN$ where PN is the ordinate of P.

(Dehli 1945)

12. If the normal at any point P of the ellipse $x^2/a^2+y^2/b^2=1$ meets the major and minor axes at G and g respectively, and if CF is the perpendicular upon this normal from the centre C of the ellipse, show that

$$PF.PG=b^2 \text{ and } PF.Pg=a^2.$$ *(Dacca 1945)*

13. Show that the length of the normal at any point of an ellipse intercepted between the curve and its major axis is $\frac{b\sqrt{(rr')}}{a}$, where r, r' are the focal distances of the point.

14. If the chord through the points whose eccentric angles are θ and ϕ on the ellipse $x^2/a^2+y^2/b^2=1$ passes through the focus, prove that $\tan\frac{\theta}{2}\tan\frac{\phi}{2}+\frac{1-e}{1+e}=1$.

(Calcutta 1945)

15. The tangent at the point α on the ellipse $x^2/a^2+y^2/b^2=1$ meets the auxiliary circle in two points which subtend a right angle at the curve; show that the eccentricity of the ellipse is $(1+\sin^2\alpha)^{-1/2}$. *(Travancore 1947)*

16. Show that the chord joining the points α, β on the

ellipse $x^2/a^2+y^2/b^2=1$ will subtend a right angle,

(i) at a vertex if $\tan\frac{\alpha}{2}\tan\frac{\beta}{2}=-\frac{a^2}{b^2}$;

(ii) at the centre if $\tan\alpha\tan\beta=-\frac{a^2}{b^2}$.

17. Show that the line $lx+my=n$ will cut the ellipse $x^2/a^2+y^2/b^2=1$ in points whose eccentric angles differ by $\pi/2$ if $a^2l^2+b^2m^2=2n^2$.

18. The tangent and the normal at a point P on the ellipse $x^2/a^2+y^2/b^2=1$ meet the major axis in T and T', so that $TT'=a$. Prove that the eccentric angle θ of P is given by $e\cos^2\theta+\cos\theta-1=0$ where e is the eccentricity of the ellipse.

19. Show that the area of a triangle inscribed in an ellipse is $\frac{1}{2}ab\,\{\sin(\beta-\gamma)+\sin(\gamma-\alpha)+\sin(\alpha-\beta)\}$ where α, β, γ are the eccentric angles of the angular points.

20. Show that the area of a triangle inscribed in an ellipse is maximum when the eccentric angles of the vertices differ from each other by $2\pi/3$.

21. If $\phi_1, \phi_2, \phi_3, \phi_4$ are the eccentric angles of the points in which an ellipse is cut by a circle, prove that

$$\phi_1+\phi_2+\phi_3+\phi_4=2n\pi.$$

22. To any point P of an ellipse $b^2x^2+a^2y^2=a^2b^2$ corresponds a point Q on the auxiliary circle. The normals to the ellipse at P and the circle at Q meet in R. Prove that the locus of R is $x^2+y^2=(a+b)^2$.

23. PQ is a chord of the ellipse $x^2/a^2+y^2/b^2=1$ parallel to a given line. Find the locus of the intersection of the normals at P and Q.

24. If the line joining two points whose eccentric angles are α, β, cut the major axis of an ellipse at a distance d from the centre, show that

$$\tan\frac{\alpha}{2}\tan\frac{\beta}{2}=\frac{d-a}{d+a}.$$

25. If p and p' be the perpendiculars from the centre and a focus of the ellipse $x^2/a^2+y^2/b^2=1$ on any tangent and r the focal distance of the point of contact, prove that $p'a=pr$.

26. Find the equation of the lines joining the origin to the points in which the lines $x \cos \alpha+y \sin \alpha=p$ meets the ellipse $x^2/a^2+y^2/b^2=1$. Hence show that if the intercepted portion of the chord subtends a right angle at the centre, it always touches a circle of radius $\dfrac{ab}{\sqrt{(a^2+b^2)}}$ concentric with the ellipse.

27. Show that the locus of the point of intersection of tangents at two points on the ellipse $x^2/a^2+y^2/b^2=1$, when the difference of their eccentric angles is $2k$, is

$$\frac{x^2}{a^2}+\frac{y^2}{b^2}=\sec^2 k.$$

(Calcutta 1947)

28. The normal at P meets the axes in G and g; show that the loci of the middle points of PG and Pg are respectively the ellipses

$$\frac{4x^2}{a^2(1-e^2)^2}+\frac{4y^2}{b^2}=1 \quad \text{and} \quad a^2x^2+b^2y^2=\tfrac{1}{4}(a^2-b^2)^2.$$

29. PQ, PR are focal chords of an ellipse and 2α, 2β, 2γ are eccentric angles of P, Q and R respectively; prove that

$$\tan \alpha \tan \beta . \tan \gamma=\cot \alpha.$$

30. A tangent to the ellipse $x^2/a^2+y^2/b^2=1$ meets the ellipse $x^2/a^2+y^2/b^2=a+b$ in the points P and Q; prove that the tangents at P and Q are at right angles.

7·19. Tangents through a given point.

Let the given point be (x_1, y_1) through which tangents are drawn to the ellipse, $\dfrac{x^2}{a^2}+\dfrac{y^2}{b^2}=1.$

Any tangent to the ellipse is

$$y=mx+\sqrt{(a^2m^2+b^2)}. \qquad ...(1)$$

If it passes through (x_1, y_1),

$$y_1=mx_1+\sqrt{(a^2m^2+b^2)}$$

or $$(y_1-mx_1)^2=a^2m^2+b^2$$

or $$m^2(x_1^2-a^2)-2mx_1y_1+y_1^2-b^2=0. \qquad ...(2)$$

This is a quadratic in m; hence two tangents can be drawn from the point (x_1, y_1) to the ellipse.

The tangents will be real, coincident or imaginary according as

$$4x_1^2y_1^2 >=< 4(x_1^2-a^2)(y_1^2-b^2)$$

or $$\frac{x_1^2}{a^2}+\frac{y_1^2}{b^2} >=< 1.$$

or according as the point (x_1, y_1) is outside, on or inside the ellipse.

7·20. Pair of tangents from an external point.

Let (h, k) be another point on the tangent drawn from the point (x_1, y_1) to the ellipse $\frac{x^2}{a^2}+\frac{y^2}{b^2}=1$.

The equation to the line joining (x_1, y_1) and (h, k) is

$$y-y_1=\frac{k-y_1}{h-x_1}(x-x_1)$$

or $$y=\frac{k-y_1}{h-x_1}.x+\frac{hy_1-kx_1}{h-x_1}. \qquad ...(1)$$

If it is a tangent to the ellipse, it must be of the form

$$y=mx+\sqrt{(am^2+b^2)}. \qquad ...(2)$$

Comparing (1) and (2),

$$m=\frac{k-y_1}{h-x_1} \text{ and } \sqrt{(a^2m^2+b^2)}=\frac{hy_1-kx_1}{h-x_1}. \qquad ...(3)$$

Eliminating m from equations (3), we get

$$a^2\left(\frac{k-y_1}{h-x_1}\right)^2+b^2=\left(\frac{hy_1-kx_1}{h-x_1}\right)^2$$

or $$a^2(k-y_1)^2+b^2(h-x_1)^2=(hy_1-kx_1)^2.$$

This is an equation in h and k, which does not involve any variable. Hence the locus of the point (h, k) is

$$(xy_1-x_1y)^2=a^2(y-y_1)^2+b^2(x-x_1)^2$$

which can be rewritten in the form

$$\left(\frac{x^2}{a^2}+\frac{y^2}{b^2}-1\right)\left(\frac{x_1^2}{a^2}+\frac{y_1^2}{b^2}-1\right)=\left(\frac{xx_1}{a^2}+\frac{yy_1}{b^2}-1\right)^2$$

or $$SS_1=T^2$$

where $$S\equiv\frac{x^2}{a^2}+\frac{y^2}{b^2}-1,\ S_1\equiv\left(\frac{x_1^2}{a^2}+\frac{y_1^2}{b^2}-1\right)$$

and $$T\equiv\frac{xx_1}{a^2}+\frac{yy_1}{b^2}-1.$$

7·21. Chord of Contact.

Let $T(x_1, y_1)$ be an external point and let tangents TP and TQ be drawn from T to the ellipse $x^2/a^2+y^2/b^2=1$ touching at P and Q whose coordinates are (p_1, q_1) and (p_2, q_2) respectively.

The line joining P and Q is the chord of contact of the point T.

The equation to the straight line PQ is given by

$$y-q_1=\frac{q_2-q_1}{p_2-p_1}(x-p_1) \qquad \ldots(1)$$

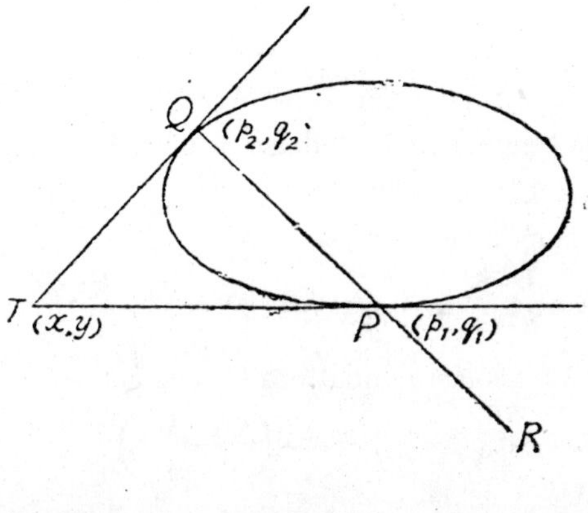

As the tangent at (p_1, q_1) passes through the point (x_1, y_1),

$$\frac{x_1 p_1}{a^2}+\frac{y_1 q_1}{b^2}=1 \qquad \ldots(2)$$

Similarly because the tangent at $Q\,(p_2, q_2)$ passes through the point (x_1, y_1),

$$\therefore \quad \frac{x_1 p_2}{a^2}+\frac{y_1 q_2}{b^2}=1 \qquad \ldots(3)$$

Subtracting (2) from (3), we get

$$\frac{y_1}{b^2}(q_2-q_1)+\frac{x_1}{a^2}(p_2-p_1)=0,$$

or

$$\frac{q_2-q_1}{p_2-p_1}=-\frac{b^2}{a^2}\cdot\frac{x_1}{y_1}$$

Substituting for $\dfrac{q_2-q_1}{p_2-p_1}$ in (1), we get the equation of the chord of contact as

$$y-q_1=-\frac{b^2}{a^2}\cdot\frac{x_1}{y_1}(x-p_1)$$

or

$$\frac{yy_1}{b^2}+\frac{xx_1}{a^2}=\frac{x_1 p_1}{a^2}+\frac{y_1 q_1}{b^2}$$

$$=1 \text{ from (2).}$$

Hence the equation of the chord of contact of the point (x_1, y_1) is

$$\frac{xx_1}{a^2}+\frac{yy_1}{b^2}=1,$$

7·22. Polar :—*To find the equation of the polar of a given point (x_1, y_1) with respect to the ellipse*

$$\frac{x_1^2}{a^2}+\frac{y_1^2}{b^2}=1.$$

Let a straight line through the given point $P(x_1, y_1)$ cut the ellipse in points Q and R. If the tangents at Q and R meet at the point T, we have to find the locus of the point T.

Let the coordinates of the point T be (h, k).

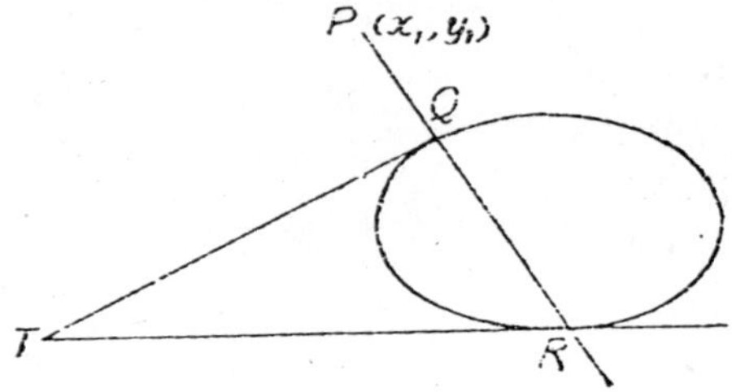

Then QR is the chord of contact of the point $T(h, k)$. Hence its equation is

$$\frac{xh}{a^2}+\frac{yk}{b^2}=1.$$

It passes through the point $P(x_1, y_1)$.

$$\therefore \quad \frac{hx_1}{a^2}+\frac{ky_1}{b^2}=1. \qquad \ldots(1)$$

But this is a relation between h and k on the basis of the given condition. Hence the locus of (h, k) is

$$\frac{xx_1}{a^2}+\frac{yy_1}{b^2}=1,$$

which is the equation of the polar of the point (x_1, y_1).

7·23. Solved Examples.

Ex. 1. *Show that the locus of the poles of normal chords of the ellipse* $\frac{x^2}{a^2}+\frac{y^2}{b^2}=1$ *is the curve*

$$\frac{a^6}{x^2}+\frac{b^6}{y^2}=(a^2-b^2)^2.$$

(Agra 1954)

Let the pole be (h, k).

Then the polar with respect to the ellipse is

$$\frac{xh}{a^2}+\frac{yk}{b^2}=1. \qquad \ldots(1)$$

But this is the same as the normal to the ellipse whose equation is

$$ax \sec\phi - by \operatorname{cosec}\phi = a^2 - b^2 \qquad \ldots(2)$$

Comparing (1) and (2),

$$\frac{\frac{h}{a^2}}{a\sec\phi} = \frac{\frac{k}{b^2}}{-b\operatorname{cosec}\phi} = \frac{1}{a^2-b^2} \qquad \ldots(3)$$

From (3), $$\cos\phi = \frac{a^3}{(a^2-b^2).h}$$

and $$\sin\phi = -\frac{b^3}{(a^2-b^2).k}.$$

Eliminating ϕ, we get

$$\frac{a^6}{(a^2-b^2)^2 h^2} + \frac{b^6}{(a^2-b^2)^2 k^2} = 1.$$

Hence the locus of (h, k) is

$$\frac{a^6}{x^2} + \frac{b^6}{y^2} = (a^2-b^2)^2.$$

Ex. 1. *From the point O (h, k) two tangents OP and OQ are drawn to the ellipse* $\frac{x^2}{a^2}+\frac{y^2}{b^2}=1$, *show that the area of the triangle OPQ is*

$$\frac{ab\left(\frac{h^2}{a^2}+\frac{k^2}{b^2}-1\right)^{3/2}}{\left(\frac{h^2}{a^2}+\frac{k^2}{b^2}\right)}$$

and the area of the quadrilateral CPOQ is

$$ab\left[\frac{h^2}{a^2}+\frac{k^2}{b^2}-1\right]^{1/2}.$$

The chord of contact of the point (h, k) is given by

$$\frac{xh}{a^2}+\frac{yk}{b^2}=1 \qquad \ldots(1)$$

The length of the chord of contact is given by

$$(x_1-x_2)\sqrt{(1+m^2)}$$

where m is the slope of the chord of contact and x_1, x_2 are the abscissae of the chord of contact with the ellipse

$$\frac{x^2}{a^2}+\frac{y^2}{b^2}=1 \qquad \ldots(2)$$

Substituting the value of y from (1) in (2), we get

$$\frac{x^2}{a^2}+\left(1-\frac{xh}{a^2}\right)^2\frac{b^2}{k^2}=1,$$

or $$x^2\left(\frac{1}{a^2}+\frac{b^2}{a^4}\frac{h^2}{k^2}\right)-\frac{2b^2}{a^2}\cdot\frac{xh}{k^2}+\frac{b^2}{k^2}-1=0$$

or $$x^2(a^2k^2+b^2h^2)-2b^2hxa^2+a^4b^2-a^4k^2=0.$$

If the roots of this equation are x_1 and x_2,

$$x_1+x_2=\frac{2b^2ha^2}{a^2k^2+b^2h^2} \qquad \ldots(3)$$

and $$x_1x_2=\frac{a^4(b^2-k^2)}{(a^2k^2+b^2h^2)} \qquad \ldots \quad \ldots \quad \ldots(4)$$

$$\therefore \quad (x_1-x_2)=\sqrt{\{(x_1+x_2)^2-4x_1x_2\}}$$

$$=\sqrt{\left\{\frac{4b^4h^2a^4}{(a^2k^2+b^2h^2)}-\frac{4a^4(b^2-k^2)}{(a^2k^2+b^2h^2)}\right\}}$$

$$=2\sqrt{\left\{\frac{b^4h^2a^4-a^4(b^2-k^2)(h^2a^2+b^2h^2)}{(a^2k^2+b^2h^2)^2}\right\}}$$

$$=2a^2\sqrt{\left\{\frac{b^2h^2k^4+a^2k^2-a^2k^2b^2}{(a^2k^2+b^2h^2)^2}\right\}}$$

$$=\frac{2a^2\times abk\sqrt{\left(\frac{h^2}{a^2}+\frac{k^2}{b^2}-1\right)}}{a^2b^2\left(\frac{h^2}{a^2}+\frac{k^2}{b^2}\right)}$$

$$=\frac{2ak}{b}\frac{\sqrt{\left(\frac{h^2}{a^2}+\frac{k^2}{b^2}-1\right)}}{\left(\frac{h^2}{a^2}+\frac{k^2}{b^2}\right)}.$$

Also m of the chord of contact is

$$-\frac{b^2}{a^2}\cdot\frac{h}{k}.$$

$\therefore$ length of the chord of contact

$$=\frac{\sqrt{\left(1+\frac{b^4h^2}{a^4k^2}\right)}\cdot\sqrt{\left(\frac{h^2}{a^2}+\frac{k^2}{b^2}-1\right)}\cdot\frac{2ak}{b}}{\left(\frac{h^2}{a^2}+\frac{k^2}{b^2}\right)}.$$

Perpendicular from (h, k) on the chord of contact is

$$\frac{\left(\frac{h^2}{a^2}+\frac{k^2}{b^2}-1\right)}{\sqrt{\left(\frac{h^2}{a^4}+\frac{k^2}{b^4}\right)}}$$

or
$$\frac{\left(\frac{h^2}{a^2}+\frac{k^2}{b^2}-1\right)}{\frac{k}{b^2}\sqrt{\left(1+\frac{b^4h^2}{a^4k^2}\right)}}.$$

Hence the area of the triangle OPQ

$$=\frac{ab\left(\frac{h^2}{a^2}+\frac{k^2}{b^2}-1\right)^{3/2}}{\left(\frac{h^2}{a^2}+\frac{k^2}{b^2}\right)}.$$

Length of the perpendicular from C on the chord of contact is

$$\frac{1}{\sqrt{\left(\frac{h^2}{a^4}+\frac{k^2}{b^4}\right)}}$$

or
$$\frac{1}{\frac{k}{b^2}\sqrt{\left(1+\frac{b^4}{a^4}\frac{h^2}{k^2}\right)}}.$$

Hence the area of the triangle CPQ is

$$\frac{ab\left(\frac{h^2}{a^2}+\frac{k^2}{b^2}-1\right)^{1/2}}{\left(\frac{h^2}{a^2}+\frac{k^2}{b^2}\right)}.$$

$\therefore$ area of quadrilateral $OCPQ$

$=$area of $\triangle\ OPQ+$area of $\triangle\ CPQ$

$$=\frac{ab\left(\frac{h^2}{a^2}+\frac{k^2}{b^2}-1\right)^{1/2}}{\left(\frac{h^2}{a^2}+\frac{k^2}{b^2}\right)}\left\{\frac{h^2}{a^2}+\frac{k^2}{b^2}-1+1\right\}$$

$$=ab\left(\frac{h^2}{a^2}+\frac{k^2}{b^2}-1\right)^{1/2}.$$

7·24. Diameter. The locus of the middle points of a system of parallel chords of an ellipse is called a diameter and the chords are called double ordinates.

7·25. To find the equation of diameter.

Let $y=mx+c$ be one of a system of parallel chords of the ellipse

$$\frac{x^2}{a^2}+\frac{y^2}{b^2}=1.$$

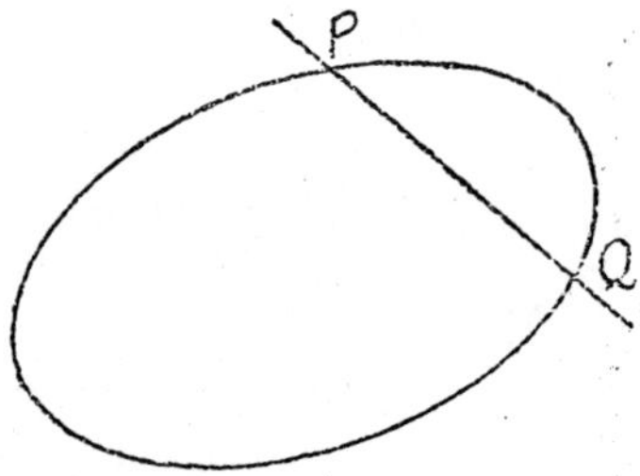

The abscissae of the points of intersection P and Q of the chord $y=mx+c$ and the ellipse

$$\frac{x^2}{a^2}+\frac{y^2}{b^2}=1$$

are given by the equation

$$\frac{x^2}{a^2}+\frac{(mx+c)^2}{b^2}=1$$

or
$$x^2\left(\frac{1}{a^2}+\frac{m^2}{b^2}\right)+\frac{2mxc}{b^2}+\frac{c^2}{b^2}=1$$

or $$x^2(b^2+a^2m^2)+2mca^2x+a^2(c^2-b^2)=0. \quad \ldots \quad \ldots(1)$$

Let x_1 and x_2 be the roots of this equation and let (h, k) be the coordinates of the middle point of PQ.

Then $$h=\frac{x_1+x_2}{2}=-\frac{mca^2}{b^2+a^2m^2} \quad \ldots(2) \text{ from (1).}$$

But the point (h, k) also lies on the chord

$$y=mx+c.$$

$$\therefore \quad k=mh+c. \quad \ldots \quad \ldots \quad \ldots(3)$$

Eliminating c between (2) and (3), we get

$$k=mh-\frac{(b^2+a^2m^2)\,h}{ma^2}$$

$$=-\frac{b^2h}{ma^2}.$$

Hence the locus of the point (h, k) is

$$y=-\frac{b^2x}{ma^2},$$

i.e. the equation of the diameter of chords of the system $y=mx+c$ is

$$y=-\frac{b^2}{a^2m}.x.$$

Corollary. If $y=m_1x$ is a diameter of the system

$$y=mx+c,$$

$$m_1=-\frac{b^2}{a^2m}$$

or $$mm_1=-\frac{b^2}{a^2}.$$

Similarly if $y=mx$ is a diameter of the system

$$y=m_1x+c,$$

$$m=-\frac{b^2}{a^2m_1}$$

or $$mm_1=-\frac{b^2}{a^2}.$$

Hence if one diameter bisects chords parallel to one another, the second diameter bisects chords parallel to the first. Two such diameters are called conjugate diameters.

Hence $y=mx$ and $y=m_1x$ are conjugate diameters and the ellipse if $mm_1=-\dfrac{b^2}{a^2}$.

7·25. Some propositions on conjugate diameters.

(1) *The eccentric angles of the extremities of two conjugate semi-diameters of an ellipse differ by a right angle.*

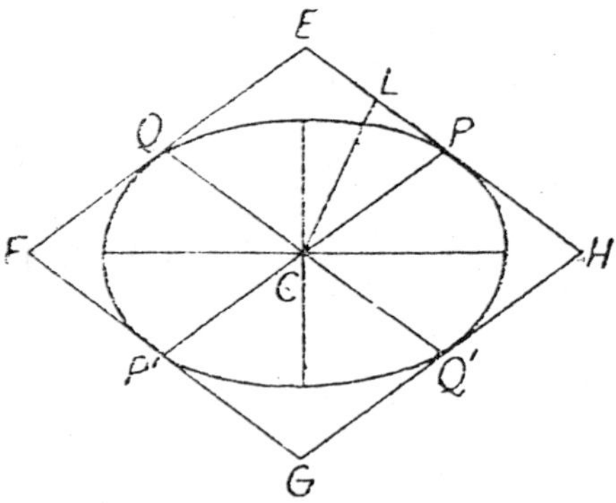

Let ϕ_1 and ϕ_2 be the eccentric angles of P and Q respectively, the extremities of the conjugate diameters CP and CQ.

The slope of CP, m is $\dfrac{b}{a}\dfrac{\sin\phi_1}{\cos\phi_1}$.

The slope of CQ, m_1 is $\dfrac{b}{a}\dfrac{\sin\phi_2}{\cos\phi_2}$.

But $$mm_1=-\frac{b^2}{a^2}.$$

$$\therefore\quad \frac{b^2}{a^2}\frac{\sin\phi_1\sin\phi_2}{\cos\phi_1\cos\phi_2}=-\frac{b^2}{a^2}$$

or $$\cos \phi_1 \cos \phi_2+\sin \phi_1 \sin \phi_2=0$$

or $$\cos (\phi_2-\phi_1)=0.$$

$$\therefore \quad \phi_2-\phi_1=\pi/2.$$

(2) *The sum of the squares of conjugate semi-diameters of an ellipse is equal to the sum of the squares of the semi-axes of the ellipse.*

Let the coordinates of P be $(a \cos \phi, b \sin \phi)$.

Then the coordinates of Q are

$$a \cos (\phi+\pi/2),\ b \sin (\phi+\pi/2).$$

$$\therefore \quad CP^2=a^2 \cos^2 \phi+b^2 \sin^2 \phi$$

$$CQ^2=(-a \sin \phi)^2+(b \cos \phi)^2$$

$$=a^2 \sin^2 \phi+b^2 \cos^2 \phi.$$

By addition, $CP^2+CQ^2=a^2+b^2$.

(3) *The tangents at the extremities of a diameter of an ellipse are parallel to the conjugate diameter.*

Equation to the tangent at the point P $(a \cos \phi, b \sin \phi)$

is $$\frac{x \cos \phi}{a}+\frac{y \sin \phi}{b}=1. \qquad \ldots(1)$$

The point Q is $(-a \sin \phi, b \cos \phi)$.

$\therefore$ the equation to CQ is

$$y=-\frac{b}{a}\frac{\cos \phi}{\sin \phi}x$$

or $$\frac{x}{a} \cos \phi+\frac{y}{b} \sin \phi=0. \qquad \ldots(2)$$

It is obvious that (1) and (2) are parallel to each other.

Similarly the tangent at P' is parallel to CQ and the diameter PP' is parallel to the tangents at Q and Q'.

(4) *The area of a parallelogram formed by the tangents at the extremities of two conjugate diameters of an ellipse is constant and is equal to the product of the axes.*

The equation of the tangent at P is

$$\frac{x}{a} \cos \phi+\frac{y}{b} \sin \phi=1. \qquad \ldots(1)$$

Length of the perpendicular CL from C on (1) is

$$\frac{1}{\sqrt{\left(\frac{\cos^2 \phi}{a^2}+\frac{\sin^2 \phi}{b^2}\right)}}.$$

Length of $CQ=\sqrt{(a^2 \sin^2 \phi+b^2 \cos^2 \phi)}$.

$\therefore$ area of the parallelogram $CPEQ$

$$=\frac{1}{\sqrt{\left(\frac{\cos^2 \phi}{a^2}+\frac{\sin^2 \phi}{b^2}\right)}}.\sqrt{(a^2 \sin^2 \phi+b^2 \cos^2 \phi)}$$

$$=ab.$$

Hence the area of the parallelogram $EFGH=4ab$.

7·26. Equi-conjugate diameters. If two conjugate diameters of an ellipse are equal, they are called equi-conjugate diameters.

To find the equation of the equi-conjugate diameters of the ellipse $\frac{x^2}{a^2}+\frac{y^2}{b^2}=1$.

Let the eccentric angles of the extremities P and Q of two conjugate diameters be ϕ and $\pi/2+\phi$; then

$$CP^2=CD^2$$

or $$a^2 \cos^2 \phi+b^2 \sin^2 \phi=a^2 \sin^2 \phi+b^2 \cos^2 \phi$$

or $$a^2 \cos 2\phi=b^2 \cos 2\phi$$

or $$\cos 2\phi\,(a^2-b^2)=0.$$

$\therefore$ $\cos 2\phi=0$, so that

$2\phi=\pi/2$ or $3\pi/2$ or $\phi=\pi/4$ or $3\pi/4$.

The equation to CP is

$$y=\frac{b}{a}\cdot\frac{\sin \phi}{\cos \phi}.x$$

$$=\frac{b}{a}x \qquad \text{as } \phi=\frac{\pi}{4}.$$

The equation to CQ is

$$y=-\frac{b}{a}\cdot\frac{\cos \phi}{\sin \phi}.x$$

$$y=-\frac{b}{a}x \qquad \text{as } \phi=\frac{\pi}{4}.$$

The other value of ϕ also gives the same equations.

Hence the equations of the equi-conjugate diameters are

$$x/a=\pm y/b.$$

7·27. Supplemental chords.

The chords joining any point on the ellipse to the extremities of a diameter are called the supplemental chords.

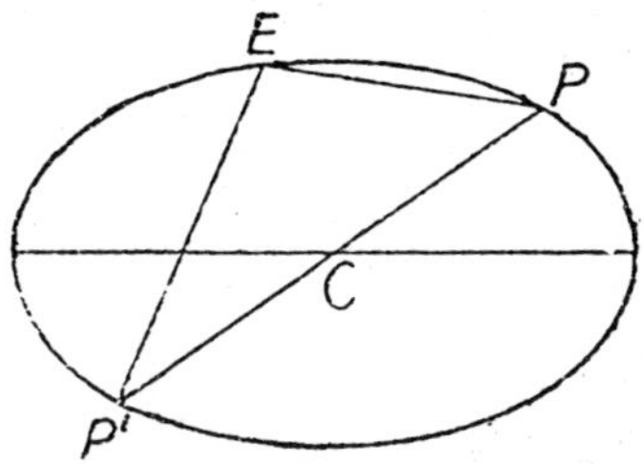

To prove that any two supplemental chords of an ellipse are parallel to a pair of conjugate diameters.

Let PCP' be a diameter and let E be a point on the ellipse.

Let the eccentric angles of P and E be ϕ and ϕ_1 respectively. Then the eccentric angle of P' is $\pi+\phi$.

Slope of EP, $m_1=\dfrac{b\sin\phi-b\sin\phi_1}{a\cos\phi-a\cos\phi_1}$.

Slope of EP', $m_2=\dfrac{-b\sin\phi-b\sin\phi_1}{-a\cos\phi-a\cos\phi_1}$.

$$\therefore\quad m_1m_2=\frac{b^2}{a^2}\cdot\frac{(\sin^2\phi-\sin^2\phi_1)}{(\cos^2\phi-\cos^2\phi_1)}$$

$$=\frac{b^2}{a^2}\frac{(1-\cos^2\phi-1+\cos^2\phi_1)}{\cos^2\phi-\cos^2\phi_1}$$

$$=-\frac{b^2}{a^2}.$$

Hence EP and EP' are parallel to a pair of conjugate diameters.

7·28. Chord bisected at the point (h, k).

The equation to the diameter through (h, k) is

$$y = \frac{k}{h} x.$$

The required chord is parallel to the conjugate diameter of $y = \frac{k}{h} x$.

Hence if the slope of this chord is m,

$$m \times \frac{k}{h} = -\frac{b^2}{a^2}.$$

$$\therefore \quad m = -\frac{b^2}{a^2} \cdot \frac{h}{k}.$$

The equation of the chord is therefore given by

$$y - k = -\frac{b^2}{a^2} \cdot \frac{h}{k} (x - h)$$

or $$\frac{k}{b^2}(y - k) + \frac{h}{a^2}(x - h) = 0.$$

7·29. Equation of an ellipse referred to a pair of conjugate diameters as axes.

Let the axes be the diameters CP and CQ. The equation referred to major axis and minor axis as axes is

$$\frac{x^2}{a^2} + \frac{y^2}{b^2} = 1. \qquad \ldots(1)$$

To get the equation referred to CP and CQ as axes, we will have to substitute expressions of the type $lx + my$ and $l_1 x + m_1 y$ for x and y respectively in (1) so that the new equation will be of the form

$$Ax^2 + Hxy + By^2 = 1 \qquad \ldots(2)$$

Axis CP bisects all chords parallel to CQ. Hence if (x_1, y_1) is a point on (2), $(x_1, -y_1)$ must also lie on it, so that

$$Ax_1^2 + Hx_1 y_1 + By_1^2 = 1 \qquad \ldots(3)$$

and $$Ax_1^2 - Hx_1y_1 + By_1^2 = 1 \qquad \ldots(4)$$

Subtracting (4) from (3),

$$2Hx_1y_1 = 0.$$

$$\therefore \quad H = 0.$$

The equation to the ellipse becomes

$$Ax^2 + By^2 = 1 \qquad \ldots(5)$$

Let the lengths of CP and CQ be a_1 and b_1 respectively, so that the coordinates of P and Q with respect to the next axes are $(a_1, 0)$ and $(0, b_1)$ respectively. These points lie on (5).

$$\therefore \quad Aa_1^2 = 1 \text{ and } Bb_1^2 = 1.$$

$$\therefore \quad A = \frac{1}{a_1^2} \text{ and } B = \frac{1}{b_1^2}$$

so that the equation to the ellipse becomes

$$\frac{x^2}{a_1^2} + \frac{y^2}{b_1^2} = 1.$$

7·30. Solved Examples.

Ex. 1. *Show that the locus of the middle points of the chords of an ellipse which subtend a right angle at the centre is*

$$\frac{x^2}{a^4} + \frac{y^2}{b^4} = \frac{a^2+b^2}{a^2b^3}\left(\frac{x^2}{a^2} + \frac{y^2}{b^2}\right)^2.$$

(Allahabad 1953)

Let (h, k) be the middle point of the chord. Then the equation of the chord is

$$\frac{k}{b^2}(y-k) + \frac{h}{a^2}(x-h) = 0,$$

or $$\frac{xh}{a^2} + \frac{yk}{b^2} = \frac{h^2}{a^2} + \frac{k^2}{b^2} \qquad \ldots(1)$$

The equation of the lines joining the centre to the points of intersection of (1) and the ellipse

$$\frac{x^2}{a^2} + \frac{y^2}{b^2} = 1 \qquad \ldots(2)$$

will be obtained by making (2) homogeneous with the help

of (1), so that the equation is

$$\frac{x^2}{a^2}+\frac{y^2}{b^2}=\frac{\left(\frac{xh}{a^2}+\frac{yk}{b^2}\right)^2}{\left(\frac{h^2}{a^2}+\frac{k^2}{b^2}\right)^2}$$

or
$$\left(\frac{x^2}{a^2}+\frac{y^2}{b^2}\right)\left(\frac{h^2}{a^2}+\frac{k^2}{b^2}\right)^2=\left(\frac{xh}{a^2}+\frac{yk}{b^2}\right)^2 \qquad \ldots(3)$$

The line (3) are at right angles to one another if

$$\left[\left(\frac{h^2}{a^2}+\frac{k^2}{b^2}\right)^2\frac{1}{a^2}-\frac{h^2}{a^4}+\left(\frac{h^2}{a^2}+\frac{k^2}{b^2}\right)^2\frac{1}{b^2}-\frac{k^2}{b^4}\right]=0$$

Hence the locus of (h, k) is

$$\frac{x^2}{a^4}+\frac{y^2}{b^4}=\left(\frac{1}{a^2}+\frac{1}{b^2}\right)\left(\frac{x^2}{a^2}+\frac{y^2}{b^2}\right)^2$$

$$=\frac{a^2+b^2}{a^2b^2}\left(\frac{x^2}{a^2}+\frac{y^2}{b^2}\right)^2.$$

Ex. 2. *CP and CD are conjugate diameters of the ellipse*

$$\frac{x^2}{a^2}+\frac{y^2}{b^2}=1.$$

Prove that the locus of the orthocentre of the triangle CPD is the curve

$$2\,(b^2y^2+a^2x^2)^3=(a^2-b^2)^2\,(b^2y^2-a^2x^2)^2.$$

(Rajputana 1956)

Let the extremities P and D have coordinates $(a\cos\phi, b\sin\phi)$ and $(-a\sin\phi, b\cos\phi)$ respectively.

Let the coordinates of the orthocentre be (h, k).

Slope of $PD=\frac{b}{a}\,\frac{(-\cos\phi+\sin\phi)}{(\cos\phi+\sin\phi)}$

Slope of the line joining (h, k) and $(0, 0)=\frac{k}{h}$.

Both these lines are perpendicular to each other.

$$\therefore\quad \frac{k}{h}\cdot\frac{b}{a}\,\frac{(-\cos\phi+\sin\phi)}{(\cos\phi+\sin\phi)}=-1 \qquad \ldots(1)$$

Slope of CD is $-\frac{b \cos \phi}{a \sin \phi}$.

Slope of line joining (h, k) and P is

$$\frac{k-b \sin \phi}{h-a \cos \phi}.$$

Both are perpendicular to each other.

$$\therefore \quad \left(\frac{k-b \sin \phi}{h-a \cos \phi}\right)\left(-\frac{b}{a} \frac{\cos \phi}{\sin \phi}\right)=-1$$

or $$(k-b \sin \phi)\, b \cos \phi=(h-a \cos \phi)\, a \sin \phi$$

or $$ha \sin \phi-bk \cos \phi=(a^2-b^2) \sin \phi \cos \phi. \qquad \ldots(2)$$

From (1), $$\frac{\cos \phi-\sin \phi}{ha}=\frac{\cos \phi+\sin \phi}{kb}$$

$$=\frac{\cos \phi-\sin \phi+\cos \phi+\sin \phi}{ka+kb}$$

$$=\frac{\cos \phi+\sin \phi-\cos \phi+\sin \phi}{kb-ha}$$

$$=\frac{\sqrt{\{(\cos \phi-\sin \phi)^2+(\cos \phi+\sin \phi)^2\}}}{\sqrt{(k^2b^2+h^2a^2)}}$$

or $$\frac{2 \cos \phi}{kb+ha}=\frac{2 \sin \phi}{kb-ha}=\sqrt{\left(\frac{2}{k^2b^2+h^2a^2}\right)} \quad \ldots \qquad \ldots(3)$$

Substituting the values of $\cos \phi$ and $\sin \phi$ from (3) in (2), we get

$$\frac{ha.(kb-ha)}{\sqrt{\{2\,(k^2b^2+h^2a^2)}}-\frac{kh\,(kb+ha)}{\sqrt{\{2\,(k^2b^2+h^2a^2)}}=\frac{(a^2-b^2)\,(k^2b^2-h^2a^2)}{2\,(k^2b^2+h^2a^2)}$$

or $$\sqrt{\{2\,(k^2b^2+h^2a^2)\}}\,[-(k^2b^2+h^2a^2)]=(a^2-b^2)\,(k^2b^2-h^2a^2)$$

or $$\{2\,(k^2b^2+h^2a^2)^3=(a^2-b^2)^2\,(k^2b^2-h^2a^2)^2.$$

Hence the locus of (h, k) is

$$2\,(b^2y^2+a^2x^2)^3=(a^2-b^2)^2\,(b^2y^2-a^2x^2)^2.$$

Ex. 3. *If $\lambda_1 \lambda'$ be the angles which any two conjguate diameters subtend at any fixed point on an ellipse, prove that*

$$\cot^2 \lambda+\cot^2 \lambda'=const. \qquad \textit{(Allahabad 1948)}$$

Let the extremities P and Q have coordinates

$$(a \cos \phi, b \sin \phi), (-a \sin \phi, b \cos \phi).$$

The coordinates of P' and Q' are $(-a \cos \phi, -b \sin \phi)$ and $(a \sin \phi, -b \cos \phi)$ respectively.

Let the coordinates of any point on the ellipse R be $(a \cos \phi_1, b \sin \phi_1)$.

The slope of $RP = \dfrac{b \sin \phi - b \sin \phi_1}{a \cos \phi - a \cos \phi_1}$

$$= \frac{b}{a} \cdot \frac{2 \cos \frac{\phi+\phi_1}{2} \sin \frac{\phi-\phi_1}{2}}{2 \sin \frac{\phi+\phi_1}{2} \sin \frac{\phi_1-\phi}{2}}$$

$$= -\frac{b}{a} \cot \frac{\phi+\phi_1}{2}.$$

The slope of $RP' = \dfrac{-b \sin \phi - b \sin \phi_1}{-a \cos \phi - a \cos \phi_1}$

$$= \frac{b}{a} \cdot \frac{2 \sin \frac{\phi+\phi_1}{2} \cos \frac{\phi-\phi_1}{2}}{2 \cos \frac{\phi+\phi_1}{2} \cos \frac{\phi-\phi_1}{2}}$$

$$= \frac{b}{a} \tan \frac{\phi+\phi_1}{2}.$$

Hence the angle λ subtended by PP' at R is given by

$$\tan \lambda = \frac{\frac{b}{a}\left[\tan \frac{\phi+\phi_1}{2} + \cos \frac{\phi+\phi_1}{2}\right]}{1 - \frac{b^2}{a^2}}$$

$$= \frac{ab}{(a^2-b^2) \sin \frac{\phi+\phi_1}{2} \cos \frac{\phi+\phi_1}{2}}$$

$$= \frac{2ab}{(a^2-b^2) \sin (\phi+\phi_1)}.$$

The slope of $RQ = \dfrac{b \cos \phi - b \sin \phi_1}{-a \sin \phi - a \cos \phi_1}$

$$= -\frac{b}{a} \frac{[\cos \phi + \cos (\pi/2 + \phi_1)]}{[\sin \phi + \sin (\pi/2 + \phi_1)]}$$

$$= -\frac{b}{a} \frac{2 \cos \left(\dfrac{\phi + \phi_1}{2} + \dfrac{\pi}{4}\right) \cos \left(\dfrac{\phi - \phi_1}{2} - \dfrac{\pi}{4}\right)}{2 \sin \left(\dfrac{\phi + \phi_1}{2} + \dfrac{\pi}{4}\right) \cos \left(\dfrac{\phi - \phi_1}{2} - \dfrac{\pi}{4}\right)}$$

$$= -\frac{b}{a} \cot \left(\frac{\phi + \phi_1}{2} + \frac{\pi}{4}\right).$$

The slope of $RQ' = \dfrac{-b \cos \phi - b \sin \phi_1}{a \sin \phi - a \cos \phi_1}$

$$= -\frac{b}{a} \frac{[\cos \phi + \sin \phi_1]}{[\sin \phi - \cos \phi_1]}$$

$$= \frac{b}{a} \frac{[\cos \phi + \cos (\pi/2 - \phi_1)]}{[\cos (\pi/2 + \phi) + \cos \phi_1]}$$

$$= \frac{b}{a} \frac{2 \cos \left(\dfrac{\phi - \phi_1}{2} + \dfrac{\pi}{4}\right) \cos \left(\dfrac{\phi + \phi_1}{2} - \dfrac{\pi}{4}\right)}{2 \cos \left(\dfrac{\phi + \phi_1}{2} + \dfrac{\pi}{4}\right) \cos \left(\dfrac{\phi - \phi_1}{2} + \dfrac{\pi}{4}\right)}$$

$$= \frac{b}{a} \cdot \frac{\sin \left(\dfrac{\phi + \phi_1}{2} + \dfrac{\pi}{4}\right)}{\cos \left(\dfrac{\phi + \phi_1}{2} + \dfrac{\pi}{4}\right)}$$

$$= \frac{b}{a} \tan \left(\frac{\phi + \phi_1}{2} + \frac{\pi}{4}\right).$$

If λ' is the angle subtended by QQ' at R.

$$\tan \lambda' = \frac{b}{a} \frac{\left[\tan \left(\dfrac{\phi + \phi_1}{2} + \dfrac{\pi}{4}\right) + \cot \left(\dfrac{\phi + \phi_1}{2} + \dfrac{\pi}{4}\right)\right]}{1 - \dfrac{b^2}{a^2}}$$

$$\tan \lambda' = \frac{ab}{a^2-b^2} \cdot \frac{1}{\sin\left(\frac{\phi+\phi_1}{2}+\frac{\pi}{4}\right)\cos\left(\frac{\phi+\phi_1}{2}+\frac{\pi}{4}\right)}$$

$$= \frac{2ab}{(a^2-b^2)\sin(\phi+\phi_1+\pi/2)}$$

$$= \frac{2ab}{(a^2-b^2)\cos(\phi+\phi_1)}.$$

$$\therefore \quad \cot^2\lambda + \cot^2\lambda' = \frac{(a^2-b^2)^2}{4a^2b^2}\left[\cos^2(\phi+\phi_1)+\sin^2(\phi+\phi_1)\right]$$

$$= \frac{(a^2-b^2)^2}{4a^2b^2}$$

$$= \text{const.}$$

7·31. Concyclic points on an ellipse.

To prove that the two straight lines through the points of intersection of an ellipse with any circle make equal angles with the axes of the ellipse.

Let the equation to the ellipse be

$$\frac{x^2}{a^2}+\frac{y^2}{b^2}=1 \qquad \ldots(1)$$

and let the equation to a circle be

$$x^2+y^2+2gx+2fy+c=0. \qquad \ldots(2)$$

Then the equation

$$\frac{x^2}{a^2}+\frac{y^2}{b^2}-1+\lambda(x^2+y^2+2gx+2fy+c)=0 \qquad \ldots(3)$$

represents a curve which passes through the points of intersection of (1) and (2). Equation (3) represents a pair of straight lines if

$$(\lambda c-1)\left(\lambda+\frac{1}{a^2}\right)\left(\lambda+\frac{1}{b^2}\right)-\left(\lambda+\frac{1}{a^2}\right)\lambda^2f^2-\left(\lambda+\frac{1}{b^2}\right)\lambda^2g^2=0. \qquad \ldots(4)$$

This is a cubic equation in λ and will give three values of λ, corresponding to each of which there will be a pair of straight lines.

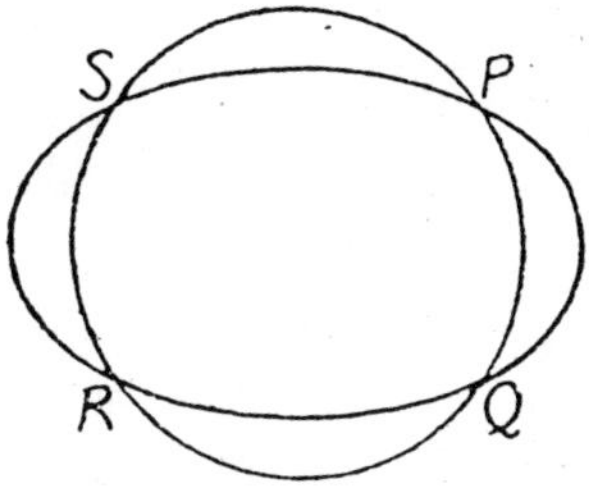

It is also clear from the figure. The pairs of straight lines are

PQ, *RS*; *PS*, *QR*; and *PR*, *QS*.

Then (3) represents straight lines they are parallel to the lines

$$\frac{x^2}{a^2}+\frac{y^2}{b^2}+\lambda(x^2+y^2)=0$$

or
$$x^2\left(\frac{1}{a^2}+\lambda\right)+y^2\left(\frac{1}{b^2}+\lambda\right)=0$$

or
$$y=\pm x\sqrt{\left\{\frac{\lambda+(1/a^2)}{\lambda+(1/b^2)}\right\}}$$

which are equally inclined to the axes.

7·32. Circle through three given points on an ellipse.

To find the centre of the circle passing through three points on an ellipse, whose eccentric angles are α, β, γ.

Let a circle cut an ellipse in the points P, Q, R and S whose eccentric angles are α, β, γ, δ.

The equation to the straight line PQ is

$$\frac{x}{a}\cos\left(\frac{\alpha+\beta}{2}\right)+\frac{y}{b}\sin\left(\frac{\alpha+\beta}{2}\right)=\cos\left(\frac{\alpha-\beta}{2}\right) \quad \ldots(1)$$

The equation to the straight line RS is

$$\frac{x}{a}\cos\left(\frac{\gamma+\delta}{2}\right)+\frac{y}{b}\sin\left(\frac{\gamma+\delta}{2}\right)=\cos\left(\frac{\gamma-\delta}{2}\right) \quad \ldots(2)$$

These lines make equal angles with the axes.

$$\therefore \quad \tan\left(\frac{\alpha+\beta}{2}\right)=-\tan\left(\frac{\gamma+\delta}{2}\right)$$

i.e. $$\frac{\alpha+\beta}{2}=n\pi-\frac{\gamma+\delta}{2}$$

or $$\alpha+\beta+\gamma+\delta=2n\pi. \qquad \ldots(3)$$

At a point where any circle

$$x^2+y^2+2gx+2fy+c=0$$

cuts the ellipse, the eccentric angle satisfies the relation

$$a^2\cos^2\phi+b^2\sin^2\phi+2ga\cos\phi+2fb\sin\phi+c=0$$

or $$(a^2-b^2)\cos^2\phi+2ga\cos\phi+c+b^2=-2fb\sin\phi$$

or $$\{(a^2-b^2)\cos^2\phi+2ga\cos\phi+c+b^2\}^2=4f^2b^2-4f^2b^2\cos^2\phi.$$

Hence $\cos\alpha+\cos\beta+\cos\gamma+\cos\delta=\dfrac{-4ga}{(a^2-b^2)}$.

Similarly putting the equation in $\sin\phi$, we get

$$\sin\alpha+\sin\beta+\sin\gamma+\sin\delta=\frac{-4fb}{(b^2-a^2)}.$$

But $$\alpha+\beta+\gamma+\delta=2n\pi.$$

$$\therefore \quad \cos\delta=\cos(\alpha+\beta+\gamma)$$

and $$\sin\delta=-\sin(\alpha+\beta+\gamma)$$

The centre of the circle is $(-g, -f)$.

Hence the centre is given by

$$x=\frac{a^2-b^2}{4a}[\cos\alpha+\cos\beta+\cos\gamma+\cos(\alpha+\beta+\gamma)]$$

$$y=\frac{b^2-a^2}{4b}[\sin\alpha+\sin\beta+\sin\gamma-\sin(\alpha+\beta+\gamma)]$$

7·33. Conormal points :—To prove that in general four normals can be drawn to an ellipse from any point, and sum of the eccentric angles of the conormal points is equal to an odd multiple of two right angles.

The equation to the normal to the ellipse

$$\frac{x^2}{a^2}+\frac{y^2}{b^2}=1$$

at the point ϕ is

$$ax \sec \phi - by \operatorname{cosec} \phi = a^2 - b^2.$$

If it passes through a point (x_1, y_1)

$$\frac{ax_1}{\cos \phi} - \frac{by_1}{\sin \phi} = a^2 - b^2$$

or
$$\frac{ax_1}{\cos^2 \frac{\phi}{2} - \sin^2 \frac{\phi}{2}} - \frac{by_1}{2 \sin \frac{\phi}{2} \cos \frac{\phi}{2}} = a^2 - b^2$$

or
$$\frac{ax_1}{1 - \tan^2 \frac{\phi}{2}} - \frac{by_1}{2 \tan \frac{\phi}{2}} = \frac{a^2 - b^2}{1 + \tan^2 \frac{\phi}{2}}$$

Let $\tan \frac{\phi}{2} = t$; then

$$\frac{ax_1}{1 - t^2} - \frac{by_1}{2t} = \frac{a^2 - b^2}{1 + t^2}$$

or
$$ax_1 - 2t(1 + t^2) - by_1(1 - t^4) = (a^2 - b^2) . 2t(1 - t^2)$$

or
$$by_1 t^4 + 2(ax_1 + a^2 - b^2) t^3 + 2(ax_1 - a^2 + b^2) t - by_1 = 0 \quad \ldots(1)$$

This gives four values of t and corresponding four values of $\tan \phi/2$. Corresponding to one value of $\tan \phi/2$, there is one point on the ellipse. Hence from one point (x_1, y_1) four normals can be drawn to an ellipse.

If the roots of equation (1) are t_1, t_2, t_3, t_4,

$$\Sigma t_1 = -\frac{2(ax_1 + a^2 - b^2)}{by_1} \quad \ldots(2)$$

$$\Sigma t_1 t_2 = 0 \quad \ldots(3)$$

$$\Sigma t_1 t_2 t_3 = -\frac{2(ax_1 - a^2 + b^2)}{by_1} \quad \ldots(4)$$

$$t_1 t_2 t_3 t_4 = -1. \quad \ldots(5)$$

Suppose corresponding to t_1, t_2, t_3, t_4 the corresponding values of the eccentric angles are ϕ_1, ϕ_2, ϕ_3 and ϕ_4.

$$\tan \frac{(\phi_1 + \phi_2 + \phi_3 + \phi_4)}{2} = \frac{\Sigma t_1 - \Sigma t_1 t_2 t_3}{1 - \Sigma t_1 t_2 + t_1 t_2 t_3 t_4}$$

$$=\frac{\Sigma t_1-\Sigma t_1t_2t_3}{1-0-1}=\infty.$$

$$\therefore \quad \frac{\phi_1+\phi_2+\phi_3+\phi_4}{2}=n\pi+\frac{\pi}{2}$$

i.e.

$$\phi_1+\phi_2+\phi_3+\phi_4=2n\pi+\pi$$
$$=(2n+1)\pi.$$

7·34. Solved Examples.

Ex. 1. *If α, β, γ be the eccentric angles of three points on the ellipse $\frac{x^2}{a^2}+\frac{y^2}{b^2}=1$ at which the normals are concurrent, then*

$$\sin(\alpha+\beta)+\sin(\beta+\gamma)+\sin(\gamma+\alpha)=0. \quad \textit{(Agra 1945)}$$

If the normal at (h, k) meet the ellipse at points α, β, γ and δ, then we know that

$$\Sigma \tan\frac{\alpha}{2}\tan\frac{\beta}{2}=0$$

and

$$\tan\frac{\alpha}{2}\tan\frac{\beta}{2}\tan\frac{\gamma}{2}\tan\frac{\delta}{2}=-1.$$

Eliminating $\tan \delta/2$ between these equations, we get

$$\tan\frac{\alpha}{2}\tan\frac{\beta}{2}+\tan\frac{\alpha}{2}\tan\frac{\gamma}{2}+\tan\frac{\beta}{2}\tan\frac{\gamma}{2}$$
$$-\cot\frac{\alpha}{2}\cot\frac{\beta}{2}-\cot\frac{\alpha}{2}\cot\frac{\gamma}{2}-\cot\frac{\beta}{2}\cot\frac{\gamma}{2}=0.$$

This gives

$$\Sigma\left(\tan\frac{\alpha}{2}\tan\frac{\beta}{2}-\cot\frac{\alpha}{2}\cot\frac{\beta}{2}\right)=0$$

or

$$\Sigma\frac{\cos^2\frac{\alpha}{2}\cos^2\frac{\beta}{2}-\sin^2\frac{\alpha}{2}\sin^2\frac{\beta}{2}}{\sin\frac{\alpha}{2}\cos\frac{\alpha}{2}\sin\frac{\beta}{2}\cos\frac{\beta}{2}}=0$$

or

$$\Sigma\frac{\cos\frac{\alpha+\beta}{2}\cos\frac{\alpha-\beta}{2}}{\sin\alpha\sin\beta}=0$$

or $$\Sigma \frac{\cos\alpha+\cos\beta}{\sin\alpha\sin\beta}=0$$

or $$\frac{\cos\alpha+\cos\beta}{\sin\alpha\sin\beta}+\frac{\cos\beta+\cos\gamma}{\sin\beta\sin\gamma}+\frac{\cos\gamma+\cos\alpha}{\sin\gamma\sin\alpha}=0$$

or $$\sin\gamma(\cos\alpha+\cos\beta)+\sin\alpha(\cos\beta+\cos\gamma)+\sin\beta(\cos\gamma+\cos\alpha)=0$$

or $$\sin(\beta+\gamma)+\sin(\gamma+\alpha)+\sin(\alpha+\beta)=0.$$

Ex. 2. *Show that in general four normals can be drawn to an ellipse from a point (h, k) and that the feet of these normals lie on a rectangular hyperbola which passes through $(0, 0)$ and (h, k).*

(Agra 1946)

The normals to the ellipse $\frac{x^2}{a^2}+\frac{y^2}{b^2}=1$ at the ends of the chord $lx+my-1=0$ and $l'x+m'y-1=0$ will be concurrent if $a^2ll'=b^2mm'=-1$.

The equation to the normal at the point (x_1, y_1) is

$$(y-y_1)\frac{x_1}{a^2}-(x-x_1)\frac{y_1}{b^2}=0.$$

It passes through the point (h, k).

$$\therefore \quad (k-y_1)\frac{x_1}{a^2}-(h-x_1)\frac{y_1}{b^2}=0.$$

This is a relation between (x_1, y_1) on the basis of the given condition.

Hence the locus of (x_1, y_1) is

$$(k-y)\frac{x}{a^2}-(h-x)\frac{y}{b^2}=0$$

or $$xy\left(\frac{1}{b^2}-\frac{1}{a^2}\right)+\frac{k}{a^2}x-\frac{h}{b^2}y=0$$

which is a rectangular hyperbola passing through the points $(0, 0)$ and (h, k). Now

$$lx+my-1=0$$

and $$l'x+m'y-1=0$$

are lines passing through the feet of the normals. Hence

$$\frac{x^2}{a^2}+\frac{y^2}{b^2}-1+\lambda\,(lx+my-1)\,(l'x+m'y-1)=0$$

is the equation to a conic passing through the feet of the normals.

This conic must represent a rectangular hyperbola as shown before, so that the coefficients of terms containing x^2, y^2 and the constant term should be zero.

$$\therefore \quad \frac{1}{a^2}+\lambda ll'=0$$

$$\frac{1}{b^2}+\lambda mm'=0$$

and

$$-1+\lambda=0.$$

Eliminating λ from the above equation,

$$a^2ll'=b^2mm'=-1.$$

Examples XV.

1. (a) For the ellipse $8x^2+12y^2=96$, find the pair of conjugate semi-diameters inclined at an angle $\tan^{-1} 7$.

   (b) If P, Q be the extremities of conjugate diameters CP, CQ of the ellipse $8x^2+12y^2=96$; find the locus of the point of intersection of the tangents at P, Q. *(C. U. 1948)*

2. (a) Show that $4x-3y+4=0$ and $x+3y-7=0$ are parallel to the conjugate diameters of $4x^2+9y^2=36$.

   (b) If the line $\frac{l}{a}x+\frac{m}{b}y=n$ cuts the ellipse $b^2x^2+a^2y^2=a^2b^2$ at the ends of conjugate diameters, prove that $l^2+m^2=2n^2$. *(C. U. 1949)*

3. Calculate the eccentric angle of the two points P (8, 3) and Q (6, -4) on the ellipse $\frac{x^2}{100}+\frac{y^2}{25}=1$ and satisfy

yourselves that the semi-diameters with P, Q as end points are conjugate to each other.

4. Show that the sum of reciprocals of the squares of any two diameters of an ellipse which are at right angles to one another is constant.

5. PCP' is a diameter of an ellipse and QCQ' is the corresponding diameter of the auxiliary circle; show that the area of the parallelogram formed by the tangents at P, P', Q, Q' is $\dfrac{8a^2b}{(a-b)\sin 2\phi}$, where ϕ is the eccentric angle of P.

6. A pair of conjugate diameters is produced to meet the directrix. Show that the orthocentre of the triangle so formed is the focus. *(Nagpur 1947)*

7. Prove that if CP and CD be two conjugate semi-diameters of the ellipse $x^2/a^2+y^2/b^2=1$, the locus of the middle point of PD is $x^2/a^2+y^2/b^2=\frac{1}{2}$. *(Aligarh 1945)*

8. If P and D are the ends of conjugate diameters of the ellipse $x^2/a^2+y^2/b^2=1$, show that the locus of the foot of the perpendicular from the centre on PD is

$$2(x^2+y^2)^2=a^2x^2+b^2y^2.$$ *(Aligarh 1955)*

9. If θ, θ' be the angles subtended by the major axis of an ellipse at the extremities of a pair of conjugate diameters, show that $\cot^2\theta+\cot^2\theta'$ is constant.

10. Find the locus of the middle points of the chords of contact of the tangents drawn to the ellipse $\dfrac{x^2}{a^2}+\dfrac{y^2}{b^2}=1$ from points on the director circle.

11. Prove that if the polar of a point P with respect to an ellipse passes through Q, then the polar of Q passes through P.

12. Tangents are drawn from a point P to an ellipse $\frac{x^2}{a^2}+\frac{y^2}{b^2}=1$; show that the locus of P when the chord of contact subtends a right angle at the centre is

$$\frac{x^2}{a^4}+\frac{y^2}{b^4}=\frac{1}{a^2}+\frac{1}{b^2}.$$

13. If p, q be the lengths of two tangents to the ellipse $\frac{x^2}{a^2}+\frac{y^2}{b^2}=1$ at right angles to one another, prove that

$$\frac{4(a^2+b^2)^3}{p^2+q^2}=-\left\{a^2+b^2+a^2b^2\left(\frac{1}{p^2}+\frac{1}{q^2}\right)\right\}^2.$$

14. Chords of the ellipse $\frac{x^2}{a^2}+\frac{y^2}{b^2}=1$; always touch the ellipse $\frac{x^2}{\alpha^2}+\frac{y^2}{\beta^2}=1$, show that the locus of their poles is

$$\frac{\alpha^2x^2}{a^4}+\frac{\beta^2y^2}{b^4}=1.$$

15. If the product of the perpendiculars from the foci of the ellipse $\frac{x^2}{a^2}+\frac{y^2}{b^2}=1$ upon the polar of P be always c^2, prove that the locus of P is the ellipse

$$b^4(c^2+a^2-b^2)x^2+c^2a^4y^2=a^4b^4.$$ *(Utkal 1948)*

16. Tangents are drawn from any point on the circle $x^2+y^2=a^2$ to the ellipse $\frac{x^2}{a^2}+\frac{y^2}{b^2}=1$. Prove that the locus of the middle points of the chords of contact is

$$\left(\frac{x^2}{a^2}+\frac{y^2}{b^2}\right)^2=\left(\frac{x^2+y^2}{a^2}\right).$$ *(Lucknow 1946)*

17. If ω is the difference of the eccentric angles of two points on an ellipse the tangents at which are at right angles, prove that $ab \sin \omega=\lambda\mu$ where λ, μ are the

semi-diameters parallel to the tangents at the points, and a, b are the semi-axes of the ellipse.

18. Show that the tangents at the ends of conjugate diameters of the ellipse $\frac{x^2}{a^2}+\frac{y^2}{b^2}=1$ intersect on the ellipse $\frac{x^2}{a^2}+\frac{y^2}{b^2}=2$. *(Agra 1952)*

19. Find the locus of the point of intersection of normals at two points of an ellipse which are the extremities of conjugate diameters. *(Lucknow 1943)*

20. Show that the locus of the middle points of normal chords of the ellipse $\frac{x^2}{a^2}+\frac{y^2}{b^2}=1$ is the curve

$$\left(\frac{x^2}{a^2}+\frac{y^2}{b^2}\right)^2\left(\frac{a^6}{x^2}+\frac{b^6}{y^2}\right)=(a^2-b^2)^2.$$

21. Show that the locus of the middle points of chords of constant length $2c$ of the ellipse $\frac{x^2}{a^2}+\frac{y^2}{b^2}=1$ is

$$\left(\frac{x^2}{a^2}+\frac{y^2}{b^2}-1\right)\left(\frac{x^2}{a^4}+\frac{y^2}{b^4}\right)+\frac{c^2}{a^2b^2}\left(\frac{x^2}{a^4}+\frac{y^2}{b^4}\right)=0.$$

(Aligrah 1954)

22. If (x_1, y_1) be the middle point of a chord of the ellipse $\frac{x^2}{a^2}+\frac{y^2}{b^2}=1$; show that its length is

$$2ab\sqrt{\left[\left(\frac{x_1^2}{a^2}+\frac{y_1^2}{b^2}\right)\left\{\left(\frac{x_1^2}{a^2}+\frac{y_1^2}{b^2}\right)^{-1}-1\right\}\right]}$$

23. The normals at four points on $\frac{x^2}{a^2}+\frac{y^2}{b^2}=1$ meet in the point (α, β). Prove that the mean position of the four points is $\left\{\frac{1}{2}\frac{a^2\alpha}{a^2-b^2}, \frac{1}{2}\frac{b^2\beta}{(b^2-a^2)}\right\}$. *(Lucknow 1944)*

24. Prove that the locus of the centre of a circle which cuts

the ellipse $\frac{x^2}{a^2}+\frac{y^2}{b^2}=1$ in the fixed point (α, β) and in two other points at the extremities of a diameter is the ellipse

$$2a^2x^2+2b^2y^2=(a^2-b^2)(\alpha x-\beta y). \qquad \textit{(Utkal 1947)}$$

25. If the normals at the four points (x_1, y_1), (x_2, y_2), (x_3, y_3) and (x_4, y_4) on the ellipse $\frac{x^2}{a^2}+\frac{y^2}{b^2}=1$ are concurrent, prove that

$$(x_1+x_2+x_3+x_4)\left(\frac{1}{x_1}+\frac{1}{x_2}+\frac{1}{x_3}+\frac{1}{x^4}\right)=4.$$

(Delhi 1947)

26. Tangents are drawn from any point on the conic $\frac{x^2}{a^2}+\frac{y^2}{b^2}=4$; so that the conic $\frac{x^2}{a^2}+\frac{y^2}{b^2}=1$; prove that the normals at the points of contact meet on the conic

$$a^2x^2+b^2y^2=\tfrac{1}{4}(a^2-b^2)^2. \qquad \textit{(Utkal 1946)}$$

27. The normals at points whose eccentric angles are α, β, γ, δ on the ellipse $\frac{x^2}{a^2}+\frac{y^2}{b^2}=1$ meet at a point. Prove that

(i) $\Sigma \sin(\alpha+\beta)=0$

(ii) $\Sigma \cos(\alpha+\beta)=0$. *(Lucknow 1951)*

28. If four normals at four points of an ellipse $\frac{x^2}{a^2}+\frac{y^2}{b^2}=1$ meet in a point and two points lie on the line $\frac{lx}{a}+\frac{my}{b}=1$, show that the other two points will lie on $\frac{x}{al}+\frac{y}{bm}=1$.

29. If the points of intersection of

$$\frac{x^2}{a^2}+\frac{y^2}{b^2}=1$$

and $$\frac{x^2}{\alpha^2}+\frac{y^2}{\beta^2}=1$$

be the end points of conjugate diameters of the former, prove that $\frac{a^2}{\alpha^2}+\frac{b^2}{\beta^2}=2.$

(*Agra 1956*)

30. The normal at a variable point P of the ellipse $b^2x^2+a^2y^2=a^2b^2$ cuts the diameter CD conjugate to CP in Q. Prove that $PQ.CD$ is constant and that the equation to the locus of Q is

$$\frac{a^2}{x^2}+\frac{b^2}{y^2}=\left(\frac{a^2-b^2}{x^2+y^2}\right)^2.$$

31. From the point $(a\cos\phi,\ b\sin\phi)$ of the ellipse $\frac{x^2}{a^2}+\frac{y^2}{b^2}=1$, three normals (other than the normal at the point) are drawn to the ellipse; prove that their feet lie on the circle,

$$x^2+y^2-\frac{b^2}{a}.x\cos\phi-\frac{a^2}{b}y\sin\phi-a^2-b^2=0.$$

32. Tangents are drawn from any point on the ellipse $\frac{x^2}{a^2}+\frac{y^2}{b^2}=1$ to the circle $x^2+y^2=r^2$. Prove that the chord of contact touches the ellipse $a^2x^2+b^2y^2=r^2$. If $\frac{1}{r^2}=\frac{1}{a^2}+\frac{1}{b^2}$, prove that the line joining the centre to the points of contact with the circle are conjugate diameters of the second ellipse.

33. Show that if (x_1, y_1) be the middle point of a chord of the ellipse $\frac{x^2}{a^2}+\frac{y^2}{b^2}=1$, (α, β) the point of intersection of the normals, and (h, k) that of the tangents at its extremities, then

$$\frac{a^2\alpha}{h}+\frac{b^2\beta}{k}=(a^2-b^2)\left(\frac{hx_1}{a^2}-\frac{ky_1}{b^2}\right).$$

34. Prove that the circle whose diameter is the chord

$$\frac{x}{a}\cos\left(\frac{\theta_1+\theta_2}{2}\right)+\frac{y}{b}\sin\left(\frac{\theta_1+\theta_2}{2}\right)=\cos\left(\frac{\theta_1-\theta_2}{2}\right)$$

cuts the ellipse $\frac{x^2}{a^2}+\frac{y^2}{b^2}=1$ in two other points whose join is the line

$$\frac{x}{a}\cos\left(\frac{\theta_1+\theta_2}{2}\right)-\frac{y}{b}\sin\left(\frac{\theta_1+\theta_2}{2}\right)=\frac{a^2+b^2}{a^2-b^2}\cos\left(\frac{\theta_1-\theta_2}{2}\right).$$

35. A parallelogram circumscribes the ellipse $\frac{x^2}{a^2}+\frac{y^2}{b^2}-1=0$ and two of its angular points are on the line $x^2-h^2=0$; prove that the other two are on the conic

$$\frac{x^2}{a^2}+y^2\left(1-\frac{a^2}{h^2}\right)b^2-1=0.$$

36. A triangle circumscribes the ellipse $\frac{x^2}{a^2}+\frac{y^2}{b^2}=1$ and its centroid lies on the axis of x at a distance c from the centre; prove that the angular points of the triangle lie on the conic

$$\left(\frac{x-3c}{a^2}\right)^2+\frac{y^2\,(a^2-9c^2)}{a^2b^2}=4.$$

CHAPTER VIII

THE HYPERBOLA

8·1. Definition. The locus of a point which moves so that its distance from a fixed point (the focus) is e times its distance from a fixed straight line (the directrix), e being a fixed quantity greater then one, is called a hyperbola.

To find the standard equation of a hyperbola.

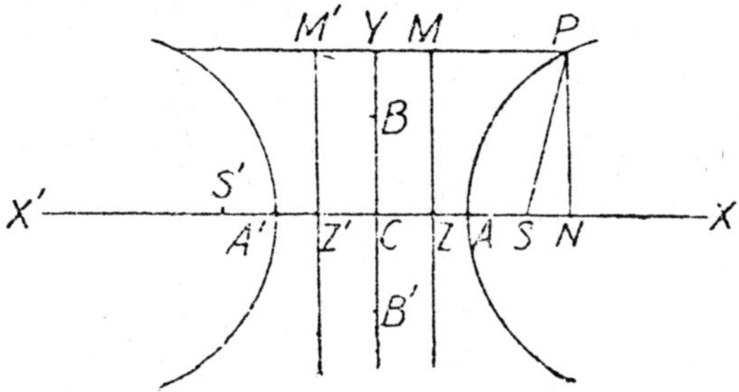

Let S be the focus and ZM the directrix. Draw SZ perpendicular from S on the directrix.

Since $e > 1$, the line SZ can be divided internally and externally in the points A and A' respectively. Then points A and A' lie on the hyperbola.

Let $AA'=2a$.

Let C the middle point of AA' be the origin, CX the x-axis and a perpendicular line CY the y-axis.

Now $$SA=e.AZ \qquad \ldots(1)$$

and $$SA'=e.A'Z. \qquad \ldots(2)$$

$$\therefore \quad SA+SA'=e\,(AZ+A'Z)$$

$$=2ae,$$

or $$2CS=2ae.$$

$$\therefore \quad CS=ae.$$

Also $$SA' - SA = e\,(A'Z - AZ)$$

or $$AA' = e\,(AA' - AZ - AZ)$$
$$= 2e\,(CA - AZ)$$

or $$2a = 2e \,.\, CZ.$$
$$\therefore \quad CZ = a/e.$$

Coordinates of S are $(ae, 0)$ and the equation to ZM is $x = a/e$.

Let P be any point (x, y) on the hyperbola.

Now $$SP = e \,.\, PM.$$
$$\sqrt{\{(x-ae)^2 + y^2\}} = e \,.\, (x - a/e)$$

or $$(x-ae)^2 + y^2 = e^2\,(x - a/e)^2$$

or $$x^2\,(e^2-1) - y^2 = a^2\,(e^2-1)$$

or $$\frac{x^2}{a^2} - \frac{y^2}{a^2\,(e^2-1)} = 1$$

or $$\frac{x^2}{a^2} - \frac{y^2}{b^2} = 1 \qquad \text{if } b^2 = a^2\,(e^2-1),$$

which is the standard form of the equation of the hyperbola.

8·2. Tracing the hyperbola.

The points (x, y), $(x, -y)$, $(-x, y)$ and $(-x, -y)$ all satisfy the equation of the hyperbola simultaneously. Hence it is symmetrical about both axes.

Also any chord of the hyperbola will be bisected at C, which is therefore the centre of the curve.

From the equation to the curve,
$$y = \pm \frac{b}{a} \sqrt{(x^2 - a^2)}.$$

If x is less than a, y is imaginary.

The curve therefore does not lie between A and A'.

If $x = \pm a$, $y = 0$ and for all values of x greater than a, y has two equal and opposite values. If $x \to \pm a$, y also $\to \pm \infty$.

The curve has therefore two infinite branches passing through A and A', one lying to the right of A and the other to the left of A'.

If $x=0$, $y=\pm b\sqrt{(-1)}$. Therefore the intersections of the curve and the y-axis are imaginary.

The line AA' is called the transverse axis. The line BB', where B, B' are the points on the y-axis distant b, from the centre, is called the conjugate axis.

As in the case of ellipse, there a is second focus S', whose coordinates are $(-ae, 0)$ and a corresponding second directrix $Z'M'$ at the same distance a/e from C as ZM.

If the length of semi latus rectum is l, it is given by

$$\frac{a^2e^2}{a^2}-\frac{l^2}{b^2}=1.$$

$$\therefore \quad l^2=b^2(e^2-1)$$

$$=\frac{b^4}{a^4} \qquad [\because \quad b^2=a^2(e^2-1)]$$

$$\therefore \quad l=\frac{b^2}{a}.$$

PNP' is a double ordinate of the point P.

8·3. Focal distances of a point.

Let P be any point (x, y) on the hyperbola, and let PM' be the perpendicular from P on $Z'M'$.

Now
$$\begin{aligned} SP &= e.PM \\ &= e.NZ \\ &= e(CN-CZ) \\ &= e\left(x-\frac{a}{e}\right) \\ &= ex-a. \end{aligned}$$

Also
$$\begin{aligned} S'P &= e.PM' \\ &= e.NZ' \\ &= e(CN+CZ') \end{aligned}$$

$$=e\left(x+\frac{a}{e}\right)$$

$$=ex+a.$$

$$\therefore \quad S'P-SP=2a.$$

Hence the difference of the focal distances of any point on the hyperbola is constant, and is equal to the transverse axis.

8·4. Propositions about the hyperbola.

The difference in the equations to the ellipse and the hyperbola is that the sign of b^2 in ellipse is positive, but in hyperbola is negative. Hence the results obtained for ellipse hold good for the hyperbola, if the sign of b^2 is changed.

(i) The equation of the tangent at any point (x_1, y_1) on the hyperbola $\frac{x^2}{a^2}-\frac{y^2}{b^2}=1$ is

$$\frac{xx_1}{a^2}-\frac{yy_1}{b^2}=1.$$

(ii) The normal at any point (x_1, y_1) on the curve is

$$\frac{x-x_1}{\frac{x_1}{a^2}}=\frac{y-y_1}{-\frac{y_1}{b^2}}.$$

(iii) The equation of the polar of any point (x_1, y_1) or to the chord of contact of tangents drawn from (x_1, y_1) is

$$\frac{xx_1}{a^2}-\frac{yy_1}{b^2}=1.$$

(iv) The straight line $y=mx+\sqrt{(a^2m^2-b^2)}$ is a tangent to the curve for all values of m.

(v) The straight line $lx+my=n$ touches the curve if

$$a^2l^2-b^2m^2=n^2.$$

(vi) The equation of the director circle is

$$x^2+y^2=a^2-b^2$$

which is imagiary if $b > a$.

(vii) The locus of the middle points of chords parallel to the diameter $y=mx$ is the diameter $y=m_1x$, where

$$mm_1=\frac{b^2}{a^2}.$$

such diameters having called conjugate diameters.

(viii) The chord whose middle point is (x_1, y_1) is given by

$$\frac{xx_1}{a^2}-\frac{yy_1}{b^2}=\frac{x_1^2}{a^2}-\frac{y_1^2}{b^2}.$$

8·5. Position of a point with respect to the hyperbola :—The point (x_1, y_1) lies within, upon, or without the hyperbola according as the expression

$$\frac{x_1^2}{a^2}-\frac{y_1^2}{b^2}-1$$

is positive, zero or negative.

8·6. Parametric Coordinates.

The coordinates of any point on the hyperbola can be expressed as

$$x=a \sec \theta, \quad y=b \tan \theta$$

or $$x=a \cosh t, y=b \sinh t.$$

The equation of the tangent at $(a \sec \theta, b \tan \theta)$ is

$$\frac{x}{a} \sec \theta-\frac{y}{b} \tan \theta=1$$

or $$\frac{x}{a}-\frac{y}{b} \sin \theta=\cos \theta.$$

The equation of the normal at this point is

$$\frac{\sec \theta}{a}(y-b \tan \theta)+\frac{\tan \theta}{b}(x-a \sec \theta)=0$$

or $$by \cot \theta+ax \cos \theta=a^2+b^2.$$

8·7. Limiting case of a hyperbola.

The equation to the hyperbola is

$$\frac{x^2}{a^2}-\frac{y^2}{a^2(e^2-1)}=1$$

or $$x^2-\frac{y^2}{e^2-1}=a^2.$$

If e is kept constant and a tends to 0, the hyperbola approximates to the pair of lines

$$x^2-\frac{y^2}{e^2-1}=0,$$

the coordinate axes bisecting the angles between them.

8·8. Polar equation of hyperbola.

The cartesian equation of the hyperbola is

$$\frac{x^2}{a^2}-\frac{y^2}{b^2}=1$$

which on conversion to polar coordinates becomes

$$\frac{1}{r^2}=\frac{\cos^2\theta}{a^2}-\frac{\sin^2\theta}{b^2} \qquad \ldots \quad \ldots \quad \ldots(1)$$

If θ is zero, $r^2=a^2$ or $r=\pm a$.

$\frac{1}{r^2}=0$ or r tends to ∞ when $\frac{\cos^2\theta}{a^2}=\frac{\sin^2\theta}{b^2}$

or $$\tan\theta=\pm\frac{b}{a}.$$

When $\frac{\sin^2\theta}{b^2}>\frac{\cos^2\theta}{a^2}$, *i.e.* $\tan\theta>\frac{b}{a}$, $\frac{1}{r^2}$ is negative and r is imaginary that is the radius vector does not cut the curve in real points.

8·9. Asymptote. Definition. A straight line which meets the conic in two points both of which are situated at an infinite distance, but which is itself not altogether at infinity is called an asymptote.

Equation of the asymptote.

The straight line $$y=mx+c \qquad \ldots(1)$$

intersects the hyperbola $$\frac{x^2}{a^2}-\frac{y^2}{b^2}=1 \qquad \ldots(2)$$

in points whose abscissae are given by

$$\frac{x^2}{a^2}-\frac{(mx+c)^2}{b^2}=1$$

or $$x^2\left(\frac{1}{a^2}-\frac{m^2}{b^2}\right)-\frac{2mcx}{b^2}-\frac{c^2}{b^2}-1=0. \quad \ldots(3)$$

If the line (1) is an asymptote, both the roots of (3) must be infinite. Hence the coefficients of x^2 and x in (3) must both be zero,

i.e. $$\frac{1}{a^2}-\frac{m^2}{b^2}=0 \quad \text{or} \quad m=\pm\frac{b}{a}$$

and $$\frac{cm}{b^2}=0 \quad \text{or} \quad c=0 \text{ (} m \text{ not being equal to 0).}$$

Substituting these values in (1), the equation of the asymptotes is

$$y=\pm\frac{b}{a}x \quad \ldots(4)$$

or $$\frac{x}{a}+\frac{y}{b}=0 \quad \text{and} \quad \frac{x}{a}-\frac{y}{b}=0. \quad \ldots(5)$$

From (4) it is evidently clear that the asymptotes are inclined at an angle $\tan^{-1}\frac{b}{a}$ to the axes.

From (5) the combined equation of the asymptotes is

$$\frac{x^2}{a^2}-\frac{y^2}{b^2}=0.$$

Alternate method—Definition. An asymptote is a straight line at a finite distance from the origin to which the tangent to a curve tends as the point of contact tends to infinity.

The tangent at the point (h, k) to the hyperbola $\frac{x^2}{a^2}-\frac{y^2}{b^2}=1$ is

$$\frac{xh}{a^2}-\frac{yk}{b^2}=1. \quad \ldots(1)$$

But (h, k) lies on the hyperbola,

$$\therefore \quad \frac{h^2}{a^2}-\frac{k^2}{b^2}=1. \qquad \ldots(2)$$

Eliminating k from (1) and (2), we get

$$\frac{xh}{a^2} \pm \frac{y}{b}\sqrt{\left(\frac{h^2}{a^2}-1\right)}=1, \qquad \ldots(3)$$

which gives the equation of two tangents to the hyperbola at two points whose abscissa is h.

Equation (3) can be written as

$$\frac{x}{a} \pm \frac{ay}{b}\sqrt{\left(\frac{1}{a^2}-\frac{1}{h^2}\right)}=\frac{a}{h}. \qquad \ldots(4)$$

If h tends to ∞, equation (4) becomes

$$\frac{x}{a} \pm \frac{y}{b}=0.$$

These are therefore the equations to the asymptotes.

8·10. Rectangular Hyperbola. The angle between the asymptotes is

$$2 \tan^{-1} \frac{b}{a}.$$

This will be a right angle if

$$\tan^{-1} \frac{b}{a}=\frac{\pi}{4}, \quad i.e. \text{ if } \frac{b}{a}=1 \quad \text{or} \quad b=a.$$

Such a hyperbola is called a rectangular hyperbola. Hence the equation of a rectangular hyperbola is

$$x^2-y^2=a^2$$

and its asymptotes are $y=\pm x$, the two lines inclined to the axes at an angle of 45°.

8·11. Solved Examples.

Ex. 1. *Find the equation to the hyperbola whose eccentricity is $\frac{5}{4}$, whose focus is $(b, 0)$ and whose directrix is $4x-3y=b$.*

Let (x, y) be a point on the hyperbola. The distance

of the point from the focus is e times its distance from the directrix.

$$\therefore \quad (x-b)^2+y^2=\left(\frac{5}{4}\right)^2\left(\frac{4x-3y-b}{\sqrt{25}}\right)^2$$

or $$x^2+y^2-2bx+b^2=\tfrac{1}{16}(4x-3y-b)^2$$

or $$16(x^2+y^2-2bx+b^2)=(4x-3y-b)^2$$

or $$7y^2+15b^2-24bx+24xy-6yb=0.$$

Ex. 2. *Trace the hyperbola*

$$2x^2-y^2+18x+27=0.$$

The equation can be written as

$$2(x^2+9x)-y^2+27=0$$

or $$2(x+9)^2-y^2+27-\tfrac{81}{2}=0$$

or $$\frac{(x+\frac{9}{2})^2}{1/2}-\frac{y^2}{1}=\frac{27}{2}$$

or $$\frac{(x+\frac{9}{2})^2}{\frac{27}{4}}-\frac{y^2}{\frac{27}{2}}=1.$$

Transferring the origin to the point $(-\frac{9}{2}, 0)$ the equation becomes

$$\frac{x^2}{\frac{27}{4}}-\frac{y^2}{\frac{27}{2}}=1.$$

Hence the centre is $(-\frac{9}{2}, 0)$.

The semi axes are $\dfrac{\sqrt{27}}{2}$ and $\sqrt{\frac{27}{2}}$,

i.e. $\dfrac{3\sqrt{3}}{2}$ and $3\sqrt{\frac{3}{2}}$ respectively

as $$b^2=a^2(e^2-1),$$

$$\tfrac{27}{2}=\tfrac{27}{4}(e^2-1) \quad \text{or} \quad e^2-1=2;$$

$$\therefore \quad e=\sqrt{3}.$$

Ex. 3. *The normal to the hyperbola meets the axes in M and N, and lines MP and NP are drawn at right angles to the axes. Prove that the locus of P is the hyperbola $a^2x^2-b^2y^2=(a^2+b^2)^2$.*

(Punjab 1954)

The equation to the normal to the hyperbola $\frac{x^2}{a^2}-\frac{y^2}{b^2}=1$ at the point $(a \sec \theta, b \tan \theta)$ is

$$ax \cos \theta+by \cot \theta=a^2+b^2.$$

It meets the axis $y=0$ in point

$$\left(\frac{a^2+b^2}{a \cos \theta}, 0\right)$$

and the axis $x=0$ in point

$$\left(0, \frac{a^2+b^2}{b \cot \theta}\right).$$

The line perpendicular to x-axis at $\left(\frac{a^2+b^2}{a \cos \theta}, 0\right)$ is

$$x=\frac{a^2+b^2}{a \cos \theta}. \qquad \ldots(1)$$

The line perpendicular to y-axis at $\left(0, \frac{a^2+b^2}{b \cot \theta}\right)$ is

$$y=\frac{a^2+b^2}{b \cot \theta}. \qquad \ldots(2)$$

From (1), $$\cos \theta=\frac{a^2+b^2}{ax}. \qquad \ldots(3)$$

From (2), $$\cot \theta=\frac{a^2+b^2}{by}. \qquad \ldots(4)$$

From (3) and (4) by division,

$$\sin \theta=\frac{by}{ax}. \qquad \ldots(5)$$

From (3) and (5) eliminating θ, we get

$$\frac{(a^2+b^2)^2}{a^2x^2}+\frac{b^2y^2}{a^2x^2}=1$$

or $$(a^2+b^2)^2+b^2y^2=a^2x^2$$

or $$a^2x^2-b^2y^2=(a^2+b^2)^2$$

which is the locus of the point of intersection of (1) and (2)

Ex. 4. *A series of hyperbolas is drawn having a common transverse axis of length 2a. Prove that the locus of a point P on each hyperbola, such that its distance from the transverse axis is equal to its distance from an asymptote, is the curve*

$$(x^2-y^2)^2=4x^2(x^2-a^2). \qquad \textit{(Agra 1961)}$$

Let the hyperbolas be $\frac{x^2}{a^2}-\frac{y^2}{\lambda^2}=1$, where λ is a variable.

Let (h, k) be a point on the hyperbola.

The equation to an asymptote is $\frac{x}{a}-\frac{y}{\lambda}=0$.

Perpendicular from (h, k) on the asymptote is given by

$$\frac{\frac{h}{a}-\frac{k}{\lambda}}{\sqrt{\left(\frac{1}{a^2}+\frac{1}{\lambda^2}\right)}}. \qquad \ldots(1)$$

The distance of (h, k) from the transverse axis is k.

$$\therefore \quad \frac{\frac{h}{a}-\frac{k}{\lambda}}{\sqrt{\left(\frac{1}{a^2}+\frac{1}{\lambda^2}\right)}}=k$$

or $$\left(\frac{h}{a}-\frac{k}{\lambda}\right)^2=k^2\left(\frac{1}{a^2}+\frac{1}{\lambda^2}\right)$$

or $$\frac{h^2}{a^2}-\frac{2kh}{a\lambda}=\frac{k^2}{a^2}$$

or $$\frac{1}{a^2}(h^2-k^2)=\frac{2kh}{a\lambda} \qquad \ldots(2)$$

But (h, k) is a point on the hyperbola.

$$\therefore \quad \frac{h^2}{a^2}-\frac{k^2}{\lambda^2}=1. \qquad \ldots(3)$$

Eliminating λ between (2) and (3), we get

$$\frac{1}{a^4}(h^2-k^2)^2=\frac{4h^2}{a^2}\left(\frac{h^2}{a^2}-1\right)$$

or $$(h^2-k^2)^2=4h^2(h^2-a^2).$$

Hence the locus of (h, k) is

$$(x^2-y^2)^2=4x^2(x^2-a^2).$$

Ex. 5. *A series of chords of the hyperbola* $\frac{x^2}{a^2}-\frac{y^2}{b^2}=1$ *are tangents to the circle described on the straight line joining the foci of the hyperbola as diameter; show that the locus of their poles with respect to the hyperbola is* $\frac{x^2}{a^4}+\frac{y^2}{b^4}=\frac{1}{a^2+b^2}$.

Let a chord of the hyperbola be

$$y=mx+c. \qquad \ldots(1)$$

The equation to the circle on the line joining the foci $(ae, 0)$, $(-ae, 0)$ as diameter is

$$(x-ae)(x+ae)+y^2=0$$

or $$x^2+y^2=a^2e^2. \qquad \ldots(2)$$

(1) touches (2) if $c=ae\sqrt{(1+m^2)}$.

Hence the equation (1) becomes

$$y=mx+ae\sqrt{(1+m^2)}. \qquad \ldots(3)$$

Let the pole of (3) be (h, k). Then the polar is

$$\frac{xh}{a^2}-\frac{yk}{b^2}=1. \qquad \ldots(4)$$

Hence equations (3) and (4) represent the same locus.

Comparing coefficients in (3) and (4),

$$\frac{\frac{h}{a^2}}{m}=\frac{\frac{k}{b^2}}{1}=\frac{1}{-ae\sqrt{(1+m^2)}}. \qquad \ldots(5)$$

From (5), $$m=\frac{b^2}{a^2}\cdot\frac{h}{k} \qquad \ldots(6)$$

and $$a^2e^2(1+m^2)=\frac{b^4}{k^2}. \qquad \ldots(7)$$

Eliminating m from (6) and (7),

$$a^2e^2\left(1+\frac{b^4}{a^4}\cdot\frac{h^2}{k^2}\right)=\frac{b^4}{k^2}$$

or $$a^2e^2\left(\frac{k^2}{b^4}+\frac{h^2}{a^4}\right)=1$$

or $$\frac{h^2}{a^4}+\frac{k^2}{b^4}=\frac{1}{a^2e^2}$$

or $$\frac{h^2}{a^4}+\frac{k^2}{b^4}=\frac{1}{a^2+b^2}.$$

Hence the locus of (h, k) is

$$\frac{x^2}{a^4}+\frac{y^2}{b^4}=\frac{1}{a^2+b^2}.$$

8·12. Conjugate hyperbola.

The hyperbola $-\frac{x^2}{a^2}+\frac{y^2}{b^2}=1$ which has its transverse and conjugate axes along the y-axis and x-axis respectively is conjugate to the hyperbola $\frac{x^2}{a^2}-\frac{y^2}{b^2}=1$ which has its transverse and conjugate axes along the x-axis and y-axis respectively. It means that hyperbolas

$$\frac{x^2}{a^2}-\frac{y^2}{b^2}=1 \qquad \ldots(1)$$

and $$-\frac{x^2}{a^2}+\frac{y^2}{b^2}=1 \qquad \ldots(2)$$

are conjugate to each other.

The asymptotes of both are

$$\frac{x^2}{a^2}-\frac{y^2}{b^2}=0 \qquad \ldots(3)$$

From equations (1), (2) and (3) it is observed that equation (3) differs from (1) by a constant and that equation (2) differs from (3) by the same constant.

In any transformation of axes the left hand members of (1), (2) and (3) will change in the same way and the right hand constants will differ in the same manner as before.

Hence the equation of the asymptotes differs from

that of the hyperbola by a constant and the equation of the conjugate hyperbola differs from that of the asymptotes by the same constant.

8·13. Properties of conjugate diameters and conjugate hyperbola.

1. *Only one of a pair of conjugate diameters of a hyperbola meets the curve in real points.*

The straight lines $y=m_1x$ and $y=m_2x$ are conjugate diameters if

$$m_1m_2=\frac{b^2}{a^2}.$$

$y=m_1x$ intersects the hyperbola in points whose abscissae are given by

$$\frac{x^2}{a^2}-\frac{m_1^2x^2}{b^2}=1$$

or
$$x^2=\frac{a^2b^2}{b^2-a^2m_1^2}.$$

These points are real if $b^2 > a^2m_1^2$

or
$$m_1 < \frac{b}{a}.$$

But as
$$m_1m_2=\frac{b^2}{a^2},$$

$$m_2 > \frac{b}{a},$$

i e. the inclination of the other diameter is greater than that of the asymptote and therefore the other diameter will intersect the hyperbola in imaginary points.

Similarly, we can prove that if $y=m_2x$ intersects the hyperbola in real points, then $y=m_1x$ intersects it in imaginary points.

2. *If a pair of diameters be conjugate with respect to a hyperbola, they will also be conjugate with respect to its conjugate hyperbola.*

The straight lines $y=m_1x$ and $y=m_2x$ are conjugate diameters of the hyperbola

$$\frac{x^2}{a^2}-\frac{y^2}{b^2}=1$$

if
$$m_1m_2=\frac{b^2}{a^2}. \qquad \ldots \quad \ldots \quad \ldots(1)$$

The equation to the conjugate hyperbola is

$$-\frac{x^2}{a^2}+\frac{y^2}{b^2}=1$$

i.e. instead of a^2, we have $-a^2$ and instead of b^2, we have $-b^2$.

Making the above changes in relation (1), we get

$$m_1m_2=\frac{-b^2}{-a^2}=\frac{b^2}{a^2}$$

which is the same as (1).

3. *Out of a pair of conjugate diameters of a hyperbola, one meets the hyperbola in real points and the other meets the conjugate hyperbola in real points.*

Let the diameters be $y=m_1x$ and $y=m_2x$, so that

$$m_1m_2=\frac{b^2}{a^2}.$$

Let $y=m_1x$ meet the hyperbola in real points, so that $m_1<\frac{b}{a}$ as before and $m_2>\frac{b}{a}$.

The straight line $y=m_2x$ meets the conjugate hyperbola in points whose abscissae are given by

$$x^2\left(\frac{m_2^2}{b^2}-\frac{1}{a^2}\right)=1$$

or
$$x^2=\frac{a^2b^2}{m_2^2a^2-b^2}.$$

Since $m_2>\frac{b}{a}$, these points are real.

4. *If CP and CD are two conjugate diameters of a hyperbola, then*

$$CP^2 - CD^2 = a^2 - b^2.$$

Let the point P be $(a \sec \theta, b \tan \theta)$.

The equation to CP is therefore

$$y = \frac{b \tan \theta}{a \sec \theta} x = \frac{b}{a} \sin \theta . x.$$

Hence $$m_1 = \frac{b}{a} \sin \theta.$$

But $$m_1 m_2 = \frac{b^2}{a^2}.$$

$$\therefore \quad m_2 = \frac{b^2}{a^2} \cdot \frac{a}{b} \cdot \frac{1}{\sin \theta}$$

$$= \frac{b}{a} \cdot \frac{1}{\sin \theta}.$$

The equation to CD is therefore

$$y = \frac{b}{a} \cdot \frac{1}{\sin \theta} . x$$

This meets the conjugate hyperbola

$$-\frac{x^2}{a^2} + \frac{y^2}{b^2} = 1$$

in points whose abscissae are

$$-\frac{x^2}{a^2} + \frac{1}{a^2 \sin \theta} . x^2 = 1$$

or $$\frac{x^2}{a^2} \left(\frac{1 - \sin^2 \theta}{\sin^2 \theta} \right) = 1$$

or $$x = \pm a \tan \theta.$$

$$\therefore \quad y = \pm b \sec \theta.$$

Hence the coordinates of D are $(a \tan \theta, b \sec \theta)$.

$$\therefore \quad CP^2 - CD^2 = a^2 \sec^2 \theta + b^2 \tan^2 \theta - a^2 \tan^2 \theta - b^2 \sec^2 \theta$$

$$= (a^2 - b^2)(\sec^2 \theta - \tan^2 \theta)$$

$$= a^2 - b^2.$$

5. *If PCP′ and DCD′ be a pair of conjugate diameters to a hyperbola, then the tangents to the hyperbola at P, P′, D, D′ form a parallelogram of constant area whose vertices lie on the asymptotes.*

Let the point P be $(a \sec\theta, b\tan\theta)$. Then we have shown that the points D, P' and D' are $(a\tan\theta, b\sec\theta)$, $(-a\sec\theta, -b\tan\theta)$ and $(-a\tan\theta, -b\sec\theta)$ respectively.

The equation to the tangent to the hyperbola at P is

$$\frac{x}{x}\sec\theta - \frac{y}{b}\tan\theta = 1 \qquad \ldots \quad \ldots \quad \ldots(1)$$

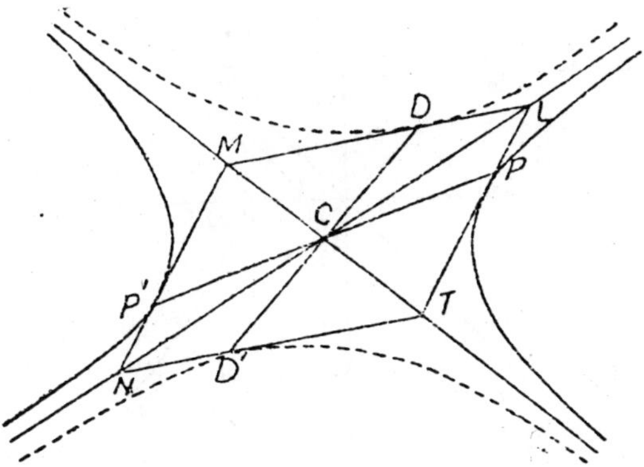

Also the tangent to the conjugate hyperbola at D is

$$\frac{x}{a}\tan\theta - \frac{y}{b}\sec\theta = -1. \qquad \ldots(2)$$

The point of intersection of (1) and (2) is $a(\sec\theta + \tan\theta)$, $b(\sec\theta + \tan\theta)$ which lies on the asymptote $\frac{x}{a} - \frac{y}{b} = 0$.

Similarly the other points M, N and T can be shown to lie on the asymptotes.

Area of the parallelogram $LMNT$

$= 4 \times$ area of the parallelogram $CPLD$

$= 4 \times CD \times$ perpendicular from C on PL

$$=4\times\sqrt{(a^2\tan^2\theta+b^2\sec^2\theta)}\times\frac{1}{\sqrt{\left(\frac{\sec^2\theta}{a^2}+\frac{\tan^2\theta}{b^2}\right)}}$$

$=4ab$ which is constant.

8·14. Equation of the hyperbola referred to its asymptotes.

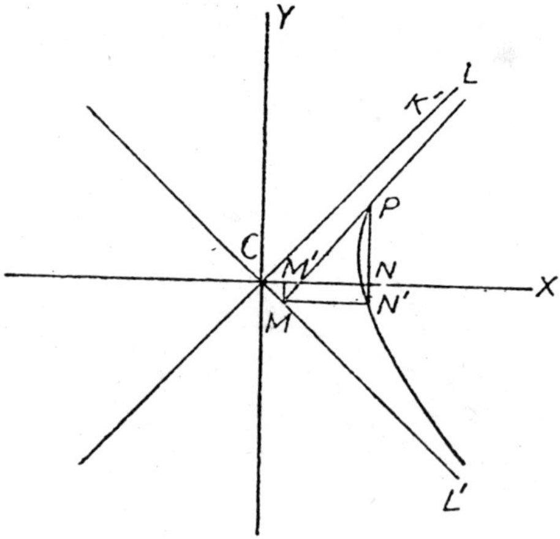

Let the asymptotes be CL and CL'.

Let α be the angle LCN; then

$$\tan\alpha=\frac{b}{a}.$$

$$\therefore\quad \sin\alpha=\frac{b}{\sqrt{(a^2+b^2)}}\quad\text{and}\quad\cos\alpha=\frac{a}{\sqrt{(a^2+b^2)}}.$$

Let $P(x, y)$ be a point on the hyperbola and let its coordinates be (p, q) when referred to asymptotes as axes so that CM is p and PM is q.

Draw MM' perp. to the transverse axis and let PN, the perpendicular through P meet the parallel to the transverse axis through M in N'.

Then $\qquad x=CN=CM'+M'N=CM'+MN'$

$$=CM\cos\alpha+PM\cos\alpha$$
$$=(p+q)\cos\alpha.$$
$$=\frac{(p+q)\,a}{\sqrt{(a^2+b^2)}} \qquad \ldots(1)$$

and $$y=PN=PN'-N'N$$
$$=PN'-MM'$$
$$=PM\sin\alpha-CM\sin\alpha$$
$$=(p-q)\sin\alpha$$
$$=\frac{(p-q)\,b}{\sqrt{(a^2+b^2)}}. \qquad \ldots(2)$$

But (x, y) lies on the hyperbola

$$\frac{x^2}{a^2}-\frac{y^2}{b^2}=1.$$

Substituting from (1) and (2),

$$\frac{(p+q)^2}{a^2+b^2}-\frac{(p-q)^2}{a^2+{}^2b}=1$$

or $$4pq=a^2+b^2$$

or $$pq=\frac{a^2+b^2}{4}.$$

Generalising the equation to the hyperbola referred to asymptotes as axes is $xy=\dfrac{a^2+b^2}{4}$,

or $$xy=c^2 \text{ where } c^2=\frac{a^2+b^2}{4}.$$

In the special case when the hyperbola is rectangular,

$$b=a \quad \text{and} \quad xy=\frac{a^2}{2}.$$

The equation to the conjugate hyperbola will be

$$xy=-\frac{a^2+b^2}{4}.$$

Note. In the case of a rectangular hyperbola $x^2-y^2=a^2$, the axes can be rotated by $-45°$ to get the asymptotes as the new axes, so that we can write

$$\frac{x}{\sqrt{2}}+\frac{y}{\sqrt{2}} \text{ for } x \text{ and } \frac{x}{\sqrt{2}}-\frac{y}{\sqrt{2}} \text{ for } y.$$

The equation to the hyperbola then becomes

$$\tfrac{1}{2}(x+y)^2-\tfrac{1}{2}(x-y)^2=a^2$$

or
$$xy=\frac{a^2}{2}.$$

8·15. Tangent, polar and normal to the hyperbola referred to asymptotes as axes.

The coordinates of any point on the hyperbola $xy=c^2$ can be taken as $(ct, c/t)$ where t is a parameter.

(i) Let $\left(ct_1, \frac{c}{t_1}\right)$ and $\left(ct_2, \frac{c}{t_2}\right)$ be any two points on the hyperbola.

The equation to the line joining them is

$$y-\frac{c}{t_1}=\frac{\frac{c}{t_2}-\frac{c}{t_1}}{ct_2-ct_1}(x-ct_1)$$

or
$$yc(t_2-t_1)-\frac{c^2}{t_1}(t_2-t_1)=\frac{xc}{t_1t_2}(t_1-t_2)-\frac{c^2t_1}{t_1t_2}(t_1-t_2)$$

or
$$y-\frac{c}{t_1}=-\frac{x}{t_1t_2}+\frac{c}{t_2}$$

or
$$yt_1t_2-ct_2=-x+ct_1$$

or
$$t_2t_1y+x=c(t_1+t_2) \qquad \ldots(1)$$

When t_2 tends to t_1, this becomes

$$t_1^2y+x=2ct_1 \qquad \ldots(2)$$

which is the equation of the tangent at t_1.

Writing x_1 for ct_1 and y_1 for $\frac{c}{t_1}$, we get the equation of the tangent at (x_1, y_1) to the hyperbola $xy=c^2$, as

$$xy_1+yx_1=2c^2. \qquad \ldots(3)$$

(ii) The polar can be found by the methods used before as

$$xy_1+yx_1=2c^2.$$

(iii) The normal at $\left(ct_1, \frac{c}{t_1}\right)$ is at right angles to the tangent at this point and passes through this point. Hence its equation is

$$y-\frac{c}{t_1}=+t_1^2(x-ct_1)$$

or $$xt_1^3-t_1y-ct_1^4+c=0. \quad \ldots(4)$$

8·16. Equation of the hyperbola referred to conjugate diameters as axes.

The equation of the hyperbola referred to its transverse and conjugate axes as the axes of reference is

$$\frac{x^2}{a^2}-\frac{y^2}{b^2}=1 \quad \ldots(1)$$

where a and b are the semi-axes.

Rotate the axes through suitable angles so that they coincide with the conjugate diameters as oblique axes.

We shall have to substitute expressions of the type $lx+my$ and $l'x+m'y$ for x and y respectively, so that the transformed equation is of the type

$$Ax^2+2Hxy+By^2=1. \quad \ldots(1)$$

Now chords parallel to CD are bisected by chord CP. Hence if one extremity of one such chord is (x_1, y_1) referred to new axes, the other extremity will be $(x_1, -y_1)$.

Both these points must satisfy equation (1). Therefore

$$Ax_1^2+2Hx_1y_1+By_1^2=1, \quad \ldots(2)$$

$$Ax_1^2-2Hx_1y_1+By_1^2=1. \quad \ldots(3)$$

Subtracting (3) from (2),

$$4Hx_1y_1=0$$

so that $$H=0 \quad [x_1 \text{ or } y_1 \text{ being not}=0].$$

Hence the equation becomes

$$Ax^2+By^2=1. \qquad \ldots(4)$$

If a' is the length of CP and l' the length of CD, then the points $(a', 0)$ and $(0, ib')$ lie on the hyperbola.

Substituting these coordinates in (4), we get

$$Aa'^2=1 \text{ and } -Bb'^2=1$$

so that

$$A=\frac{1}{a'^2} \quad \text{and} \quad B=-\frac{1}{b'^2}.$$

The equation of the hyperbola then becomes

$$\frac{x^2}{a'^2}-\frac{y^2}{b'^2}=1.$$

8·17. Solved examples.

Ex. 1. *If e and e' be the eccentricities of a hyperbola and its conjugate, prove that $1/e^2+1/e'^2=1$.* *(Agra 1941)*

The equation to the hyperbola is

$$\frac{x^2}{a^2}-\frac{y^2}{b^2}=1$$

where

$$b^2=a^2(e^2-1). \qquad \ldots(1)$$

The conjugate hyperbola is $-\dfrac{x^2}{a^2}+\dfrac{y^2}{b^2}=1.$

If e' is its eccentricity, the conjugate axis now being a,

$$a^2=+b^2(e'^2-1). \qquad \ldots(2)$$

Multiplying (1) and (2), we get

$$a^2b^2=a^2b^2(e^2-1)(e'^2-1)$$

or

$$1=e^2e'^2-e'^2-e^2+1$$

or

$$\frac{1}{e^2}+\frac{1}{e'^2}=1.$$

Ex. 2. *Show that the tangent at any point of a hyperbola cuts off a triangle of constant area from the asymptotes, and that the portion of it intercepted between the asymptotes, is bisected at the point of contact.* *(Agra 1961)*

The equation of the hyperbola referred to its asymptotes as axes is $xy=c^2$.

The equation of the tangent at any point (x_1, y_1) is

$$xy_1+yx_1=2c^2.$$

It meets the x-axis which is an asymptote at $\left(\frac{2c^2}{y_1}, 0\right)$ and the y-axis at $\left(0, \frac{2c^2}{x_1}\right)$.

The middle point is $\left(\frac{c^2}{y_1}, \frac{c^2}{x_1}\right)$.

But (x_1, y_1) is a point on the hyperbola.

$$\therefore \quad x_1 y_1=c^2.$$

Therefore the coordinates of the middle point are (x_1, y_1) which is the given point.

If the angle between the asymptotes is 2α, the area of the triangle between the asymptotes

$$=\frac{1}{2} \cdot \frac{2c^2}{y_1} \cdot \frac{2c^2}{x_1} \sin 2\alpha$$

$$=2x_1 y_1 \sin 2\alpha$$

$$=2c^2 . 2 \sin \alpha \cos \alpha$$

$$=4c^2 . \frac{ab}{a^2+b^2}$$

as $\sin \alpha=\frac{b}{\sqrt{(a^2+b^2)}}$ and $\cos \alpha=\frac{a}{\sqrt{(a^2+b^2)}}$.

$\therefore$ the required area $=ab$ as $c=\frac{a^2+b^2}{4}$

$$=\text{a constant.}$$

Ex. 3. *Find the equation of the hyperbola whose asymptotes are the straight lines $x+2y+1=0$ and $2x+y+3=0$ and which passes through the point $(1, 2)$.* *(Jammu and Kashmir 1957)*

The equation of the hyperbola differs from the equation of the asymptotes by a constant. Hence the equation of the hyperbola is $(x+2y+1)(2x+y+3)+c=0$.

But it passes through the point (1, 2).

$$\therefore \quad (1+4+1)(2+2+3)+c=0.$$

$$\therefore \quad c=-42.$$

Hence the required equation is

$$(x+2y+1)(2x+y+3)-42=0$$

or $$2x^2+5xy+2y^2+5x+7y+3-42=0$$

or $$2x^2+5xy+2y^2+5x+7y-39=0.$$

Ex. 4. *Find the asymptotes of $xy-3x-2y=0$. What is the equation of the conjugate hyperbola?* *(Jammu and Kashmir 1955)*

The equation of the asymptotes of $xy-3x-2y=0$ is

$$xy-3x-2y+c=0.$$

But this must represent two straight lines. Therefore

$$2\left(-\tfrac{3}{2}\right)(-1)\left(\tfrac{1}{2}\right)-c\left(\tfrac{1}{2}\right)^2=0 \quad \text{or} \quad c=6.$$

Hence the equation of the asymptotes is

$$xy-3x-2y+6=0.$$

The difference in the constant between the equation of the asymptotes and the conjugate hyperbola is the same as that between the equation of the asymptotes and the hyperbola.

Hence the equation of the conjugate hyperbola is

$$xy-3x-2y+12=0.$$

Ex. 5. *If a rectangular hyperbola circumscribe a triangle, prove that it also passes through the orthocentre of the triangle.*

(Agra 1960)

Let the vertices of the triangle PQR be $\left(ct_1, \frac{c}{t_1}\right)$, $\left(ct_2, \frac{c}{t_2}\right)$ and $\left(ct_3, \frac{c}{t_3}\right)$. The equation to the straight line PQ is

$$y-\frac{c}{t_1}=\frac{\frac{c}{t_2}-\frac{c}{t_1}}{ct_2-ct_1}(x-ct_1)$$

or $$y-\frac{c}{t_1}=-\frac{1}{t_1t_2}(x-ct_1). \quad \ldots(1)$$

Similarly the equation to QR is

$$y-\frac{c}{t_2}=-\frac{1}{t_2t_3}(x-ct_2). \quad \ldots(2)$$

The equation to the line through $R\left(ct_3, \frac{c}{t_3}\right)$ perpendicular to PQ is

$$y-\frac{c}{t_3}=t_1t_2(x-ct_3). \quad \ldots(3)$$

The equation to the line through P perpendicular to QR is

$$y-\frac{c}{t_1}=t_2t_3(x-ct_1). \quad \ldots(4)$$

The point of intersection of (3) and (4), the orthocentre of the triangle, is $\left(-\frac{c}{t_1t_2t_3}, -ct_1t_2t_3\right)$. This point clearly satisfies the equation to the hyperbola.

Ex. 6. *A rectangular hyperbola whose centre is C is cut by any circle of radius r in the four points P, Q, R, S; prove that*

$$CP^2+CQ^2+CR^2+CS^2=4r^2. \quad \textit{(Agra 1954)}$$

Let the rectangular hyperbola be

$$xy=c^2 \quad \ldots(1)$$

and let the circle be

$$(x-f)^2+(y-g)^2=r^2. \quad \ldots(2)$$

The abscissae of the points of intersection of (1) and (2) are given by the equation

$$(x-f)^2+\left(\frac{c^2}{x}-g\right)^2=r^2$$

or $$x^2(x-f)^2+(c^2-gx)^2=x^2r^2$$

or $$x^4-2fx^3+x^2(f^2+g^2-r^2)-2gc^2x+c^4=0.$$

If x_1, x_2, x_3, x_4 are the roots of this equation, then

$$x_1+x_2+x_3+x_4=2f, \qquad \ldots(1)$$

$$x_1x_2+x_2x_3+x_3x_4+x_4x_1+x_1x_3+x_2x_4=f^2+g^2-r^2, \qquad \ldots(2)$$

$$x_1x_2x_3+x_2x_3x_4+x_3x_4x_1+x_4x_1x_2=2gc^2, \qquad \ldots(3)$$

$$x_1x_2x_3x_4=c^4. \qquad \ldots(4)$$

Similarly the ordinates of the points of intersection are given by the equation

$$y^4-2y^3g+y^2(f^2+g^2-r^2)-2fc^2y+c^4=0.$$

If y_1, y_2, y_3 and y_4 are the roots of this equation, then

$$y_1+y_2+y_3+y_4=2g \qquad \ldots(5)$$

and $$y_1y_2+y_2y_3+y_3y_4+y_4y_1+y_1y_3+y_2y_4=f^2+g^2-r^2. \ldots(6)$$

Now $CP^2+CQ^2+CR^2+CS^2=\Sigma x_1^2+\Sigma y_1^2$

$$=(\Sigma x_1)^2-2\Sigma x_1x_2+(\Sigma y_1)^2-2\Sigma y_1y_2$$

$$=4f^2-2(f^2+g^2-r^2)+4g^2-2(f^2+g^2-r^2)$$

[from (1), (2), (5) and (6)]

$$=4r^2.$$

Ex. 7. *If the normals at (x_1, y_1), (x_2, y_2), (x_3, y_3) and (x_4, y_4) on the rectangular hyperbola $xy=c^2$ meet in the point (α, β), prove that*

$$\alpha=x_1+x_2+x_3+x_4,$$

$$\beta=y_1+y_2+y_3+y_4$$

and $$x_1x_2x_3x_4=y_1y_2y_3y_4=-c^4. \qquad \textit{(Lucknow 1947)}$$

The equation to the normal at the point $\left(ct, \frac{c}{t}\right)$ is given by

$$xt^3-ty-ct^4+c=0.$$

As it passes through (α, β),

$$ct^4-\alpha t^3+t\beta-c=0.$$

It is a biquadratic in t giving four roots.

If t_1, t_2, t_3, t_4 be the roots,

$$t_1+t_2+t_3+t_4=\frac{\alpha}{c}, \qquad \ldots(1)$$

$$t_1t_2+t_2t_3+t_3t_4+t_4t_1+t_1t_3+t_2t_4=0, \qquad ...(2)$$

$$t_1t_2t_3+t_2t_3t_4+t_3t_4t_1+t_4t_1t_2=-\frac{\beta}{c}, \qquad ...(3)$$

$$t_1t_2t_3t_4=-1. \qquad ...(4)$$

From (1), $\quad ct_1+ct_2+ct_3+ct_4=\alpha;$

$$\therefore \quad x_1+x_2+x_3+x_4=\alpha.$$

Dividing (3) by (4), we get

$$\frac{1}{t_4}+\frac{1}{t_1}+\frac{1}{t_2}+\frac{1}{t_3}=\frac{\beta}{c}$$

or

$$\frac{c}{t_1}+\frac{c}{t_2}+\frac{c}{t_3}+\frac{c}{t_4}=\beta;$$

$$\therefore \quad y_1+y_2+y_3+y_4=\beta.$$

Also $\quad x_1x_2x_3x_4=c^4t_1t_2t_3t_4=-c^4. \qquad$ [from (4)]

Similarly $\quad y_1y_2y_3y_4=\dfrac{c^4}{t_1t_2t_3t_4}=-c^4.$

Ex. 8. *Show that the locus of the middle points of normal chords of the rectangular hyperbola $x^2-y^2=a^2$ is*

$$(y^2-x^2)^3=4a^2x^2y^2. \qquad \textit{(Agra 1942)}$$

The normal at the point $a \sec\theta$, $a\tan\theta$ is given by

$$ax\cos\theta+ay\cot\theta=2a^2$$

or

$$x\cos\theta+y\cot\theta=2a. \qquad ...(1)$$

Let the middle point of the chord be (h, k). Then the equation to the chord is

$$xh-yk=h^2-k^2. \qquad ...(2)$$

As (1) and (2) are the same,

$$\frac{\cos\theta}{h}=\frac{\cot\theta}{-k}=\frac{2a}{h^2-k^2}. \qquad ...(3)$$

From (3), $\quad \cos\theta=\dfrac{2ah}{h^2-k^2}$

and $\quad \sin\theta=-\dfrac{h}{k}.$

Eliminating θ, we get

$$\frac{4a^2h^2}{(h^2-k^2)^2}+\frac{h^2}{k^2}=1.$$

or
$$\frac{4a^2h^2}{(k^2-h^2)^2}=1-\frac{h^2}{k^2}$$

$$=\frac{k^2-h^2}{k^2}$$

or
$$(k^2-h^2)^3=4a^2h^2k^2.$$

Hence the locus of (h, k) is

$$(y^2-x^2)^3=4a^2x^2y^2$$

Examples XVI

1. Find the equation of the axes of the hyperbola
$$6x^2-7xy-3y^2=1.$$
[**Hint.** First asymptotes and then the axes].

2. Find the equation to the hyperbola whose directrix is $2x+y=1$, focus $(1, 2)$ and eccentricity $\sqrt{3}$.

3. Find the centre of the hyperbola
$$2x^2+y^2-3xy-5x+4y+6=0.$$
[**Hint.** Find asymptotes and their point of intersection.]

4. An ellipse and a hyperbola have the same principal axes. Show that the polar of any point on either curve with respect to the other touches the first curve.

(Calcutta 1952)

5. Prove that the asymptotes of the hyperbola
$xy=ax+by$ are $y=a$, $x=b$.

6. Find the equation of the hyperbola whose asymptotes are $2x-y-3=0$ and $3x+y-7=0$ and which passes through $(1, 1)$.

7. Find the equation to the hyperbola whose asymptotes are parallel to $2x+3y=0$ and $3x+2y=0$, whose centre is at $(1, 2)$ and which passes through $(5, 3)$.

(Calcutta 1948)

8. Find the equation of the hyperbola conjugate to

$$x^2+3xy+2y^2+2x+3y+2=0.$$

9. Show that the locus of the poles of normal chords of the hyperbola $\dfrac{x^2}{a^2}-\dfrac{y^2}{b^2}=1$ is

$$x^2y^2(a^2+b^2)^2=a^6y^2-b^6x^2.$$

10. A circle with fixed centre $(3h, 2k)$ and of variable radius cuts the rectangular hyperbola $x^2-y^2=9a^2$ at the points A, B, C, D; prove that the locus of the centroid of the triangle ABC is given by

$$(x-2h)^2-(y-2k)^2=a^2.$$

11. Show that the normal to the rectangular hyperbola $xy=c^2$ at the point t meets the curve again at the point t' such that $t^3t'=-1$.

12. A circle cuts the rectangular hyperbola $xy=1$ in the points (x_r, y_r), $r=1, 2, 3, 4$. Prove that

$$x_1x_2x_3x_4=y_1y_2y_3y_4=1. \qquad \textit{(Benares 1948)}$$

13. Find the equation of the chord of the hyperbola whose middle point is (x_1, y_1) and deduce that the locus of the middle points of chords passing through the fixed point (h, k) is a hyperbola whose centre is the point $(\frac{1}{2}h, \frac{1}{2}k)$.

14. Show that the locus of the pole of any straight line with respect to a coaxal system of circles is a hyperbola one of whose asymptotes is perpendicular to the given line, and the other is parallel to the radical axis of the system. *(Mysore 1945)*

15. Show that if a rectangular hyperbola cuts a circle in four points, the centre of mean position of the four points is mid-way between the centres of two curves.

16. If a hyperbola be rectangular and its equation be $xy=c^2$, prove that the locus of the middle points of the

chords of constant length $2d$ is

$$(x^2+y^2)(xy-c^2)=d^2xy. \qquad \textit{(Nagpur 1945)}$$

17. From any point P of a rectangular hyperbola, centre O, perpendiculars PM, PN are drawn to its principal axes. Show that the tangent at P is perpendicular to MN and that the distance of O from the tangent varies inversely as OP. *(Travancore 1946)*

18. Show that the locus of poles with respect to the parabola $y^2=4ax$ of tangents to the hyperbola $x^2-y^2=a^2$ is the ellipse $4x^2+y^2=4a^2$.

19. A circle cuts a rectangular hyperbola in A, B, C, D. If H is the orthocentre of triangle ABC, show that H and D are the extremities of a diameter of the rectangular hyperbola. *(Andhra 1947)*

20. The normals at the three points A, B, C on a rectangular hyperbola intersect at a point S on the curve. Prove that the centre of the hyperbola is the centroid of the triangle ABC.

21. Prove that if the normals at P, Q, R, S on a rectangular hyperbola intersect in a point, then the circle PQR passes through the other extremity of the diameter through S. *(Utkal 1947)*

22. If four points be taken on a rectangular hyperbola such that the chord joining any two is perpendicular to the chord joining the other two, and if α, β, γ, δ be the inclinations to either asymptote of the straight lines joining these points to the centre, prove that

$$\tan\alpha\tan\beta\tan\gamma\tan\delta=1.$$

23. If the tangent and normal to a rectangular hyperbola cut off intercepts a_1 and a_2 on one axis and b_1 and b_2 on the other, show that

$$a_1a_2+b_1b_2=0.$$

24. At any point P on the hyperbola $\frac{x^2}{a^2}-\frac{y^2}{b^2}=1$ a tangent is drawn to meet the asymptotes in Q and R. If C be the centre of the hyperbola, show that the locus of the circle circumscribing the triangle CQR is

$$4(a^2x^2-b^2y^2)=(a^2+b^2)^2. \qquad \textit{(Agra 1958)}$$

25. Two lines $x-\alpha=0$, $y-\beta=0$ are conjugate with respect to the hyperbola $xy=c^2$ (that is to say each line passes through the pole of the other). Prove that (α, β) is on the hyperbola $xy-2c^2=0$.

26. Prove that the locus of the middle points of the chords of the hyperbola

$$\frac{x^2}{a^2}-\frac{y^2}{b^2}=1,$$

substending a right angle at the centre, is the curve

$$\frac{x^2}{a^4}+\frac{y^2}{b^4}-\frac{a^2-b^2}{a^2b^2}\left(\frac{x^2}{a^2}-\frac{y^2}{b^2}\right)^2=0.$$

CHAPTER IX

TRACING A CONIC

9·1. Conic sections.

A conic section has been defined as the locus of a point P which moves so that its distance from a fixed point S, the focus, has a constant ratio e with its perpendicular distance from a fixed straight line called the directrix.

When S does not lie on the directrix, the locus is an ellipse, parabola or a hyperbola if $e <=> 1$.

The ellipse $\frac{x^2}{a^2}+\frac{y^2}{b^2}=1$ becomes a circle if $a^2=b^2$ and therefore $e=0$ The distance of the directrix from the centre $\frac{a}{e}$ becomes infinite. Hence the circle is a particular case of an ellipse whose eccentricity e is zero.

If S lies on the directrix,

$$\sin PSM=\frac{PM}{SP}=\frac{1}{e}.$$

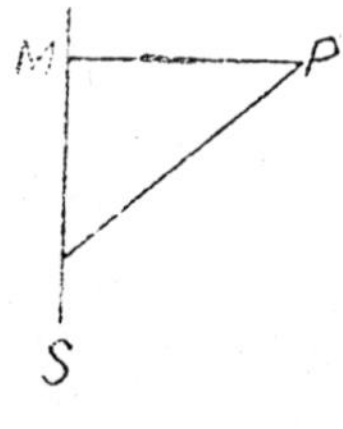

The locus of P in this case is two straight lines passing through S and inclined to the directrix SM at an angle whose sine is $1/e$.

If $e < 1$, the angle becomes imaginary and the locus is two imaginary straight lines intersecting at point S.

If $e > 1$ the locus is two real straight lines passing through S.

If $e=1$, the two lines coincide and the locus is a straight line at right angles to the directrix.

If the directrix is at infinity, we have seen that the locus is a circle, but if the focus also lies on it, it represents two lines intersecting at infinity. Hence it represents a pair of parallel straight lines.

Thus a conic section may be an ellipse, parabola, hyperbola, circle, and two straight lines, real or imaginary, coincident or parallel.

The term conic section has originated from the fact that all conics can be obtained by cutting sections from a double right-circular cone by a plane in various positions.

Let d be the semi-vertical angle of the cone.

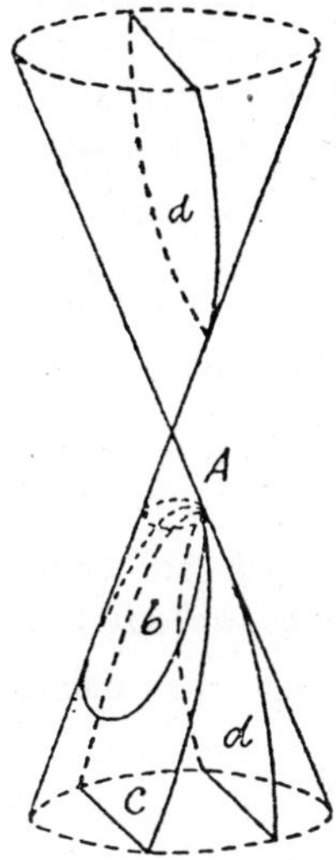

(i) Suppose the intersecting plane does not pass through the vertex of the cone and let the intersecting plane be inclined to the axis of the cone by an angle ϕ.

Then if

$\phi=\pi/2$, the section is a circle as given a in the figure.

ϕ lies between d and $\pi/2$, the section is an ellipse as given b.

$\phi=d$, the section is a parabola as given c.

$\phi < d$, the section is a hyperbola as given d.

(ii) Suppose the intersecting plane passes through the vertex of the cone.

If $\phi < d$, the section is two straight lines through the vertex.

If $\phi=d$, the section is two coincident straight lines.

If $\phi > d$, the section is two imaginary straight lines meeting at the vertex of the cone.

9·2. General equation of the second degree always represents a conic section.

Let the general equation of second degree be

$$ax^2+2hxy+by^2+2gx+2fy+c=0. \quad \ldots(1)$$

Suppose the axes are rectangular and they are turned through an angle θ, so that x and y become $x\cos\theta-y\sin\theta$ and $x\sin\theta+y\cos\theta$ respectively.

The equation (1) then becomes

$$x^2(a\cos^2\theta+2h\cos\theta\sin\theta+b\sin^2\theta)$$
$$+2xy\{h(\cos^2\theta-\sin^2\theta)-(a-b)\sin\theta\cos\theta\}$$
$$+y^2(a\sin^2\theta-2h\cos\theta\sin\theta+b\cos^2\theta)$$
$$+2x(g\cos\theta+f\sin\theta)+2y(f\cos\theta-g\sin\theta)+c=0. \quad \ldots(2)$$

If angle θ is chosen so that the term containing xy vanishes,

$$\tan 2\theta=\frac{2h}{a-b}$$

and the equation reduces to

$$Ax^2+By^2+2Gx+2Fy+C=0$$

which shows that the expression $ax^2+2hxy+by^2$ changes into Ax^2+By^2.

It is obvious that

$$a+b=A+B \text{ and } ab-h^2=AB.$$

(i) If $ab-h^2=0$, then either $A=0$ or $B=0$.

If A=0, the equation (2) reduces to

$$By^2+2Gx+2Fy+C=0$$

or
$$\left(y^2+\frac{2F}{B}y\right)+\frac{2G}{B}x+\frac{C}{B}=0$$

or
$$\left(y+\frac{F}{B}\right)^2+\frac{2G}{B}\left(x+\frac{C}{2G}-\frac{F^2}{2BG}\right)=0.$$

Transferring the origin to the point $\left(-\frac{C}{2G}+\frac{F^2}{2BG}, \frac{-F}{B}\right)$

the equation reduces to $y^2+\frac{2G}{B}x=0$ which is a parabola, whose axis is parallel to the x-axis.

Similarly if $B=0$, it will reduce to a parabola whose axis is parallel to the y-axis.

If $A=0$ and also $G=0$ the equation reduces to

$$By^2+2Fy+C=0$$

or
$$y+\frac{F}{B}=\pm\sqrt{\left(\frac{F^2}{B^2}-\frac{C}{B}\right)}$$

which are two parallel straight lines real or imaginary according as

$$F^2 > \text{or} < BC.$$

(ii) If $ab-h^2$ is not equal to 0, then both A and B are different from zero.

The equation (2) can be written as

$$A\left(x^2+\frac{2G}{A}x\right)+B\left(y^2+\frac{2F}{B}y\right)+C=0,$$

or
$$A\left(x+\frac{G}{A}\right)^2+B\left(y+\frac{F}{B}\right)^2=\frac{G^2}{A}+\frac{F^2}{B}-C.$$

Transferring the origin to the point $\left(-\frac{G}{A}, -\frac{F}{B}\right)$, the equation becomes

$$Ax^2+By^2=\frac{G^2}{A}+\frac{F^2}{B}-C=D \text{ (say)},$$

or
$$\frac{x^2}{\frac{D}{A}}+\frac{y^2}{\frac{D}{B}}=1.$$

If $\frac{D}{A}$ and $\frac{D}{B}$ are both positive, the equation represents an ellipse.

If one of $\frac{D}{A}$ and $\frac{D}{B}$ be negative, it represents a hyperbola.

If both $\frac{D}{A}$ and $\frac{D}{B}$ are negative, it is an imaginary ellipse.

If D is zero, the equation represents two straight lines, real or imaginary, according as A and B have opposite or same signs.

If A and B have opposite signs and are equal, the conic represents a rectangular hyperbola.

9·3. Centre of the conic section.

(i) Origin, centre of the conic.

Let the conic be

$$ax^2+2hxy+by^2+2gx+2fy+c=0. \quad \ldots(1)$$

The centre bisects all chords of the conic. Hence if one extremity of the chord is (x_1, y_1), and origin the centre, the other extremity must be $(-x_1, -y_1)$.

Now obviously (x_1, y_1) and $(-x_1, -y_1)$ being the extremities of the chord lie on the conic.

Therefore $ax_1^2+2hx_1y_1+by_1^2+2gx_1+2fy_1+c=0 \quad \ldots(2)$

and $ax_1^2+2hx_1y_1+by_1^2-2gx_1-2fy_1+c=0. \quad \ldots(3)$

Subtracting (3) from (2), we get

$$gx_1+fy_1=0.$$

As x_1 and y_1 are positive quantities the above result can be possible only when g and f are separately equal to zero.

Hence the equation to a conic whose centre is the origin is given by

$$ax^2+2hxy+by^2+c=0. \quad \ldots(4)$$

(ii) *Coordinates of the centre, and equation referred to centre as origin.*

Suppose the centre of the conic represented by equation (1) is (x_1, y_1).

Transferring the origin to the centre (x_1, y_1), the equation (1) becomes

$$a(x+x_1)^2+b(y+y_1)^2+2h(x+x_1)(y+y_1)+2g(x+x_1)$$
$$+2f(y+y_1)+c=0$$

or $$ax^2+2hxy+by^2+2x(ax_1+hy_1+g)+2y(hx_1+by_1+f)$$
$$+ax_1^2+by_1^2+2hx_1y_1+2gx_1+2fy_1+c=0 \quad \ldots(5)$$

As the centre of the above equation is the origin, it must not contain terms containing x and y.

$$\therefore \quad ax_1+hy_1+g=0 \quad \ldots(6)$$

and $$hx_1+by_1+f=0 \quad \ldots(7)$$

Solving the above equations for x_1 and y_1.

$$x_1=\frac{fh-bg}{ab-h^2} \text{ and } y_1=\frac{gh-af}{ab-h^2}.$$

The constant term in (5),

$$ax_1^2+2hx_1y_1+by_1^2+2gx_1+2fy_1+c$$
$$=a(ax_1+hy_1+g)+b(hx_1+by_1+f)+gx_1+fy_1+c$$
$$=gx_1+fy_1+c \text{ from (6) and (7)}$$
$$=\frac{g(fh-bg)+f(gh-af)}{ab-h^2}+c$$
$$=\frac{abc+2fgh-af^2-bg^2-ch^2}{ab-h^2}$$
$$=\frac{\Delta}{ab-h^2}.$$

The equation to the conic referred to centre as origin becomes

$$ax^2+2hxy+by^2+\frac{\Delta}{ab-h^2}=0.$$

Note :—(i) The equations (6) and (7) can be obtained by differentiating partially the original equation with respect to x and y respectively. Hence the coordinates of the centre are given by equations

$$\frac{\partial F}{\partial x}=0 \quad \text{and} \quad \frac{\partial F}{\partial y}=0$$

where the equation to the conic is given by $F=0$.

(ii) If $ab-h^2=0$, the coordinates of the centre become infinite. The conic is a parabola. In a parabola the centre is at infinity.

If $\frac{a}{h}=\frac{h}{b}=\frac{g}{f}$, both the equations (6) and (7) become the same and the centre is indeterminate. In this case there is a line of centres given by

$$ax_1+hy_1+g_1=0$$

and the conic represents a pair of parallel straignt lines, both being parallel to the line of centres.

(iii) The equation $ax^2+2hxy+by^2+2gx+2fy+c=0$ becomes $ax^2+2hxy+by^2+gx_1+fy_1+c=0$ ...(1)
when referred to centre (x_1, y_1) as origin, when x_1 and y_1 are given by

$$ax_1+hy_1+g=0 \qquad \text{...(2)}$$

and

$$hx_1+by_1+f=0. \qquad \text{...(3)}$$

If

$$gx_1+fy_1+c=0, \qquad \text{...(4)}$$

then the equation (1) represents a pair of straight lines.

Eliminating x_1, y_1 from (2), (3) and (4), we get

$$\begin{vmatrix} a & h & g \\ h & b & f \\ g & f & c \end{vmatrix}=0$$

or

$$\Delta=0$$

as the condition for the general equation to represent two straight lines.

9·4. Asymptotes. Let the equation to the conic be

$$ax^2+2hxy+by^2+2gx+2fy+c=0. \qquad \ldots(1)$$

We have seen that equation to the asymptotes differs from the equation to the conic only by a constant; it can be written as

$$ax^2+2hxy+by^2+2gx+2fy+c+\lambda=0. \qquad \ldots(2)$$

But this equation must represent a pair of straight lines. Therefore

$$\begin{vmatrix} a & h & g \\ h & b & f \\ g & f & c+\lambda \end{vmatrix}=0$$

or
$$\begin{vmatrix} a & h & g \\ h & b & f \\ g & f & c \end{vmatrix}+\begin{vmatrix} a & h & 0 \\ h & b & 0 \\ g & f & \lambda \end{vmatrix}=0$$

or
$$\Delta+\lambda\,(ab-h^2)=0$$

or
$$\lambda=-\frac{\Delta}{ab-h^2}.$$

Hence the equation of the asymptotes is

$$ax^2+2hxy+by^2+2gx+2fy+c-\frac{\Delta}{ab-h^2}=0.$$

Note. The difference in the equation of the hyperbola and the asymptote is the same constant term, which is the difference between the asymptotes and the conjugate hyperbola. Hence the equation of the conjugate hyperbola is

$$ax^2+2hxy+by^2+2gx+2fy+c-\frac{2\Delta}{ab-h^2}=0.$$

9·5. Oblique Axes. If the axes are inclined at an angle ω, we transform to rectangular axes through the same origin and suppose the equation

$$ax^2+2hxy+by^2+2gx+2fy+c=0 \qquad \ldots(1)$$

changes into

$$A'x^2+2H'xy+B'y^2+2G'x+2F'y+C=0. \qquad \ldots(2)$$

The above represents

a parabola if $A'B'=H'^2$,

an ellipse if $A'B' > H'^2$,

a hyperbola if $A'B' < H'^2$,

a rectangular hyperbola if $A'+B'=0$.

By the theory of invariants,

$$\frac{ab-h^2}{\sin^2\omega}=\frac{A'B'-H'^2}{\sin^2\pi/2}=A'B'-H'^2$$

and
$$\frac{a+b-2h\cos\omega}{\sin^2\omega}=\frac{A'+B'-2H'\cos\pi/2}{\sin^2\pi/2}=A'+B'.$$

Hence equation (1) represents

a parabola if $ab=h^2$,

an ellipse if $ab > h^2$,

a hyperbola if $ab < h^2$,

a rectangular hyperbola if $a+b-2h\cos\omega=0$.

9·6. Summary of results.

The equation $ax^2+2hxy+by^2+2gx+2fy+c=0$ represents

(i) a parabola if $h^2=ab$ and $\Delta\neq 0$,

(ii) an ellipse if $ab > h^2$,

(iii) a hyperbola if $ab < h^2$ and $\Delta\neq 0$.

(iv) a rectangular hyperbola if $a+b=0$ and $\Delta\neq 0$.

If the axes are oblique, this condition becomes

$$a+b-2h\cos\omega=0,\ \Delta\neq 0.$$

(v) a circle if $a=b$ and $h=0$.

If the axes are oblique the condition becomes

$$a:b:h=1:1:\cos\omega.$$

(vi) a pair of straight lines if

$$\Delta=abc+2fgh-af^2-bg^2-ch^2=0.$$

(vii) a pair of parallel straight lines if

$$\Delta=0 \text{ and } h^2=ab.$$

9·7. Lengths and directions of the axes.

The conic can be reduced to $ax^2+2hxy+by^2=k$ when referred to its centre as origin.

Let this conic be cut by the circle $x^2+y^2=r^2$, whose centre is also the origin.

The equation to the pair of straight lines joining the origin to the points of intersection of the conic and the circle is given by

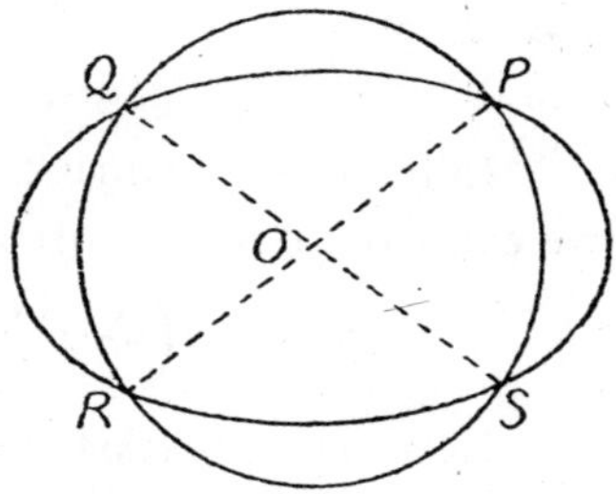

$$ax^2+2hxy+by^2=k\frac{(x^2+y^2)}{r^2}$$

or $$\left(a-\frac{k}{r^2}\right)x^2+2hxy+\left(b-\frac{k}{r^2}\right)y^2=0. \quad \ldots(1)$$

If these straight lines are coincident, they will coincide either along the minor or major axis of the conic as given

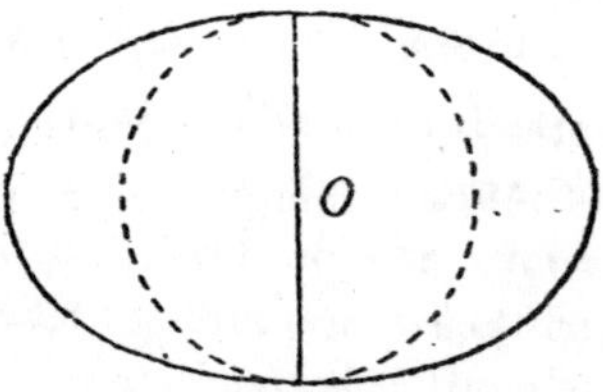

(i)

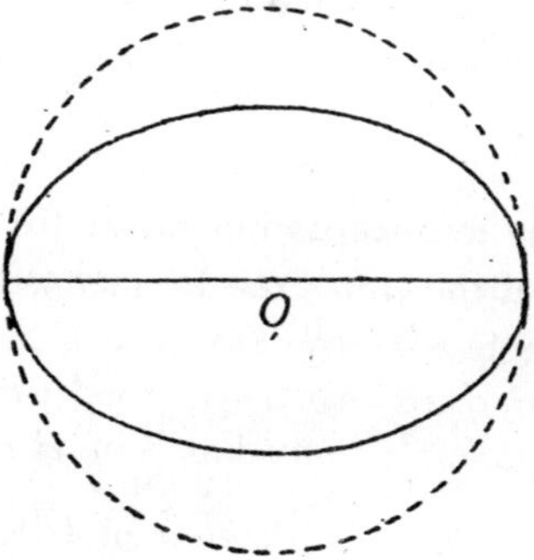

(ii)

in figures (i) and (ii). The condition for (1) to represent coincident straight lines is

$$\left(a-\frac{k}{r^2}\right)\left(b-\frac{k}{r^2}\right)=h^2, \quad \ldots(2)$$

which is a biquadratic in r and will give two roots r_1^2 and r_2^2 representing the lengths of the semi-axes of the conic.

If r_1^2 and r_2^2 are both positive, the conic is an ellipse, the greater of r_1 and r_2 representing the length of the major axis. If one of the roots r_1^2 and r_2^2 is negative, the conic is a hyperbola, the negative root representing the length of the conjugate axis.

In both cases equation (1) will be a perfect square and we can write it as

$$\left\{\left(a-\frac{k}{r^2}\right)x+hy\right\}^2=0.$$

Hence the equations of the axes are

$$\left(a-\frac{k}{r_1^2}\right)x+hy=0$$

and

$$\left(a-\frac{k}{r_2^2}\right)x+hy=0. \quad \ldots(A)$$

Otherwise :

If the coordinate axes are turned through an angle θ given by

$$\tan 2\theta=\frac{2h}{a-b}, \quad \ldots(1)$$

the term containing xy in the equation of the conic vanishes and the new axes are inclined to the coordinate axes at an angle θ.

(1) can be written as

$$\frac{2\tan\theta}{1-\tan^2\theta}=\frac{2h}{a-b}=\lambda \text{ (say)}$$

or

$$\lambda\tan^2\theta+2\tan\theta-\lambda=0. \quad \ldots(2)$$

This is a quadratic equation in $\tan\theta$. Let θ_1 and θ_2 be the two values of θ which satisfy it; then

$$\tan\theta_1 \tan\theta_2 = -1 ;$$
$$\therefore \quad \theta_1 - \theta_2 = \pi/2.$$

The polar equation of the conic is

$$r^2 (a\cos^2\theta + 2h\cos\theta\sin\theta + b\sin^2\theta) = 1$$

or
$$r^2 = \frac{1+\tan^2\theta}{a+2h\tan\theta+b\tan^2\theta}. \quad \ldots \quad \ldots \quad \ldots (3)$$

The values of r obtained by substituting the values of $\tan\theta$ from (2) will give the lengths of the axes.

The equations of the axes are

$$y = x\tan\theta_1$$

and
$$y = x\tan\theta_2.$$

9·8. Eccentricity, foci and directrices.

Let r_1 denote the bigger root and r_2 the smaller. Then the eccentricity of the conic is given by

$$r_2^2 = r_1^2 (1 - e^2)$$

or
$$e = \sqrt{\left(1 - \frac{r_2^2}{r_1^2}\right)}.$$

The foci are points on the major axis at a distance er_1 from the centre (x_1, y_1). The direction tangent $\tan\theta$ of the major axis from equation (A) of the last article is

$$\frac{-\left(a - \frac{k}{r_1^2}\right)}{h}.$$

Hence $\cos\theta$ and $\sin\theta$ can be found.

Also
$$er_1 = \sqrt{(r_1^2 - r_2^2)}.$$

$\therefore$ the foci are points

$$\{x_1 \pm \sqrt{(r_1^2 - r_2^2)}.\cos\theta,\ y_1 \pm \sqrt{(r_1^2 - r_2^2)}.\sin\theta\}.$$

The directrices are perpendicular to the major axis at a distance $\frac{r_1}{e}$ or $\frac{r_1^2}{\sqrt{(r_1^2 - r_2^2)}}$ from the centre.

Hence their equations are

$$(x-x_1)\cos\theta+(y-y_1)\sin\theta=\pm\frac{r_1^2}{\sqrt{(r_1^2-r_2^2)}}.$$

In the case of hyperbola one of r_1^2 and r_2^2 will be negative and the same formulae will hold.

9·9. Tracing a parabola.

It has been proved back that if PV is a diameter of a parabola and QV is an ordinate through any point Q on the parabola, parallel to the tangent at P, then

$$QV^2=4a\operatorname{cosec}^2\theta . PV, \qquad \ldots(1)$$

θ being the angle between the diameter PV and the tangent at P.

If QM is the perpendicular to the tangent at P and QL perpendicular to PV, then

$$QM=PV\sin\theta \text{ and } QL=QV\sin\theta,$$

so that (1) changes into

$$QL^2=4a\operatorname{cosec}\theta . QM.$$

Hence the square of the perpendicular from any point Q on the parabola from any diameter, varies as the perpendicular from Q on any tangent at the end of the diameter.

Hence, if $px+qy+r=0$ be the equation of any diameter and $p'x+q'y+r'=0$ be the equation of the tangent at its end, the equation to the parabola is

$$(px+qy+r)^2=\lambda(p'x+q'y+r'), \qquad \ldots(2)$$

where λ is a constant.

If the equation to a parabola is in the form (2), then

$$px+qy+r=0. \qquad \ldots(3)$$

is a diameter of the parabola and the axis of the parabola is parallel to it.

The line $\qquad p'x+q'y+r'=0 \qquad \ldots(4)$

is the tangent at the end of the diameter (3).

If the lines (3) and (4) are perpendicular to each other, then (3) will be the axis of the parabola and (4) the tangent at the vertex.

The equation $ax^2+2hxy+by^2+2gx+2fy+c=0$ represents a parabola if $h^2=ab$.

Putting $a=\alpha^2$ and $b=\beta^2$ and $h=\alpha\beta$, the above equation becomes

$$(\alpha x+\beta y)^2=-(2gx+2fy+c) \qquad \ldots(5)$$

This is of the same form as (2).

Hence $\alpha x+\beta y=0$ is a diameter, and the axis of the parabola is parallel to it.

Let it be $\qquad \alpha x+\beta y+k=0.$

Equation (5) then becomes

$$(\alpha x+\beta y+k)^2=2x(k\alpha-g)+2y(k\beta-f)+k^2-c. \qquad \ldots(6)$$

Choose k such that the lines

$$\alpha x+\beta y+k=0 \qquad \ldots(7)$$

and

$$2x(k\alpha-g)+2y(k\beta-f)+k^2-c=0 \qquad \ldots(8)$$

may be perpendicular to each other.

The condition for this is that

$$2\alpha(k\alpha-g)+2\beta(k\beta-f)=0.$$

$$\therefore \quad k=\frac{\alpha g+\beta f}{\alpha^2+\beta^2}.$$

The lines (7) and (8) are the axis and the tangent at the vertex respectively.

Equation (6) then becomes

$$(\alpha x+\beta y+k)^2=\frac{2(\alpha f-\beta g)}{\alpha^2+\beta^2}(\beta x-\alpha y)+k^2-c$$

which can be written as

$$\left\{\frac{\alpha x+\beta y+k}{\sqrt{(\alpha^2+\beta^2)}}\right\}^2=\frac{2(\alpha f-\beta g)}{(\alpha^2+\beta^2)^{3/2}}\left\{\frac{\beta x-\alpha y+\dfrac{(\alpha^2+\beta^2)(k^2-c)}{2(\alpha f-\beta g)}}{\sqrt{(\alpha^2+\beta^2)}}\right\} \qquad \ldots(9)$$

Now take the lines (7) and (8) as the new axes of x and y respectively. Then the equation (9) changes into

$$y^2=4px$$

where $$4p=\frac{2(\alpha f-\beta g)}{(\alpha^2+\beta^2)^{3/2}}$$

is the latus rectum of the parabola.

The vertex of the parabola is obtained by solving (7) and (8) for x and y.

Otherwise.

As before the equation to the conic can be reduced to

$$(\alpha x+\beta y)^2+2gx+2fy+c=0. \qquad \ldots(2)$$

Rotate the axes so that the line $\alpha x+\beta y=0$ becomes the new axis of x.

We have to turn the axes through an angle θ where

$$\tan\theta=-\frac{\alpha}{\beta}.$$

For x we have to substitute

$$x\cos\theta-y\sin\theta, \quad i.e. \quad \frac{\beta x+\alpha y}{\sqrt{(\alpha^2+\beta^2)}}$$

and for y $$x\sin\theta+y\cos\theta, \quad i.e. \quad \frac{-\alpha x+\beta y}{\sqrt{(\alpha^2+\beta^2)}}.$$

The equation (2), then becomes

$$y^2(\alpha^2+\beta^2)+\frac{2}{\sqrt{(\alpha^2+\beta^2)}}[g(\beta x+\alpha y)+f(\beta y-\alpha x)]+c=0$$

or $$y^2+2y.\frac{\alpha g+\beta f}{(\alpha^2+\beta^2)^{3/2}}=2x.\frac{\alpha f-\beta g}{(\alpha^2+\beta^2)^{3/2}}-\frac{c}{\alpha^2+\beta^2}$$

or $$(y-k)^2=2\frac{(\alpha f-\beta g)}{(\alpha^2+\beta^2)^{3/2}}(x-h), \qquad \ldots(3)$$

where $$k=-\frac{\alpha g+\beta f}{(\alpha^2+\beta^2)^{3/2}} \qquad \ldots(4).$$

and $$h=\frac{\sqrt{(\alpha^2+\beta^2)}}{2(\alpha f-\beta g)}\left[c-\frac{(\alpha g+\beta f)^2}{(\alpha^2+\beta^2)^2}\right]. \quad \ldots(5)$$

The equation (3) is a parabola whose latus rectum is $\frac{2(\alpha f-\beta g)}{(\alpha^2+\beta^2)^{3/2}}$, whose axis is parallel to the new axis of x, and whose vertex referred to the new axes is (h, k).

The axis of the parabola is parallel to the new axis of x; hence it makes an angle θ with the old axis of x. Therefore the perpendicular on it from the origin makes an angle $90+\theta$ with the old x-axis.

Also the length of this perpendicular is k.

Hence the axis of the parabola referred to old axes is

$$x\cos(90+\theta)+y\sin(90+\theta)=k$$

or $$-x\sin\theta+y\cos\theta=k$$

or $$\alpha x+\beta y=k\sqrt{(\alpha^2+\beta^2)}=-\frac{\alpha g+\beta f}{\alpha^2+\beta^2}. \quad \ldots(6)$$

The vertex is the point in which the line (6) meets the curve given by (2).

Hence solving (2) and (6), we get the coordinates of the vertex referred to original axes.

9·10. Focus of a parabola.

We have seen that the latus rectum of the parabola is given by $4p=\frac{2(\alpha f-\beta g)}{(\alpha^2+\beta^2)^{3/2}}$ and the equations of the axis and the tangent at the vertex are

$$\alpha x+\beta y+k=0 \quad \ldots(1)$$

and $$2x(k\alpha-g)+2y(k\beta-f)+k^2-c=0. \quad \ldots(2)$$

The vertex (x_1, y_1) can be obtained by solving (1) and (2).

The focus is a point on the axis at a distance p from the vertex. Hence its coordinates are

$$(x_1+p\cos\theta, y_1+p\sin\theta) \text{ where } \tan\theta=-\alpha/\beta,$$

so that the coordinates of the focus are

$$x_1+\frac{\beta p}{\sqrt{(\alpha^2+\beta^2)}},\ y_1-\frac{\alpha p}{\sqrt{(\alpha^2+\beta^2)}}.$$

9·11. Solved examples.

Ex. 1. *Trace the conic*

$$16x^2-24xy+9y^2+77x-64y+95=0. \quad \textit{(Benaras 1960)}$$

The given equation can be written as

$$(4x-3y+k)^2+(77-8k)\,x+2\,(3k-32)\,y+95-k^2=0 \quad \ldots(1)$$

Choose k such that the lines

$$4x-3y+k=0 \quad \ldots(2)$$

and

$$(77-8k)\,x+2\,(3k-32)\,y+95-k^2=0 \quad \ldots(3)$$

are perpendicular to each other, so that

$$4\,(77-8k)-3.2\,(2k-32)=0$$

or

$$k=10.$$

The equation (1) becomes

$$(4x-3y+10)^2=3x+4y+5$$

or

$$\left(\frac{4x-3y+10}{5}\right)^2=\frac{1}{5}\left(\frac{3x+4y+5}{5}\right). \quad \ldots(4)$$

The axis of the parabola is

$$4x-3y+10=0 \quad \ldots(5)$$

and the tangent at the vertex is

$$3x+4y+5=0. \quad \ldots(6)$$

The latus rectum is $\frac{1}{5}$.

The coordinates of the vertex are $(-\frac{11}{5}, \frac{2}{5})$, obtained by solving (5) and (6).

The x-axis meets the parabola in points given by $16x^2+77x+95=0$ which gives imaginary values of x.

The y-axis meets the parabola in points whose ordinates are given by $9y^2-64y+95=0$ or $y=\frac{19}{9}$ and 5.

With the new axes the equation to the parabola becomes $y^2=\frac{1}{5}x$.

The figure of the parabola is as shown in the diagram below :—

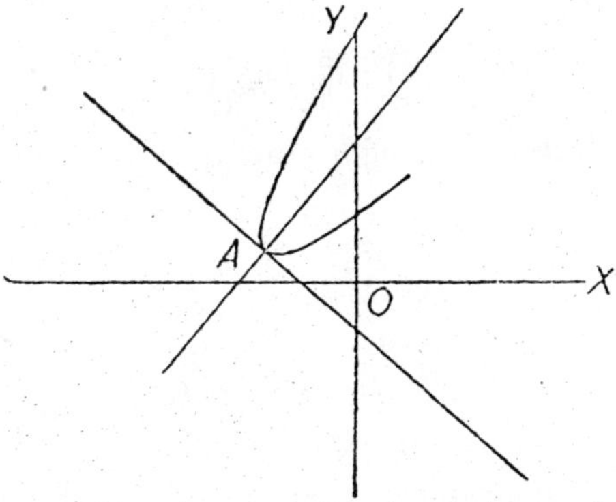

The focus is a point on the axis at a distance $\frac{1}{20}$ ahead of the vertex. Hence its coordinates are

$$\left[-\frac{11}{5}+\frac{1}{20}\cos\theta,\ \frac{2}{5}+\frac{1}{20}\sin\theta\right]$$

or $$\left[-\frac{11}{5}+\frac{1}{20}\cdot\frac{3}{5},\ \frac{2}{5}+\frac{1}{20}\cdot\frac{4}{5}\right] \qquad \text{as } \tan\theta=\frac{4}{3}$$

or $$\left[-\frac{217}{100},\ \frac{11}{25}\right].$$

2. *Trace the curve*

$$17x^2+12xy+8y^2-46x-28y+33=0.$$

(Bombay 1959)

Here $h=6$, $a=17$ and $b=8$.

Since $h^2 < ab$, the curve is an ellipse.

The coordinates of the centre are given by the equations

$$17x+6y-23=0 \qquad \ldots(1)$$

and $$6x+8y-14=0. \qquad \ldots(2)$$

Solving them, we get the centre (1, 1). The equation to the curve referred to centre as origin reduces to

$$17x^2+12xy+8y^2-23-14+33=0$$

or $$17x^2+12xy+8y^2-4=0. \qquad \ldots(3)$$

The straight lines joining the origin to the points of intersection of (3) and the circle $x^2+y^2=r^2$ are given by

$$x^2\left(17-\frac{4}{r^2}\right)+12xy+y^2\left(8-\frac{4}{r^2}\right)=0. \qquad \ldots(4)$$

The semi-axes of length r are given by the equation

$$\left(17-\frac{4}{r^2}\right)\left(8-\frac{4}{r^2}\right)=36$$

or
$$100r^4-100r^2+16=0$$

or
$$25r^4-25r^2+4=0$$

or
$$(5r^2-4)(5r^2-1)=0\ ;$$

$$\therefore \quad r_1{}^2=\tfrac{4}{5} \text{ and } r_2{}^2=\tfrac{1}{5}.$$

The equations to the major and minor axes are given by

$$\left(17-\frac{4}{r^2}\right)x+6y=0,$$

where $r^2=\frac{4}{5}$ for major axis and $r^2=\frac{1}{5}$ for the minor axis.

On substituting these values of r the equations become

$$2x+y=0 \text{ and } x-2y=0.$$

These equations when referred to the old origin become

$$2(x-1)+y-1=0$$

or
$$2x+y-3=0$$

and
$$x-2y+1=0.$$

The ellipse is cut by the x-axis in points whose abscissae are given by

$$17x^2-46x+33=0.$$

The roots of this equation are imaginary.

The ordinates of the points of intersection of y-axis with the ellipse are given by

$$8y^2-28y+33=0$$

the roots of which are again imaginary.

The ellipse does not cut the axes in real points.

The diagram is shown below :—

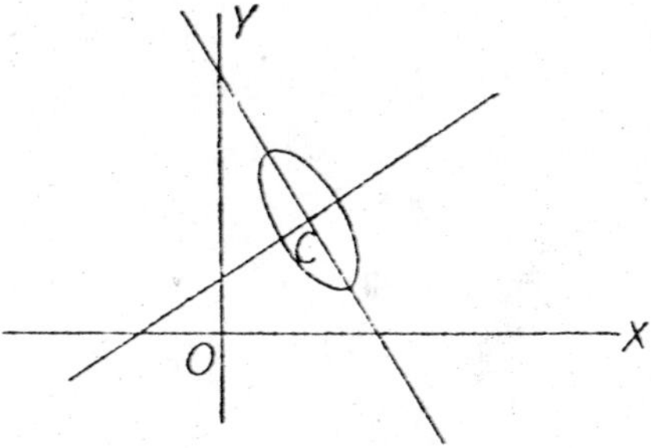

3. *Trace the conic* $8x^2-24xy+15y^2+48x-48y=0$. *Find its eccentricity.* *(Mysore 1945)*

Here $a=8$, $b=15$, $h=-12$.

Since $h^2 > ab$, the curve is a hyperbola.

The coordinates of centre are given by

$$2x-3y+6=0 \qquad \ldots(1)$$

or

$$4x-5y+8=0. \qquad \ldots(2)$$

Solving (1) and (2), the centre is (3, 4).

The equation of the conic referred to centre as origin is given by

$$8x^2-24xy+15y^2+24.3-24.4=0$$

or

$$8x^2-24xy+15y^2-24=0. \qquad \ldots(3)$$

The equation of the straight lines joining the origin to the points of intersection of (3) and $x^2+y^2=r^2$ is given by

$$\left(8-\frac{24}{r^2}\right)x^2-24xy+\left(15-\frac{24}{r^2}\right)=0. \qquad \ldots(4)$$

The semi-axes of length r are given by the equation

$$\left(8-\frac{24}{r^2}\right)\left(15-\frac{24}{r^2}\right)=144$$

or
$$r^4+23r^2-24=0$$

or
$$(r^2+24)(r^2-1)=0;$$

$$\therefore \quad r_1^2=1 \text{ and } r_2^2=-24.$$

The length of the transverse axis is 2.

The length of the conjugate axis is $4\sqrt{6}$.

Eccentricity is given by

$$24=1(e^2-1)$$

or
$$e^2=25; \quad \therefore \quad e=5.$$

The axes are given by

$$\left(8-\frac{24}{r^2}\right)x-12y=0,$$

where $r^2=1$ for transverse axis

and $r^2=-24$ for conjugate axis.

Substituting these values, the axes are

$$4x+3y=0 \text{ and } 3x-4y=0.$$

When referred to the old origin the equations to the axes are

$$4(x-3)+3(y-4)=0$$

or
$$4x+3y-24=0,$$

and
$$3(x-3)-4(y-4)=0$$

or
$$3x-4y+7=0.$$

The gradient of the transvese axis is $-\frac{4}{3}$ and the foci are points on it distant er_1, *i.e.* 5 from the centre (3, 4).

Hence the foci are

$$[3 \pm 5.\tfrac{3}{5}, 4 \pm 5.(-\tfrac{4}{5})],$$

i.e.
$$[6, 0] \text{ and } [0, 8].$$

The shape of the hyperbola is as shown below in the

diagram :—

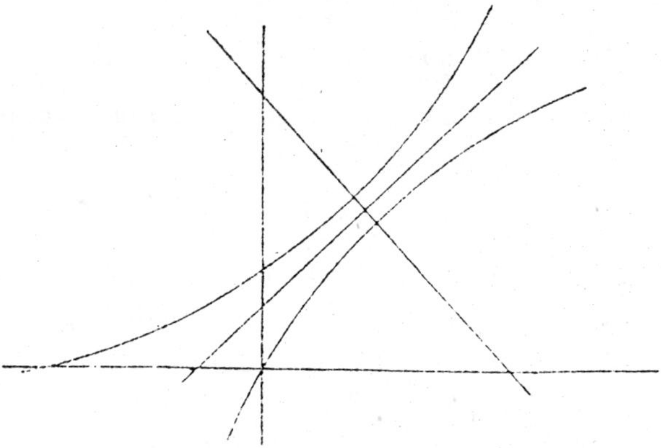

It clearly passes through the origin.

It is cut by the y-axis, where $15y^2-48y=0$,

i.e. $$y=0 \quad \text{or} \quad \tfrac{16}{5}.$$

It is cut by the x-axis at points given by

$$8x^2+48x=0,$$

i.e. $$x=0 \quad \text{or} \quad -6.$$

The directrices are perpendicular to the transverse axis at a distance $\frac{r_1}{c}$ or $\frac{1}{5}$ from the centre.

Hence their equations are

$$(x-3)\tfrac{3}{5}+(y-4)(-\tfrac{4}{5})=\pm\tfrac{1}{5}$$

or $$\tfrac{3}{5}x-\tfrac{9}{5}-\frac{4y}{5}+\tfrac{16}{5}=\pm\tfrac{1}{5}$$

or $$\tfrac{3}{5}x-\frac{4y}{5}+\tfrac{6}{5}=0 \text{ and } \tfrac{3}{5}x-\frac{4y}{5}+\tfrac{8}{5}=0,$$

i.e. $$3x-4y+6=0 \text{ and } 3x-4y+8=0.$$

Ex. 4. *If $ax^2+2hxy+by^2+2gx+2fy+c=0$ represents a rectangular hyperbola, show that the equation referred to the asym-*

ptotes as axes is

$$2\,(h^2-ab)^{3/2}\,xy-\Delta=0$$

where Δ denotes the discrminant. *(Lucknow 1944)*

The equation of the conic when referred to centre as origin is

$$ax^2+2hxy+by^2+\frac{\Delta}{ab-h^2}=0.$$

If the axes are rotated, so that they coincide with the asymptotes, the above origin will change into

$$2h_1xy+\frac{\Delta}{ab-h^2}=0.$$

By invariants,

$$h^2-ab=h_1^{\,2}$$

or $$h_1=\pm\sqrt{(h^2-ab)},$$

The required equation therefore becomes

$$2\,(h^2-ab)^{3/2}\,xy\pm\Delta=0.$$

5. *If (x_1, y_1) be the centre of the hyperbola $f(x, y)=0$, show that the asymptotes are given by the equation*

$$f(x, y)=f(x_1, y_1)$$

and the equation of the conjugate hyperbola is $f(x, y)=2f(x_1, y_1)$

$$f(x, y)\equiv ax^2+2hxy+by^2+2gx+2fy+c=0.$$

The coordinates of the centre (x_1, y_1) are given by the equations

$$ax_1+hy_1+g=0 \qquad \ldots(1)$$

and $$hx_1+by_1+f=0. \qquad \ldots(2)$$

Now $$f(x_1, y_1)=ax_1^{\,2}+2hx_1y_1+by_1^{\,2}+2gx_1+2fy_1+c$$

$$=x_1(ax_1+hy_1+g)+y_1(hx_1+by_1+f)$$

$$+gx_1+fy_1+c$$

$$=gx_1+fy_1+c=p \text{ (say)} \qquad \ldots(3)$$

Then $$gx_1+fy_1+c-p=0. \qquad \ldots(4)$$

We know that the equation to the asymptotes of the conic is given by

$$f(x, y) - \frac{\Delta}{ab-h^2} = 0.$$

Eliminating (x_1, y_1) from (1), (2) and (4), we get

$$\begin{vmatrix} a & h & g \\ h & b & f \\ g & f & c-p \end{vmatrix} = 0$$

or
$$\begin{vmatrix} a & h & g \\ h & b & f \\ g & f & c \end{vmatrix} + \begin{vmatrix} a & h & 0 \\ h & b & 0 \\ g & f & -p \end{vmatrix} = 0$$

or
$$\Delta - p(ab-h^2) = 0.$$

$$\therefore \quad p = \frac{\Delta}{ab-h^2}.$$

Hence the equation of the asymptotes is

$$f(x, y) = p = f(x_1, y_1).$$

The equation of the conjugate hyperbola is

$$f(x, y) - \frac{2\Delta}{ab-h^2} = 0$$

or
$$f(x, y) = 2p = 2f(x_1, y_1).$$

6. *Obtain the focus, directrix and axis of the perbola*

$$\sqrt{(ax)} + \sqrt{(by)} = 1.$$

If its axis passes through a fixed point, show that the locus of the focus is a rectangular hyperbola. *(Lucknow Hon's 1944)*

$$\sqrt{(ax)} + \sqrt{(by)} = 1$$

or
$$ax + by + 2\sqrt{(abxy)} = 1$$

or
$$ax + by - 1 = -2\sqrt{(abxy)}$$

or
$$(ax + by - 1)^2 = 4abxy$$

or
$$a^2x^2 + 2abxy + b^2y^2 - 2ax - 2by + 1 = 4abxy$$

or
$$(ax - by)^2 = 2ax + 2by - 1.$$

Let us introduce a constant k such that

$$(ax-by+k)^2=2xa\,(1+k)+2yb\,(1-k)+k^2-1$$

and
$$ax-by+k=0 \qquad \ldots(1)$$

is perpendicular to

$$2ax\,(1+k)+2by\,(1-k)+k^2-1=0 \qquad \ldots(2)$$

so that
$$a^2\,(1+k)-b^2\,(1-k)=0$$

or
$$k\,(a^2+b^2)=-(a^2-b^2).$$

$$\therefore \quad k=-\frac{(a^2-b^2)}{a^2+b^2}.$$

The equation to the axis is

$$ax-by=\frac{a^2-b^2}{a^2+b^2}.$$

The equation to the parabola can be written as

$$\left(\frac{ax-by+k}{\sqrt{(a^2+b^2)}}\right)^2=\frac{1}{a^2+b^2}\left[\frac{2ax.2b^2}{a^2+b^2}+2by.\frac{2a^2}{a^2+b^2}-\frac{4a^2b^2}{(a^2+b^2)^2}\right]$$

or
$$\left\{\frac{ax-by-\dfrac{a^2-b^2}{a^2+b^2}}{\sqrt{(a^2+b^2)}}\right\}^2=\frac{4ab}{(a^2+b^2)^2}\left[bx+by-\frac{ab}{(a^2+b^2)}\right]$$

or
$$\left\{\frac{ax-by-\dfrac{a^2-b^2}{a^2+b^2}}{\sqrt{(a^2+b^2)}}\right\}^2=\frac{4ab\sqrt{(a^2+b^2)}}{(a^2+b^2)^2}\left\{\frac{bx+ay-\dfrac{ab}{(a^2+b^2)}}{\sqrt{(a^2+b^2)}}\right\}$$

$$=\frac{4ab}{(a^2+b^2)^{3/2}}\left\{\frac{bx+ay-\dfrac{ab}{a^2+b^2}}{\sqrt{(a^2+b^2)}}\right\}$$

The latus rectum is $\dfrac{4ab}{(a^2+b^2)^{3/2}}$.

The vertex is obtained by solving (1) and (2) and is found to be $\left[\dfrac{a^3}{(a^2+b^2)^2}, \dfrac{b^3}{(a^2+b^2)^2}\right]$.

The gradient of the axis is $\dfrac{a}{b}$.

$\therefore$ the coordinates of the focus are

$$\left[\frac{a^3}{(a^2+b^2)^2}+\frac{ab}{(a^2+b^2)^{3/2}}\cdot\frac{b}{\sqrt{(a^2+b^2)}},\ \frac{b^3}{(a^2+b^2)^2}+\frac{ab}{(a^2+b^2)^{3/2}}\cdot\frac{a}{\sqrt{(a^2+b^2)}}\right]$$

or
$$\left[\frac{a}{(a^2+b^2)},\frac{b}{(a^2+b^2)}\right].$$

The point through which the directrix passes is as far behind the vertex as the focus is ahead of it. Hence its coordinates are

$$\left[\frac{a^3}{(a^2+b^2)^2}-\frac{ab}{(a^2+b^2)^{3/2}}\cdot\frac{b}{\sqrt{(a^2+b^2)}},\ \frac{b^3}{(a^2+b^2)^2}-\frac{ab}{(a^2+b^2)^{3/2}}\cdot\frac{a}{\sqrt{(a^2+b^2)}}\right]$$

or
$$\left[\frac{a\,(a^2-b^2)}{(a^2+b^2)^2},\frac{b\,(b^2-a^2)}{(a^2+b^2)^2}\right]. \qquad \ldots(4)$$

The directrix is a line parallel to (2) whose equation becomes
$$bx+ay-\frac{ab}{(a^2+b^2)}=0.$$

Hence its equation is $bx+ay+\lambda=0$.

But it passes through the point given by (4).

$$\therefore\quad \frac{ab\,(a^2-b^2)}{(a^2+b^2)^2}+\frac{ab\,(b^2-a^2)}{(a^2+b^2)^2}+\lambda=0$$

or
$$\lambda=0.$$

$\therefore$ the equation to the directrix is $bx+ay=0$.

Suppose the axis passes through a fixed point (h, k); then

$$ah-bk=\frac{a^2-b^2}{a^2+b^2} \text{ or } \frac{ah}{(a^2+b^2)}-\frac{bk}{(a^2+b^2)}=\frac{a^2}{(a^2+b^2)^2}-\frac{b^2}{(a^2+b^2)^2}$$

or
$$hx-ky=x^2-y^2$$

when (x, y) are the coordinates of the focus.

Hence the locus is $x^2-y^2-hx+ky=0$ which is a rectangular hyperbola.

Examples XVII

1. Show that the product of the semi-axis of the conic whose equation is $x^2-xy+2y^2-2x-6y+7=0$ is $\frac{2}{\sqrt{7}}$ and that the combined equation of its axes is
$$x^2-y^2-2xy+8y-8=0.$$ *(Poona 1959)*

2. Find the real foci and the corresponding directrices of the conic $x^2+xy+y^2+4x-7y+7=0$. *(Poona 1960)*

3. Find the centre, lengths and equations of the principal axes of the conic $6x^2+4xy+9y^2-16x-22y+9=0$. *(Bombay 1959)*

4. Find the equation of the axes of the hyperbola
$$6x^2-7xy-3y^2=1.$$ *(Calcutta 1948)*

5. Find the centre of the hyperbola
$$2x^2+y^2-3xy-5x+4y+6=0.$$ *(Calcutta 1949)*

6. Find the nature of the conic represented by the equation $x^2+2xy+y^2-2x-1=0$ and trace it. *(Benaras 1958)*

7. Trace the conic $16x^2+24xy+9y^2-5x-10y+1=0$ and find its focus. *(Benaras 1951)*

8. Trace the conic $x^2+y^2+xy+x-4y+1=0$. *(Jammu and Kashmir 1950)*

9. Trace the curve $14x^2-4xy+11y^2-44x-58y+71=0$. *(Agra 1951)*

10. Trace completely the conic
$$3x^2+8xy-3y^2-40x-20y+50=0$$
and find its eccentricity. *(Jammu and Kashmir 1951)*

11. Find the equation of the conic whose asymptotes are the lines $2x+3y-5=0$ and $5x+3y-8=0$ and which passes through the points $(1, -1)$. *(Agra 1945)*

12. Trace the conic $25x^2-120xy+144y^2-2x-29y-1=0$. *(Jammu and Kashmir 1955)*

13. Find the positions and magnitudes of the axes of the conic whose equation is $x^2-5xy+y^2+8x-20y+15=0$. *(Utkal 1946)*

14. Trace $(x+2y)^2+12x-6y=0$ and find its focus and the equation of the latus rectum. *(Aligarh 1944)*

15. Trace the conic $x^2-3xy+y^2+10x-10y+21=0$. *(Patna Hon's. 1947)*

16. Trace the conic $x^2+y^2-4xy-2x-20y-11=0$. *(Jammu and Kashmir 1956)*

17. Trace the conic $34x^2+24xy+41y^2+48x+14y-108=0$ and show that its eccentricity is $\frac{1}{\sqrt{2}}$. *(Travancore 1947)*

18. Find the positions and lengths of the axes of the conic $x^2-4xy-2y^2+10x+4y=0$, explaining the steps taken. Trace the conic. *(Punjab 1944)*

19. Find the axis, vertex, latus rectum and the focus of the parabola $16x^2-24xy+9y^2-104x-172y+44=0$. *(Agra 1944)*

20. Determine the foci and directrices of the conic whose equation is $x^2+12xy-4y^2-6x+4y+9=0$. *(Dehli 1946)*

21. In the conic $x^2-4xy+4y^2+3x+4y-1=0$, find (i) equations to the axes, (ii) coordinates of the vertex, and (iii) the length of the latus rectum. *(Jammu and Kashmir 1959)*

22. Find the equation to the hyperbola which has

$$3x-4y+7=0 \quad \text{and} \quad 4x+3y+1=0$$

for asymptotes and passes through the origin. Find its centre and trace the curve. *(Andhra 1946)*

23. Obtain the equation to the rectangular hyperbola which has for one of its asymptotes the line $x+2y-5=0$ and touches the line $17x+16y=0$ at the origin. Reduce the equation to its standard form. *(Travancore 1946)*

24. Show that the semi-axes of the conic

$$ax^2+2hxy+by^2+2gx+2fy+c=0$$

are given by

$$(ab-h^2)^3 r^4+\Delta(a+b)(ab-h^2) r^2+\Delta^2=0.$$

25. Trace the conic $17x^2-12xy+8y^2+46x-28y+17=0$.

(Aligarh 1949, Agra 1961)

26. Show that the lengths of the semi-axes of the conic

$$ax^2+2hxy+by^2=1$$

are the roots of the equation

$$(ab-h^2) r^4-(a+b) r^2+1=0.$$

27. Trace the conic

$$36x^2+24xy+29y^2-72x+126y+81=0.$$ *(Punjab 1952)*

28. Show that the latus rectum of the parabola

$$(a^2+b^2)(x^2+y^2)=(bx+ay-ab)^2 \text{ is } \frac{2ab}{(a^2+b^2)^{1/2}}.$$

29. Show that the squares of lengths of semi-axes of the conic $ax^2+2hxy+by^2=1$ are the reciprocals of the roots of $x^2-(a+b)x+(ab-h^2)=0$. *(Punjab 1954)*

30. Show that the eccentricity of the conic $ax^2+2hxy+by^2=1$ is given by the equation

$$e^4(ab-h^2)+\{(a-b)^2+4h^2\}(e^2-1)=0.$$

31. Trace the parabola $9x^2+26xy+16y^2-2x+14y+1=0$ and find the coordinates of its focus and the equation to its directrix. *(Agra 1960)*

32. Trace the conic $x^2+4xy+y^2-2x+2y-6=0$. *(Agra 1959)*

33. Trace the curve $8x^2-4xy+5y^2-16x-14y+17=0$.
Find the coordinates of its foci and show that its axes lie along $2x-y-1=0$ and $2x+4y-11=0$. *(Agra 1954)*

34. Prove that the equation of the asymptotes of the conic

$$ax^2+2hxy+by^2+2gx+2fy+c=0$$

is

$$bX^2-2hXY+aY^2=0$$

where $X=ax+hy+g$ and $Y=hx+by+f$.

35. Show that the curve given by the equations

$$x=at^2+bt+c \quad \text{and} \quad y=a't^2+b't+c'$$

is a parabola whose latus rectum is $\dfrac{(a'b-ab')^2}{(a^2+a'^2)^{3/2}}$.

36. The locus of the vertices of the rectangular hyperbolas $x^2-y^2+2\lambda xy-a^2=0$ for different values of λ, is the curve whose equation is $(x^2+y^2)^2-a^2(x^2-y^2)=0$.

37. Trace the curve $4(2y-x-3)^2-9(2x+y-1)^2=80$. Find its eccentricity and the equation to its asymptotes.

(Agra 1932)

38. Prove that the lengths of semi-axes of the conic $ax^2+2hxy+ay^2=d$ are $\sqrt{\left(\dfrac{d}{a+h}\right)}$ and $\sqrt{\left(\dfrac{d}{a-h}\right)}$ respectively and that their equation is $x^2-y^2=0$.

CHAPTER X

POLAR EQUATION OF A CONIC

10·1. The polar equation of a conic can be obtained from the cartesian equation by putting

$$x=r\cos\theta \quad \text{and} \quad y=r\sin\theta.$$

However, the equation can also be obtained directly in the polar form if the focus is the pole.

Polar equation of a conic :—

Let the focus S be the pole and let ZM be the directrix. Let SZ the perpendicular from S on the directrix ZM be the initial line.

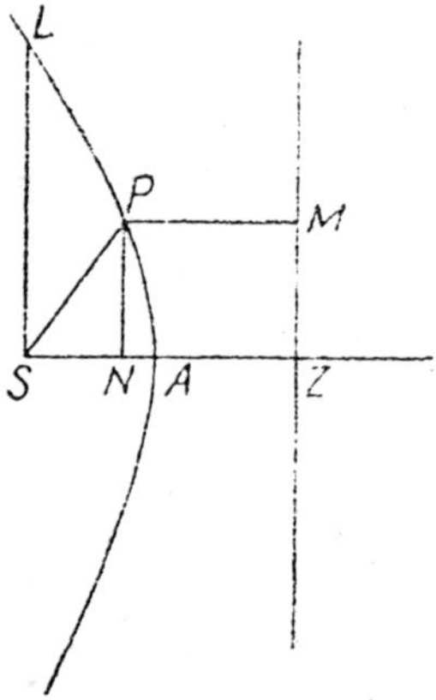

Let P be a point (r, θ) on the conic and let SL be the latus rectum.

If e be the eccentricity of the conic, then

$$SP=e.PM, \qquad \ldots(1)$$

PM being the perpendicular on the directrix.

Now $\quad PM=NZ=SZ-SN.$

But $\quad SL=e.SZ$

or $\quad SZ=\dfrac{l}{e}$, if $SL=l.$

Hence from (1),

$$SP=r=e\,(SZ-SN)$$

$$=e\left(\frac{l}{e}-r\cos\theta\right)$$

$$=l-er\cos\theta$$

or $$r(1+e\cos\theta)=l$$

or $$\frac{l}{r}=1+e\cos\theta,$$

which is a relation in (r, θ), a point on the conic and is consequently the equation to the conic.

Corollary 1. Sometimes the directrix is taken on the left side of the conic as shown in the figure, initial line being positive in the direction SZ'. In this case the vectorial angle of P is measured in the clock-wise direction, so that the equation of this conic can be obtained from the previous equation by putting $180-\theta$ for θ, and the equation becomes

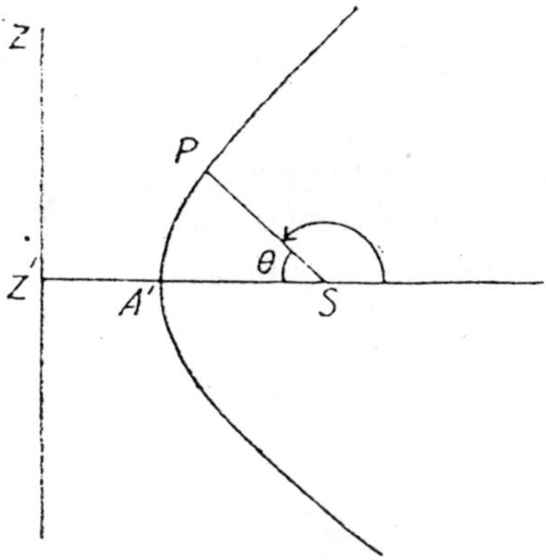

$$\frac{l}{r}=1-e\cos\theta.$$

Corollary 2. If the axis SZ of the conic is inclined at an angle α to the initial line, the equation of the conic becomes

$$\frac{l}{r}=1+e\cos(\theta-\alpha),$$

because the angle between SA and SP in this case becomes $\theta-\alpha$.

10 2. Equation of the directrices :—

(i) Let $P(r, \theta)$ be the coordinates of any point P on the directrix. Then

$$SZ = r\cos\theta.$$

But $SZ = \dfrac{l}{e}$, when l is the latus rectum.

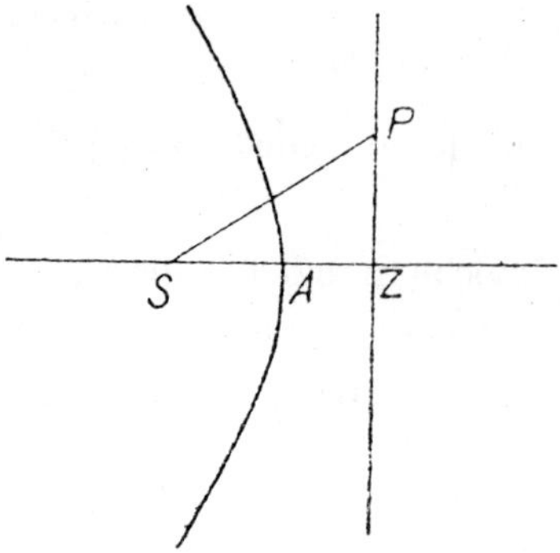

$$\therefore \quad \frac{l}{e} = r\cos\theta$$

or $$\frac{l}{r} = e\cos\theta.$$

This being a relation between r and θ, the coordinates of any point on the locus, is the equation to the locus, which is a straight line here.

The equation to the directrix of the conic $\dfrac{l}{r} = 1 + e\cos(\theta - \alpha)$ is

$$\frac{l}{r} = e\cos(\theta - \alpha).$$

(ii) If the equation to the directrix corresponding to the focus other than the pole is to be found, let us take a point $P'(r, \theta)$ on this directrix. Let the conic be a hyperbola as shown in the figure Then

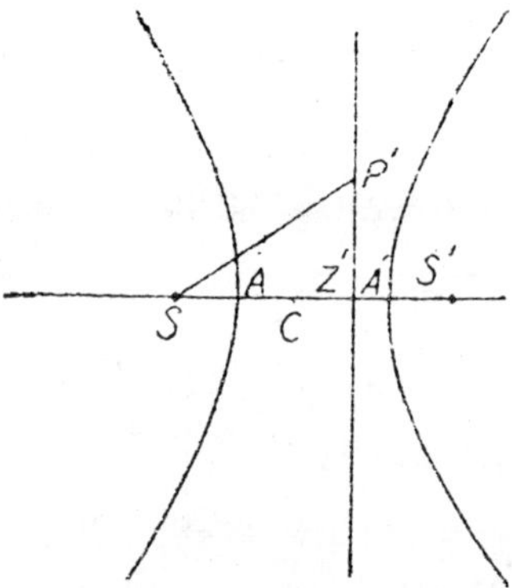

$$r\cos\theta = SZ'$$
$$= SC + CZ'$$
$$= ae + \frac{a}{e}$$

$$=\frac{a\,(e^2+1)}{e}.$$

But
$$\frac{a}{e}=\frac{l}{e\,(e^2-1)},$$

as
$$l=\frac{b^2}{a}=\frac{a^2\,(e^2-1)}{a}=a\,(e^2-1).$$

Hence the equation of the other directrix is

$$r\cos\theta=\frac{l\,(e^2+1)}{e\,(l^2-1)}$$

or
$$\frac{l}{r}=\frac{e\,(e^2-1)}{(1+e^2)}\cos\theta.$$

If the conic is an ellipse as shown in the figure,

$$r\cos(\pi-\theta)=SZ'$$
$$=SC+CZ'$$
$$=ae+\frac{a}{e}$$
$$=\frac{a\,(e^2+1)}{e}.$$

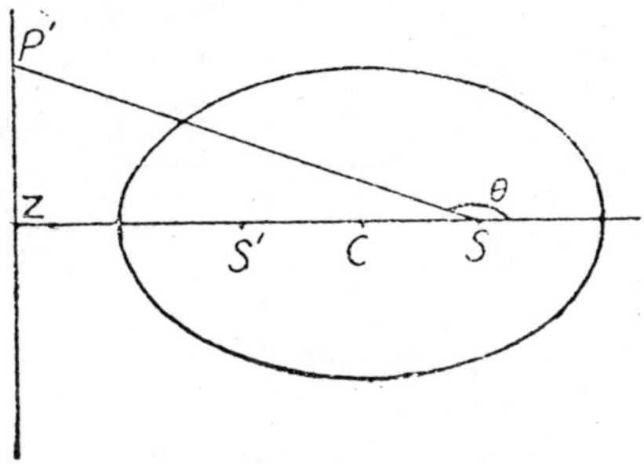

But in this case,

$$l=\frac{b^2}{a}=\frac{a^2\,(1-e^2)}{a}$$
$$=a\,(1-e^2);$$

$$\therefore\quad -r\cos\theta=\frac{l\,(e^2+1)}{e\,(1-e^2)}$$

or
$$\frac{l}{r}=\frac{e\,(e^2-1)}{(e^2+1)}\cos\theta.$$

Hence the equation is the same.

10·3. Tracing the conic $\frac{1}{r}=1+e\cos\theta$.

(i) If $e=0$, the equation becomes $r=l$ which is a circle with its centre at the pole and radius equal to l.

(ii) If $e=1$, the equation becomes

$$\frac{l}{r}=1+\cos\theta.$$

When $\theta=0$, $r=\frac{l}{2}$.

When θ increases, $1+e\cos\theta$ decreases and therefore r increases until r becomes l when $\theta=\frac{\pi}{2}$.

When θ tends to π, $1+e\cos\theta$ tends to 0 and r tends to infinity. The curve is symmetrical about the initial line. Hence the shape of the curve is a parabola as shown below—

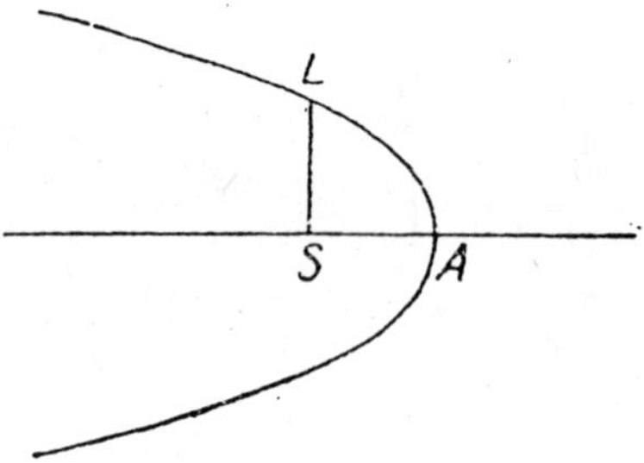

(iii) If $e<1$, then $SA=r=\frac{l}{1+e}$ when $\theta=0$.

As $\cos\theta$ is maximum when $\theta=0$, this is the minimum value of r.

When $\theta=\pi$, $\cos\theta$ is minimum ; hence $SA'=r=\frac{l}{1-e}$ is the maximum value of r.

When $\theta=\frac{\pi}{2}$, $r=l$.

The curve being symmetrical about the initial line, the

shape of the curve is an ellipse as shown in the figure.

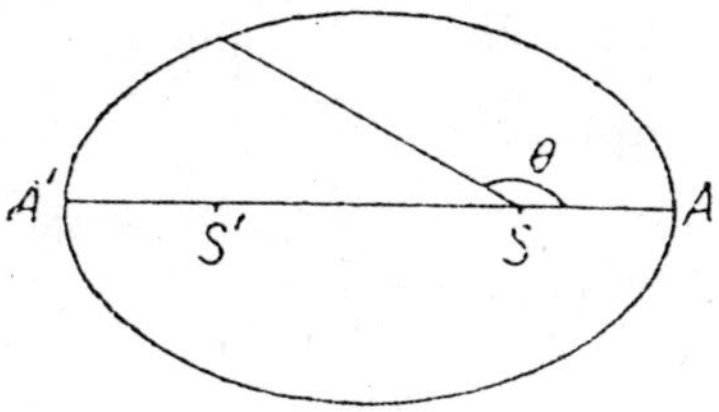

(iv) If $e > 1$, $r=\dfrac{l}{1+e}$ when $\theta=0$ which is again the minimum value SA of r.

When $\theta=\dfrac{\pi}{2}$, $r=l$.

When θ increases beyond $\dfrac{\pi}{2}$, r increases still further until r tends to infinity.

When θ tends to $\cos^{-1}\left(-\dfrac{1}{e}\right)=\alpha$, suppose, this part of the curve is AK.

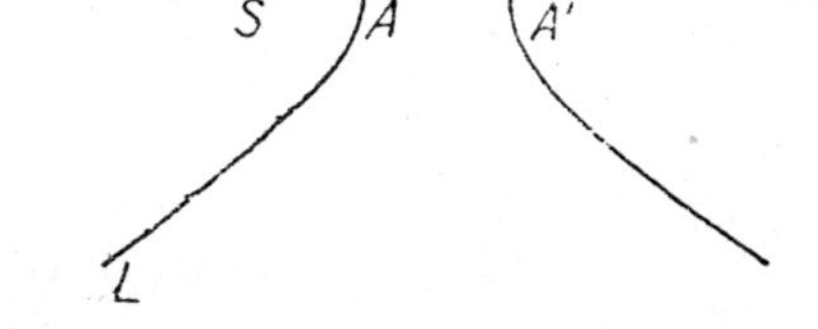

When θ increases from α to π, r decreases from infinity to $-\dfrac{l}{e-1}$ and is negative.

When $r=\dfrac{l}{e-1}$, the value is greater than SA and is SA'. When $\theta=\alpha$ from the left side, r tends to $-\infty$.

Hence the shape of the curve described is $A'K'$.

For values of θ between π and 2π, the curve can be drawn by symmetry and the portions described are similarly

AL and $A'L'$. $A'L'$ is described for θ lying between π and $2\pi-\alpha$ and AL for θ lying between $2\pi-\alpha$ and 2π.

10·4. Asymptotes :—

The asymptotes are two lines through the centre equally inclined to the initial line at angles α, where

$$\cos\alpha=-\frac{1}{e}.$$

The coordinates of the centre are $(ae, 0)$. The perpendicular ST from S on the asymptote makes an angle $-(\pi/2-\alpha)$ with the initial line. The length of the perpendicular is

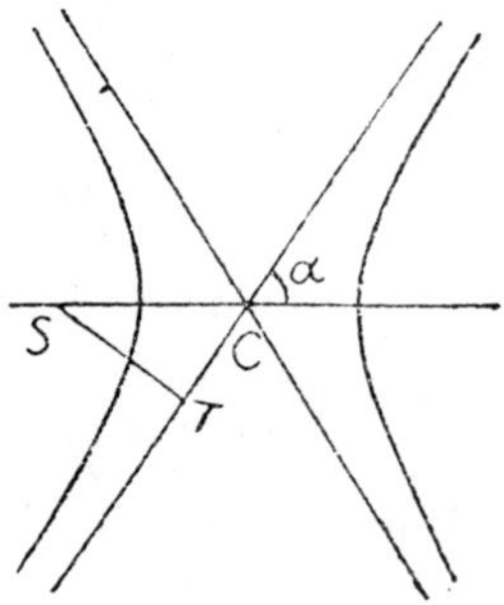

$$SC\sin\alpha=ae\sin\alpha$$

$$=\frac{ae.\sqrt{(e^2-1)}}{e}$$

$$=a\sqrt{(e^2-1)}.$$

The equations of the asymptotes are therefore

$$r\cos\left(\theta-\alpha+\frac{\pi}{2}\right)=a\sqrt{(e^2-1)}$$

and $r\cos\left(\theta+\alpha-\frac{\pi}{2}\right)=a\sqrt{(e^2-1)}.$

These equations can be written as

$$-r\sin(\theta-\alpha)=a\sqrt{(e^2-1)}$$

and $$r\sin(\theta+\alpha)=a\sqrt{(e^2-1)},$$

or $$\frac{l}{r}=-\sqrt{(e^2-1)}\sin(\theta-\alpha),$$

and $$\frac{l}{r}=+\sqrt{(e^2-1)}\sin(\theta+\alpha).$$

Representing them by one equation, we have

$$\frac{l}{r}=\mp\sqrt{(e^2-1)}(\sin\theta\cos\alpha\mp\cos\theta\sin\alpha)$$

$$=\mp\sqrt{(e^2-1)}\left(-\frac{\sin\theta}{e}\mp\cos\theta\,\frac{\sqrt{(e^2-1)}}{e}\right)$$

$$=\frac{\sqrt{(e^2-1)}}{e}\{\sqrt{(e^2-1)}\cos\theta\pm\sin\theta\}.$$

10·5. Equation of a chord of the conic.

Let the conic be $\frac{l}{r}=1+e\cos\theta.$ ...(1)

Let the chord join the points whose vectorial angles are α and β.

As the chord is a straight line, let its polar equation be

$$\frac{l}{r}=A\cos\theta+B\cos\left\{\theta-\frac{\alpha+\beta}{2}\right\} \qquad ...(2)$$

It passes through points whose vectorial angles are α and β; hence

$$\frac{l}{r_1}=A\cos\alpha+B\cos\left(\alpha-\frac{\alpha+\beta}{2}\right)$$

or
$$\frac{l}{r_1}=A\cos\alpha+B\cos\left(\frac{\alpha-\beta}{2}\right). \qquad ...(3)$$

and
$$\frac{l}{r_2}=A\cos\beta+B\cos\left(\frac{\alpha-\beta}{2}\right). \qquad ...(4)$$

The points (r_1, α) and (r_2, β) also lie on the conic (1).

Therefore
$$\frac{l}{r_1}=1+e\cos\alpha$$

and
$$\frac{l}{r_2}=1+e\cos\beta \qquad ...(6)$$

Equating the values of r_1 and r_2 from (3), (5) and (4), (6) respectively, we get

$$A\cos\alpha+B\cos\left(\frac{\alpha-\beta}{2}\right)=1+e\cos\alpha. \qquad ...(7)$$

$$A\cos\beta+B\cos\left(\frac{\alpha-\beta}{2}\right)=1+e\cos\beta. \qquad ...(8)$$

Subtracting (8) from (7), we get

$$A=e.$$

Substituting for A in (7), we get

$$B=\sec\left(\frac{\alpha-\beta}{2}\right).$$

Hence the equation of the chord joining the points α and β is

$$\frac{l}{r}=e\cos\theta+\sec\left(\frac{\alpha-\beta}{2}\right)\cos\left\{\theta-\frac{\alpha+\beta}{2}\right\}. \qquad \ldots(9)$$

Corollary. If the points have vectorial angles $\alpha-\beta$ and $\alpha+\beta$, the equation of the chord joining them is

$$\frac{l}{r}=e\cos\theta+\sec\beta\cos(\theta-\alpha).$$

10·6. Equation of the tangent.

The equation of the chord joining the points α and β is

$$\frac{l}{r}=e\cos\theta+\sec\left(\frac{\alpha-\beta}{2}\right)\cos\left(\theta-\frac{\alpha+\beta}{2}\right).$$

Taking the limit when $\beta\to\alpha$ the equation becomes

$$\frac{l}{r}=e\cos\theta+\cos(\theta-\alpha)$$

which is the equation of the tangent to the conic at the point α.

Corollary. The asymptote is the limiting case of a tangent when r tends to ∞. We know that r tends ∞ when $\cos\alpha=-\dfrac{1}{e}$. Substituting this value of α in the equation of the tangent, we get

$$\frac{l}{r}=e\cos\theta+\cos\left\{\theta-\cos^{-1}\left(-\frac{1}{e}\right)\right\}$$

which is the equation of the asymptotes. Since $\cos^{-1}\left(-\dfrac{1}{e}\right)$ has two values lying between θ and 2π, threre are two asymptotes.

The equation of the asymptotes becomes

$$\frac{l}{r}=e\cos\theta+\cos\theta\left(-\frac{1}{e}\right)\pm\sin\theta\,\frac{\sqrt{(e^2-1)}}{e},$$

or

$$\frac{l}{r}=\frac{\sqrt{(e^2-1)}}{e}\{\sqrt{(e^2-1)}\cos\theta\pm\sin\theta\}.$$

10·7. Equation of the normal.

The equation of the tangent to the conic at the point α is

$$\frac{l}{r}=e\cos\theta+\cos(\theta-\alpha)$$

$$=(e+\cos\alpha)\cos\theta+\sin\theta\sin\alpha,$$

or

$$l=(e+\cos\alpha)\,r\cos\theta+\sin\alpha.r\sin\theta$$

or

$$l=(e+\cos\alpha)\,x+\sin\alpha.y, \qquad \ldots(1)$$

changing into cartesians.

Any line perpendicular to (1) is

$$k=(e+\cos\alpha)\,y-x\sin\alpha$$

or

$$k=r\sin\theta\,(e+\cos\alpha)-r\cos\theta.\sin\alpha.$$

$$=e.r\sin\theta+r\sin(\theta-\alpha).$$

As it passes through the point (r_1, α) where r_1 is given by

$$\frac{l}{r_1}=1+e\cos\alpha,$$

we have

$$k=er_1\sin\alpha$$

$$=\frac{e\sin\alpha.l}{(1+e\cos\alpha)}.$$

Hence the equation of the normal at the point α is

$$\frac{e\sin\alpha.l}{1+e\cos\alpha}=er\sin\theta+r\sin(\theta-\alpha)$$

or

$$\frac{l}{r}\cdot\frac{e\sin\alpha}{1+e\cos\alpha}=e\sin\theta+\sin(\theta-\alpha).$$

10·8. Equation of the polar.

We know that the polar of a point (r_1, θ_1) is the chord of contact of tangents drawn to the conic from (r_1, θ_1).

Let the vectorial angles of the points of contact be α and β. Then the equation of the chord joining α and β is given by

$$\frac{l}{r}=e\cos\theta+\sec\left(\frac{\beta-\alpha}{2}\right)\cos\left\{\theta-\frac{\alpha+\beta}{2}\right\}. \quad \ldots(1)$$

But the tangents at α and β pass through the point (r_1, θ_1).

Hence
$$\frac{l}{r_1}=e\cos\theta_1+\cos(\theta_1-\alpha) \quad \ldots(2)$$

and
$$\frac{l}{r_1}=e\cos\theta_1+\cos(\theta_1-\beta). \quad \ldots(3)$$

From (2) and (3), we get

$$\cos(\theta_1-\alpha)=\cos(\theta_1-\beta)$$

so that
$$\theta_1-\alpha=-(\theta_1-\beta) \text{ as } \alpha\neq\beta$$

or
$$\theta_1=\frac{\alpha+\beta}{2} \quad \ldots(4)$$

From (2) and (4), we get

$$\frac{l}{r_1}=e\cos\theta_1+\cos\frac{\beta-\alpha}{2}. \quad \ldots(5)$$

Substituting for $\frac{\beta-\alpha}{2}$ and $\frac{\alpha+\beta}{2}$ from (4) and (5) in (1), we get

$$\frac{l}{r}=e\cos\theta+\frac{1}{\left(\frac{l}{r_1}-e\cos\theta_1\right)}\cos(\theta-\theta_1)$$

or
$$\left(\frac{l}{r}-e\cos\theta\right)\left(\frac{l}{r_1}-e\cos\theta_1\right)=\cos(\theta-\theta_1)$$

which is the equation of the polar of of (r_1, θ_1).

10·9. Solved Examples.

Ex. 1. *If PSP' and QSQ' are two perpendicular focal chords of a conic, then*

(a) $$\frac{1}{SP}+\frac{1}{SP'}=\frac{2}{l}$$

where l is the latus rectum. (Agra 1947)

(b) $$\frac{1}{PS.SP'}+\frac{1}{QS.SQ'}=\text{constant.}$$ (Agra 1949)

(a) Let the conic be $l/r=1+e\cos\theta$ and let the vectorial angle of P be α. Then the vectorial angle of P' is $\pi+\alpha$. Now SP is the radius vector corresponding to a point P whose vectorial angle is α and this point lies on the conic.

$$\therefore \quad \frac{l}{SP}=1+e\cos\alpha$$

Similarly $$\frac{l}{SP'}=1+e\cos(\pi+\alpha).$$

$$\therefore \quad \frac{l}{SP}+\frac{l}{SP'}=1+e\cos\alpha+1+e\cos(\pi+\alpha)$$
$$=1+e\cos\alpha+1-e\cos\alpha$$
$$=2,$$

or $$\frac{1}{SP}+\frac{1}{SP'}=\frac{2}{l}.$$

(b) QSQ' is a chord perpendicular to PSP'. Hence the vectorial angles of Q and Q' are $\frac{\pi}{2}+\alpha$ and $\frac{3\pi}{2}+\alpha$ respectively.

$$\therefore \quad \frac{l}{SQ}=1+e\cos\left(\frac{\pi}{2}+\alpha\right)$$
$$=1-e\sin\alpha$$

and $$\frac{l}{SQ'}=1+e\cos\left(\frac{3\pi}{2}+\alpha\right)$$
$$=1+e\sin\alpha.$$

$$\therefore \quad \frac{l^2}{SP.SP'}+\frac{l^2}{SQ.SQ'}=1-e^2\cos^2\alpha+1-e^2\sin^2\alpha$$

$$=2-e^2$$

or $$\frac{1}{SP.SP'}+\frac{1}{SQ.SQ'}=\frac{2-e^2}{l^2}$$

$$=\text{constant.}$$

Ex. 2. *The tangent at the point P on a conic cuts a directrix in the point Q. Show that PQ subtends a right angle at the corresponding focus.* *(Poona 1950)*

Let the equation of the conic be $l/r=1+e\cos\theta$ and let the vectorial angle of P be α.

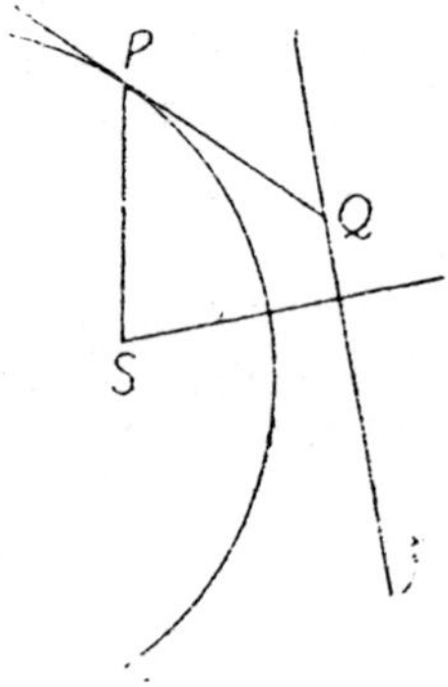

Then the equation of the tangent at P is $\dfrac{l}{r}=e\cos\theta+\cos(\theta-\alpha)$. ...(1)

The equation of the directrix is

$$\frac{l}{r}=e\cos\theta. \qquad ...(2)$$

The tangent at P meets the directrix at Q whose vectorial angle is given from (1) and (2) by

$$\cos(\theta-\alpha)=0 \quad \text{or} \quad \theta=\pi/2+\alpha.$$

Hence the angle PSQ is $\pi/2$.

Ex. 3. *Find the condition that the line $l/r=A\cos\theta+B\sin\theta$ may be a tangent to the conic $l/r=1+e\cos\theta$.* *(Punjab 1951)*

Suppose the given line

$$\frac{l}{r}=A\cos\theta+B\sin\theta \qquad ...(1)$$

is a tangent to the conic at the point α.

Then the equation of the tangent at α is also

$$\frac{l}{r}=e\cos\theta+\cos(\theta-\alpha)$$

or $$\frac{l}{r}=(\cos\alpha+e)\cos\theta+\sin\theta\sin\alpha. \qquad ...(2)$$

Now (1) and (2) represent the same line. Hence comparing coefficients, we get

$$A=\cos\alpha+e \qquad ...(3)$$

and $$B=\sin\alpha. \qquad ...(4)$$

Eliminating α between (3) and (4), we get

$$(A-e)^2+B^2=1$$

which is the required condition.

Ex. 4. *PSP' is a focal chord of a conic. Prove that the angle between the tangents at P and P' is* $\tan^{-1}\frac{2e\sin\alpha}{1-e^2}$ *where* α *is the angle between the chord and the major axis.* (*Agra 1947*)

Taking major axis as the initial line, let the equation of the conic be $\frac{l}{r}=1+e\cos\theta$.

Let α be the vectorial angle of P.

Then the vectorial angle of P' is $\pi+\alpha$.

The equation of the tangent at P is

$$\frac{l}{r}=e\cos\theta+\cos(\theta-\alpha)$$

or $$\frac{l}{r}=(\cos\alpha+e)\cos\theta+\sin\theta\sin\alpha.$$

Transforming the above equation to cartesian coordinates, we get

$$l=(\cos\alpha+e)\,x+y\sin\alpha. \qquad ...(1)$$

The direction tangent of this line is

$$-\frac{(\cos\alpha+e)}{\sin\alpha}=m_1 \qquad \text{(suppose)}.$$

Similarly the equation of the tangent at P' is

$$\frac{l}{r}=e\cos\theta+\cos(\theta-\pi-\alpha)$$

or $$\frac{l}{r}=e\cos\theta-\cos(\theta-\alpha)$$

or $$\frac{l}{r}=(e-\cos\alpha)\cos\theta-\sin\theta\sin\alpha.$$

Transforming to cartesian coordinates, it becomes

$$l=(e-\cos\alpha)\,x-y\sin\alpha.$$

The direction tangent of this line is

$$\frac{e-\cos\alpha}{\sin\alpha}=m_2 \qquad \text{(suppose).}$$

The angle ϕ between the tangents is given by

$$\tan\phi=\frac{m_2-m_1}{1+m_1m_2}$$

$$=\frac{\dfrac{e-\cos\alpha+\cos\alpha+e}{\sin\alpha}}{1-\dfrac{e^2-\cos^2\alpha}{\sin^2\alpha}}$$

$$=\frac{2e\sin\alpha}{1-e^2}.$$

$$\therefore\quad \phi=\tan^{-1}\frac{2e\sin\alpha}{1-e^2}.$$

Ex. 5. *S is the focus and P, Q are two points on a conic Σ such that the angle PSQ is constant. Prove that the locus of the intersection of tangents at P and Q is a conic Σ_1, and that the line PQ touches another conic Σ_2. Show that all the three conics Σ, Σ_1 and Σ_2 have S as a common focus and that the corresponding directrix is also common.* (*Poona 1959*)

Let the vectorial angles of P and Q be $\beta+\alpha$ and $\beta-\alpha$.

Then the tangents at P and Q are

$$\frac{l}{r}=e\cos\theta+\cos(\theta-\beta-\alpha) \qquad \ldots(1)$$

and $$\frac{l}{r}=e\cos\theta+\cos(\theta-\beta+\alpha). \qquad \ldots(2)$$

The vectorial angle of the point of intersection of (1) and (2) is given by

$$\cos(\theta-\beta-\alpha)=\cos(\theta-\beta+\alpha)$$

or $$\theta-\beta-\alpha=-(\theta-\beta+\alpha)$$

or $$\theta=\beta.$$

Substituting in (1), we get

$$\frac{l}{r}=e\cos\theta+\cos\alpha$$

or $$\frac{l\sec\alpha}{r}=1+(e\sec\alpha)\cos\theta,$$

which is the locus of point of intersection of tangents at P and Q. This is clearly a conic having the same focus.

The directrix of this conic is

$$\frac{l\sec\alpha}{r}=e\sec\alpha\cos\theta.$$

or $$\frac{l}{r}=e\cos\theta,$$

which is the same as the directrix of the original conic.

The equation of the line PQ is

$$\frac{l}{r}=e\cos\theta+\sec\alpha\cos(\theta-\beta)$$

or $$\frac{l\cos\alpha}{r}=(e\cos\alpha)\cos\theta+\cos(\theta-\beta),$$

which is a tangent to the conic

$$\frac{l\cos\alpha}{r}=1+e\cos\alpha\cos\theta$$

at the point β.

The above conic also has the same focus and directrix as the original conic.

Ex. 6. *Obtain the polar equation to the circle circumscribing the triangle formed by the tangents to the parabola*

$$\frac{2a}{r}=1+\cos\theta$$

at the points 2α, 2β, 2γ and show that it passes through the focus of the parabola. *(Benaras 1960)*

The tangents at the points α, β, γ are

$$\frac{2a}{r}=\cos\theta+\cos(\theta-\alpha),$$

$$\frac{2a}{r}=\cos\theta+\cos(\theta-\beta),$$

and
$$\frac{2a}{r}=\cos\theta+\cos(\theta-\gamma).$$

The vertices of the triangle are the points of intersection of the tangents which are given by

$$\theta=\frac{\alpha+\beta}{2} \text{ and } \frac{2a}{r}=\cos\frac{\beta+\alpha}{2}+\cos\frac{\beta-\alpha}{2}$$

or
$$\theta=\frac{\alpha+\beta}{2} \text{ and } r=a\sec\frac{\alpha}{2}\sec\frac{\beta}{2}, \quad \ldots(1)$$

$$\theta=\frac{\beta+\gamma}{2} \text{ and } r=a\sec\frac{\beta}{2}\sec\frac{\gamma}{2}, \quad \ldots(2)$$

and
$$\theta=\frac{\gamma+\alpha}{2} \text{ and } r=a\sec\frac{\gamma}{2}\sec\frac{\alpha}{2}. \quad \ldots(3)$$

All these three points satisfy the equation

$$r=a\sec\frac{\alpha}{2}\sec\frac{\beta}{2}\sec\frac{\gamma}{2}\cos\left(\theta-\frac{\alpha+\beta+\gamma}{2}\right)$$

which is obviously a circle passing through the origin, which is the focus of the parabola.

Ex. 7. *A conic is described having the same focus and eccentricity as the conic $\frac{l}{r}=1+e\cos\theta$, and the two conics tonch at the point $\theta=\alpha$; prove that the length of its latus rectum is*

$$\frac{2l(1-e^2)}{e^2+2e\cos\alpha+1}.$$

(Karnatac 1958, Aligarh 1953)

Let the conic having the same focus and eccentricity as the conic

$$\frac{l}{r}=1+e\cos\theta \qquad \ldots(1)$$

be

$$\frac{k}{r}=1+e\cos(\theta-\lambda), \qquad \ldots(2)$$

where k is its semi latus rectum.

These conics touch at the point α.

Hence the tangents to these conics at the point α must be the same.

The equations of tangents at α are

$$\frac{l}{r}=e\cos\theta+\cos(\theta-\alpha) \qquad \ldots(3)$$

and

$$\frac{k}{r}=e\cos(\theta-\lambda)+\cos(\theta-\alpha). \qquad \ldots(4)$$

The above equations can be written as

$$\frac{l}{r}=(\cos\alpha+e)\cos\theta+\sin\alpha\sin\theta \qquad \ldots(5)$$

and

$$\frac{k}{r}=(e\cos\lambda+\cos\alpha)\cos\theta+(e\sin\lambda+\sin\alpha)\sin\theta. \qquad \ldots(6)$$

Both represent the same line ; hence comparing coefficients, we get

$$\frac{l}{k}=\frac{\cos\alpha+e}{\cos\alpha+e\cos\lambda}=\frac{\sin\alpha}{\sin\alpha+e\sin\alpha}. \qquad \ldots(7)$$

Considering first and second of the above equations, we get

$$el\cos\lambda=(k-l)\cos\alpha+ek. \qquad \ldots(8)$$

The first and third equations give

$$el\sin\lambda=(k-l)\sin\alpha. \qquad \ldots(9)$$

Eliminating λ from (8) and (9), we get

$$e^2l^2=(k-l)^2+e^2k^2+2ek(k-l)\cos\alpha$$

or $$(k-l)^2+e^2(k^2-l^2)+2ek(k-l)\cos\alpha=0$$

or $$k-l+e^2(k+l)+2ek\cos\alpha=0$$

or $$k(1+e^2+2e\cos\alpha)=l(1-e^2)$$

or $$k=\frac{l(1-e^2)}{e^2+2e\cos\alpha+1}.$$

Hence the latus rectum is $\dfrac{2l(1-e^2)}{e^2+2e\cos\alpha+1}$.

Ex. 8. *If the normals at α, β, γ on $\dfrac{l}{r}=1+\cos\theta$ meet in the point (ρ, ϕ), show that $2\phi=\alpha+\beta+\gamma$.* (*Agra 1942*)

The equation to the normal at the point δ is given by

$$\frac{l\sin\delta}{r(1+\cos\delta)}=\sin\theta+\sin(\theta-\delta).$$

This passes through the point (ρ, ϕ),

$$\therefore \quad \frac{l\sin\delta}{\rho(1+\cos\delta)}=\sin\phi+\sin(\phi-\delta)$$

or $$\frac{l}{\rho}\tan\frac{\delta}{2}=\sin\phi(1+\cos\delta)-\cos\phi\sin\delta$$

or $$\frac{l.\tan\delta/2}{1+\cos\delta}=\rho\sin\phi-\rho\cos\phi\frac{\sin\delta}{1+\cos\delta}$$

or $$l\tan\frac{\delta}{2}\sec^2\frac{\delta}{2}=2\rho\sin\phi-2\rho\cos\phi\tan\frac{\delta}{2}.$$

Putting $\tan\dfrac{\delta}{2}=t$ in the above equation, we get

$$lt(1+t^2)=2\rho\sin\phi-2\rho t\cos\phi$$

or $$lt^3+(l+2\rho\cos\phi)t-2\rho\sin\phi=0. \qquad \ldots(1)$$

This is a cubic equation in t. Let the three roots be t_1, t_2, t_3, where

$$t_1=\tan\frac{\alpha}{2}, \quad t_2=\tan\frac{\beta}{2} \text{ and } t_3=\tan\frac{\gamma}{2}.$$

Hence

$$t_1+t_2+t_3=\tan\frac{\alpha}{2}+\tan\frac{\beta}{2}+\tan\frac{\gamma}{2}=0. \qquad \ldots(2)$$

$$t_1t_2+t_2t_3+t_3t_1=\tan\frac{\alpha}{2}\tan\frac{\beta}{2}+\tan\frac{\beta}{2}\tan\frac{\gamma}{2}+\tan\frac{\gamma}{2}\tan\frac{\alpha}{2}$$

$$=\frac{l+2\rho\cos\phi}{l} \qquad \ldots(3)$$

and $$t_1t_2t_3=\tan\frac{\alpha}{2}\tan\frac{\beta}{2}\tan\frac{\gamma}{2}=\frac{2\rho\sin\phi}{l} \qquad \ldots(4)$$

Now $\tan\left(\frac{\alpha}{2}+\frac{\beta}{2}+\frac{\gamma}{2}\right)$

$$=\frac{\tan\frac{\alpha}{2}+\tan\frac{\beta}{2}+\tan\frac{\gamma}{2}-\tan\frac{\alpha}{2}\tan\frac{\beta}{2}\tan\frac{\gamma}{2}}{1-\left(\tan\frac{\alpha}{2}\tan\frac{\beta}{2}+\tan\frac{\beta}{2}\tan\frac{\gamma}{2}+\tan\frac{\gamma}{2}\tan\frac{\alpha}{2}\right)}$$

$$=\frac{-\frac{2\rho\sin\phi}{l}}{1-\frac{l+2\rho\cos\phi}{l}}$$

$$=\tan\phi.$$

$$\therefore \quad \frac{\alpha}{2}+\frac{\beta}{2}+\frac{\gamma}{2}=\phi$$

or $$\alpha+\beta+\gamma=2\phi.$$

Examples XVIII

1. Show that the equations

$$\frac{l}{r}=1+e\cos\theta \quad \text{and} \quad \frac{l}{r}=-1+e\cos\theta$$

represent the same conic. *(Agra 1949)*

2. In any conic prove that the sum of the reciprocals of two perpendicular focal chords is constant. *(Benares 1945)*

3. A circle of given radius passing through the focus of a

given conic intersects in A, B, C, D; sohw that

$SA.SB.SC.SD$ is constant. *(Allahabad 1943)*

4. A circle passing through the focus of a conic whose latus rectum is $2l$ meets the conic in four points whose distances from the focus are r_1, r_2, r_3 and r_4: prove that

$$\frac{1}{r_1}+\frac{1}{r_2}+\frac{1}{r_3}+\frac{1}{r_4}=\frac{2}{l}.$$ *(Utkal 1946)*

5. If two conics have a common focus, show that two of their common chords will pass through the point of intersection their directrices.

6. Prove that the locus of the middle points of focal chords of a conic section is a conic section. *(Agra 1948)*

7. PSQ, PHR are two chords of an ellipse through the foci S, H; show that $\frac{PS}{SQ}+\frac{PH}{PR}$ is independent of the position of P. *(Aligarh 1949, Agra 1945)*

8. A straight line drawn through the common focus S of a number of conics meets them in points $P_1, P_2..$ On it is taken a point Q such that the reciprocal of SQ is equal to the sum of the reciprocals of $SP_1, SP_2...$ Prove that the locus of Q is a conic section whose focus is S and the reciprocal of whose latus rectum is equal to the sum of the reciprocals of the latera recta of the given conics. *(Agra 1956)*

9. Find the locus of the pole of a chord of the conic $\frac{l}{r}=1+e\cos\theta$ which subtends a constant angle 2α at the focus. *(Agra 1951, Punjab 1950)*

10. Show that the equation of the locus of the point of intersection of two tangents to $\frac{l}{r}=1+e\cos\theta$ which are at

right angles to one another is

$$r^2 (e^2-1)-2lex \cos\theta+2l^2=0.$$

(Agra 1954, Aligarh 55, Kashmir 50)

11. Two parabolas have a common focus and axes inclined at an angle α. Prove that the locus of the point of intersection of two perpendicular tangents one to each of the parabolas is a conic. *(Agra 1952)*

12. If the tangent at any point of an ellipse makes an angle α with its major axis and an angle β with the focal radius to the point of contact, show that

$$e \cos\alpha=\cos\beta.$$

13. A focal chord SP of an ellipse is inclined at an angle α to the major axis. Show that the perpendicular from the focus on the tangent at P makes an angle

$$\tan^{-1}\left\{\frac{\sin\alpha}{e+\cos\alpha}\right\}$$

with the axis. *(Travancore 1946)*

14. The tangents at P and Q to a conic having S as a focus meet in T. Prove that

(i) $ST^2=SP.SQ$ when the conic is a parabola; and

(ii) $$\frac{1}{SP.SQ}-\frac{1}{ST^2}=\frac{1}{b^2}\sin^2\frac{PSQ}{2}$$

when the conic is central and b is its semi minor axis.

(Agra 1955)

15. Prove that two conics

$$\frac{l_1}{r}=1-e_1\cos\theta$$

and $$\frac{l_2}{r}=1-e_2\cos(\theta-\alpha)$$

will touch one another, if

$$l_1^2(1-e_2^2)+l_2^2(1-e_1^2)=2l_1l_2(1-e_1e_2\cos\alpha).$$

(Agra 1960)

16. If a chord of a conic section subtend a constant angle 2α at the focus, prove that the locus of the point where it meets the internal bisector of the angle 2α, is the conic section

$$\frac{l \cos \alpha}{r}=1-e \cos \alpha \cos \theta.$$

(Agra 1961)

17. If the tangents at any two points P and Q of a conic meet in a point T, and if the straight line PQ meets the directrix corresponding to S in a point K, prove that the angle KST is a right angle. *(Punjab 1945)*

18. Show that the equation to a circle circumscribing the triangle formed by the three tangents to the parabola $\frac{2a}{r}=1-e \cos \theta$ drawn at the points α, β, γ is

$$r=a \operatorname{cosec} \frac{\alpha}{2} \operatorname{cosec} \frac{\beta}{2} \operatorname{cosec} \frac{\gamma}{2} \sin\left(\frac{\alpha+\beta+\gamma}{2}-\theta\right).$$

(Punjab 1952)

19. If A, B, C be any three points on a parabola, and the tangents at these points form a triangle $A'B'C'$, show that $SA.SB.SC=SA'.SB'.SC'$, S being the focus of the parabola.

20. P, Q, R are three points on the conic $\frac{l}{r}=1+e \cos \theta$, the focus S being the pole. SP and SR meet the tangent at Q in M and N, so that $SM=SN=l$. Prove that PR touches the conic $\frac{l}{r}=1+2e \cos \theta$.

(Agra 1940, Lucknow 1942)

21. If POP' be a chord of a conic through a fixed point O, prove that $\tan \frac{P'SO}{2} \tan \frac{PSO}{2}=$constant, S being the focus of the conic.

22. Two equal ellipses of eccentricity e are placed with their axes at right angles and they have one focus S in common. If PQ be a common tangent, show that the angle PSQ is equal to $2 \sin^{-1}\left(\frac{e}{\sqrt{2}}\right)$.

(Punjab 1954, Agra 1958)

23. Two conics have a common focus about which one is turned. Show that two of their common chords will touch conics having the fixed focus for focus.

(Sagar 1949)

24. Show that the locus of the point of intersection of two tangents to a parabola $\frac{l}{r}=1+\cos\theta$ which cut one another at a constant angle α is the hyperbola

$$\frac{l}{r}=\cos\alpha+\cos\theta.$$

25. If SM and SN be perpendiculars from the focus S on the tangent and normal at any point on the conic and SY perpendicular to MN, obtain in polar coordinates the locus of Y. *(Delhi 1945)*

26. If the normal at L, one of the extremities of the latus rectum of the conic $\frac{l}{r}=1+e\cos\theta$, meets the curve again at Q, show that $SQ=\frac{l\,(1+3e^2+e^4)}{1+e^2-e^4}$. *(Andhra 1946)*

27. Prove that the equation to the locus of the foot of the perpendicular from the focus on the conic $\frac{l}{r}=1+e\cos\theta$ on a tangent to it is

$$r^2(e^2-1)-2ler\cos\theta+l^2=0.$$

Discuss the particular case when $e=1$. *(Nagpur 1945)*

28. If ellipses of latera recta l, l_1 and eccentricities e, e_1 have a common focus and touch each other, show that the cosine of the angle between their axes is

$$\frac{\{(l-l_1)^2-(e^2l_1^2+e_1^2l^2)\}}{2ee_1ll_1}.$$

29. If α, β, γ are the vectorial angles of three points on the conic $\frac{2a}{r}=1-\cos\theta$, the normals at which are concurrent, show that $\Sigma \cot\frac{\alpha}{2}=0$.

If the conic be $\frac{l}{r}=1+e\cos\theta$, prove that

$$\frac{\Sigma \tan\frac{\beta}{2}\tan\frac{\gamma}{2}}{\Sigma \cot\frac{\beta}{2}\cot\frac{\gamma}{2}}=\left(\frac{1+e}{1-e}\right)^2.$$

30. If the normals at three points of the parabola

$$r=a\operatorname{cosec}^2\frac{\theta}{2}$$

whose vectorial angles are α, β, γ meet in a point whose vectorial angle is δ, prove that $2\delta=\alpha+\beta+\gamma-\pi$.

31. Show that if the normals at the four points whose vectorial angles are α, β, γ, δ on the conic

$$\frac{l}{r}=1+e\cos\theta,$$

meet in a point (ρ, ϕ), then

$$\alpha+\beta+\gamma+\delta-2\phi=(2n+1)\pi.$$

32. Show that the equation to the circle which touches the curve $\frac{l}{r}=1+e\cos\theta$ at the point α and passes through the focus is $r(1+e\cos\alpha)^2=l\cos(\theta-\alpha)+el\cos(\theta-2\alpha)$.

33. An ellipse and a parabola have a common focus S and intersect in two real points P and Q of which P is the vertex of the parabola. If e be the eccentricity of the ellipse and α the angle which SP makes with the major axis, prove that $\frac{SQ}{SP}=1+\frac{4e^2 \sin^2 \alpha}{(1-e \cos \alpha)^2}$.

34. Two chords QP, PR of a conic subtend equal angles at the focus; prove that the chord QR and the tangent at P intersect on the directrix.

35. Show that the equation of pair of tangents drawn to the conic $\frac{l}{r}=1+e \cos \theta$ from the point (r', θ') is

$$(S^2-1)(S'^2-1)=P^2$$

where $S \equiv \frac{l}{r}-e \cos \theta,\ S'=\frac{l}{r'}-e \cos \theta'$

and $P \equiv \left(\frac{l}{r}-e \cos \theta\right)\left(\frac{l}{r'}-e \cos \theta'\right)-\cos(\theta-\theta')$.

Hence or otherwise prove that the asymptotes of the conic are $\frac{el}{r}=(e^2-1) \cos \theta \pm \sqrt{(e^2-1)} . \sin \theta$.

(Poona 1952)

CHAPTER XI

THE GENERAL CONIC

11·1. Conic through five points.

We know that a conic is represented by the general equation of second degree, which is

$$ax^2+2hxy+by^2+2gx+2fy+c=0 \qquad \ldots \quad \ldots (1)$$

This equation contains five independent constants as is seen by dividing the equation by any one of the constants a, b, c, f, g or h.

If the conic passes through five given points, we have five independent equations to determine the five constants. Hence the general conic can be made to satisfy five independent conditions.

The equations will not be independent if four of the given points lie on a straight line. Suppose the line passing through the four given collinear points is the x-axis and let the coordinates of these points be $(0, 0)$, $(p, 0)$, $(q, 0)$, $(r, 0)$.

As these points lie on the conic, they satisfy its equation (1), and we have

$$c=0 \qquad \ldots (2)$$

$$ap^2+2gp=0 \qquad \ldots (3)$$

$$aq^2+2gq=0 \qquad \ldots (4)$$

and

$$ar^2+2gr=0. \qquad \ldots (4)$$

Since p and q are not equal to zero and are unequal, a and g must be separately zero. Equation (5) is therefore also satisfied and the conic becomes

$$2hxy+by^2+2fy=0.$$

If the fifth point on the conic is (s, t), we have

$$2hst+bt^2+2ft=0$$

or
$$t\,(2hs+bt+2f)=0.$$

Since t is not equal to zero, the equation
$$2hs+bt+2f=0$$
which is not sufficient to determine the ratio $h:b:f$.

Therefore, an infinite number of conics can be made to pass through five points, four of which are collinear.

If only three given points are collinear, the conic will obviously be a pair of straight lines of which one passes through three given points and the other through the remaining two points.

11·2. Tangent to the conic.

Let (x_1, y_1) be the point at which the equation to the tangent to the conic
$$ax^2+2hxy+by^2+2gx+2fy+c=0 \qquad \ldots(1)$$
is to be found.

Let (x_2, y_2) be another point on the conic. Both these points lie on the conic (1). Hence
$$ax_1{}^2+2hx_1y_1+by_1{}^2+2gx_1+2fy_1+c=0 \qquad \ldots(2)$$
and
$$ax_2{}^2+2hx_2y_2+by_2{}^3+2gx_2+2fy_2+c=0. \qquad \ldots(3)$$

Subtracting (3) from (2), we get
$$a\,(x_1{}^2-x_2{}^2)+2h\,(x_1y_1-x_2y_2)+b\,(y_1{}^2-y_2{}^2)+2g\,(x_1-x_2)$$
$$+2f\,(y_1-y_2)=0 \ldots(4)$$

But $2\,(x_1y_1-x_2y_2)=(x_1+x_2)\,(y_1-y_2)+(x_1-x_2)\,(y_1+y_2)$ so that (4) becomes
$$(x_1-x_2)\,\{a\,(x_1+x_2)+h\,(y_1+y_2)+2g\}$$
$$+(y_1-y_2)\,\{b\,(y_1+y_2)+h\,(x_1+x_2)+2f\}=0$$

or
$$\frac{y_2-y_1}{x_2-x_1}=-\frac{a\,(x_1+x_2)+h\,(y_1+y_2)+2g}{h\,(y_1+y_2)+h\,(x_1+x_2)+2f} \qquad \ldots(5)$$

The equation to the line joining (x_1, y_1) and (x_2, y_2) is
$$y-y_1=\frac{y_2-y_1}{x_2-x_1}\,(x-x_1)$$

which on account of (5) becomes

$$y-y_1=-\frac{a(x_1+x_2)+b(y_1+y_2)+2g}{b(y_1+y_2)+h(x_1+x_2)+2f}(x-x_1).$$

Taking the limit when x_2 tends to x_1 and y_2 tends to y_1, the equation to the tangent at (x_1, y_1) becomes

$$y-y_1=-\frac{ax_1+hy_1+g}{hx_1+by_1+f}(x-x_1)$$

or
$$(ax_1+hy_1+g)x+(hx_1+by_1+f)y$$
$$=ax_1^2+2hx_1y_1+by_1^2+gx_1+fy_1$$
$$=-gx_1-fy_1-c \text{ by (2)}.$$

Hence the equation to the tangent to the conic at the point (x_1, y_1) is

$$axx_1+h(xy_1+x_1y)+byy_1+g(x+x_1)+f(y+y_1)+c=0.$$

Thus the equation to the tangent to the conic can be written from the equation of the conic by substituting xx_1 for x^2, yy_1 for y^2, xy_1+x_1y for $2xy$, $x+x_1$ for $2x$ and $y+y_1$ for $2y$.

Cor. 1. If the conic passes through the origin, the tangent at the origin is $gx+fy=0$ which shows that the equation to the tangent at the origin is obtained by equating the terms of the lowest degree in the equation of the conic, to zero.

Cor. 2. The equation to the polar of the point (x_1, y_1) with respect to the conic is the same as the equation to the tangent. It can be shown exactly in the same way as done before in the case of particular conics.

11·3. Diameter.

We have to find the locus of the middle point of a chord of the conic

$$ax^2+2hxy+by^2+2gx+2fy+c=0. \qquad \ldots(1)$$

Let the chord be $\quad y=mx+k \qquad \ldots(2)$

and let (p, q) be its middle point.

The abscissae of the points of intersection of (1) and (2) are given by

$$ax^2+2hx(mx+k)+b(mx+k)^2+2gx+2f(mx+k)+c=0$$

or $$x^2(a+2hm+bm^2)+2x(hk+bmk+g+mf)+bk^2+2fk+c=0 \quad \ldots(3)$$

If x_1 and x_2 are the roots of this equation,

$$x_1+x_2=-\frac{2(hk+bmk+g+mf)}{a+2hm+bm^2}. \quad \ldots(4)$$

As (p, q) is the middle point of the chord,

$$p=\frac{x_1+x_2}{2}=\frac{-(hk+bmk+g+mf)}{a+2hm+bm^2}. \quad \ldots(5)$$

Also (p, q) lies on the chord.

$$\therefore \quad q=mp+k. \quad \ldots(6)$$

Eliminating k which is the variable between (5) and (6), we get

$$p(a+2hm+bm^2)=-(h+bm)(q-mp)-g-fm \quad \ldots(7)$$

which is a relation between p and q.

The locus of (p, q) is therefore

$$x(a+2hm+bm^2)=-(h+bm)(y-mx)-g-fm$$

or $$x(a+hm)+y(h+bm)+g+fm=0.$$

If this is parallel to the straight line

$$y=m_1x,$$

$$m_1=-\frac{a+hm}{h+bm}$$

or $$a+h(m+m_1)+bmm_1=0 \quad \ldots(8)$$

which is the condition that the straight lines $y=mx$ and $y=m_1x$ be parallel to conjugate diameters of the conic.

Note. The equation of the chord whose middle point is (p, q) can be obtained thus :—

Let the equation of the chord be

$$\frac{x-p}{l}=\frac{y-q}{m}=r. \quad \ldots(1)$$

The points of intersection of (1) and the conic are obtained by substituting the values of x and y from (1) in the equation of the conic, so that

$$r^2(al^2+2hlm+bm^2)+2\{(ap+hq+g)l+(hp+bq+f)m\}r$$
$$+(ap^2+2hpq+bq^2+2gp+2fq+c)=0. \quad \ldots(2)$$

The roots of this equation must be equal and opposite as (p, q) is the middle point of the chord, so that the coefficient of r in (2) should be zero, which gives

$$(ap+hq+g)l+(hp+bq+f)m=0. \quad \ldots(3)$$

Eliminating l and m between (1) and (3), we get

$$(x-p)(ap+hq+g)+(y-q)(hp+bq+f)=0$$

or $$axp+h(xq+py)+byq+gx+fy=ap^2+2hpq+bq^2+gp+fq.$$

Adding $gp+fq+c$ to both sides, we get

$$axp+h(xq+py)+byq+g(x+p)+f(y+q)+c$$
$$=ap^2+2hpq+bq^2+2gp+2fq+c$$

or $$T=S_1$$

where $T=axp+h(xq+py)+byq+g(x+p)+f(y+q)+c$ and S_1 is obtained by substituting (p, q) for x and y in the equation to the conic.

11·4. Intersection of two conics.

Let the equations to the conics be

$$S \equiv ax^2+2hxy+by^2+2gx+2fy+c=0 \quad \ldots(1)$$

and $$S_1 \equiv a_1x^2+2h_1xy+b_1y^2+2g_1x+2f_1y+c=0. \quad \ldots(2)$$

Eliminating x between (1) and (2), we get a fourth degree equation in y, which will have four roots real or imaginary. If x^2 is eliminated between (1) and (2), we get a linear equation in x which will give four values of x corresponding to each value of y.

Hence there are four points of intersection.

11·5. Conic through the point of intersection of two conics.

Let the equations to the conics be the same as given in previous section.

Then $$S+\lambda S_1=0 \qquad \ldots(1)$$
is the equation to the conic passing through their points of intersection.

Since S and S_1 are equations of second degree in x and y, (1) is also of second degree in x and y, and therefore represents a conic.

Also (1) is satisfied when both S and S_1 are zero; it is satisfied by points common to S and S_1 whether real or imaginary.

Hence (1) is a conic which passes through the intersection of S and S_1.

Cor. If $S=0$ and $S_1=0$ are rectangular hyperbolas, then
$$a+b=0 \text{ and } a_1+b_1=0.$$

Now in the conic $S+\lambda S_1=0$, the sum of the coefficients of x^2 and y^2 is
$$(a+\lambda a_1)+(b+\lambda b_1)$$
$$=(a+b)+\lambda(a_1+b_1)$$
$$=0,$$
so that $S+\lambda S_1=0$ is also a rectangular hyperbola.

Hence all conics which pass through the intersection of two rectangular hyperbolas are themselves rectangular hyperbolas.

11·6. Straight lines passing through the intersection of two conics.

As before $S+\lambda S_1=0$ is the conic passing through the intersection of $S=0$ and $S_1=0$.

This equation can be written as
$$(a+\lambda a_1)x^2+2(h+\lambda h_1)xy+(b+\lambda b_1)y^2$$
$$+2(g+\lambda g_1)x+2(f+\lambda f_1)y+c+\lambda c_1=0.$$

If it represents a pair of lines

$$(a+\lambda a_1)(b+\lambda b_1)(c+\lambda c_1)+2(f+\lambda f_1)(g+\lambda g_1)(h+\lambda h_1)$$
$$-(a+\lambda a_1)(f+\lambda f_1)^2-(b+\lambda b_1)(g+\lambda g_1)^2$$
$$-(c+\lambda c_1)(h+\lambda h_1)^2=0. \qquad \ldots(1)$$

This is a cubic equation in λ and will give a pair of straight lines corresponding to each value of λ when substituted in $S+\lambda S_1=0$.

Hence there will be three pairs of straight lines.

Since a cubic equation has at least one real root, one pair of real straight lines at least can always be drawn through the intersection of two conics.

11·7. Conic through the intersection of a conic and two given straight lines.

Let $S\equiv ax^2+2hxy+by^2+2gx+2fy+c=0$ be the equation to the conic, and let $u=0$ and $v=0$, where

$$u\equiv lx+my+n$$

and

$$v\equiv l_1x+m_1y+n_1$$

be the equations to two given straight lines.

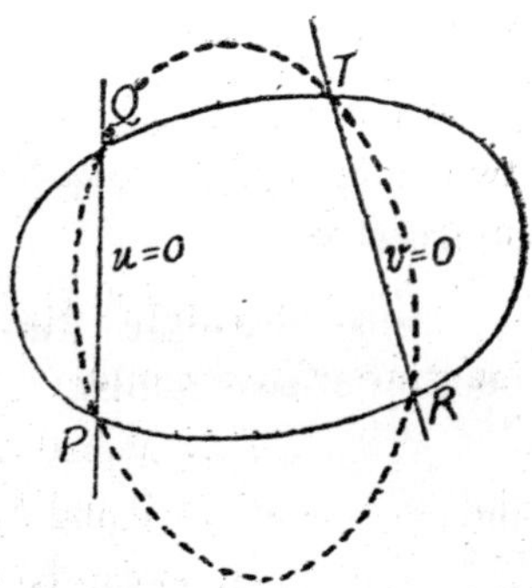

Let the straight line $u=0$ meet the conic in points P and Q and let $v=0$ meet it in points R and T.

The equation to any conic which passes through the points P, Q, R and T is

$$S=\lambda uv, \qquad \ldots(1)$$

for points common to $S=0$ and $u=0$ satisfy it and similarly the points common to $S=0$ and $v=0$ also satisfy it for all values of λ.

To determine λ one more condition is required, and therefore in all there will be five conditions to find the conic completely.

11·8. Particular cases of the equation S=λuv.

(i) Let the straight lines $u=0$ and $v=0$ intersect the conic as shown in the figure so that the points P and R coincide.

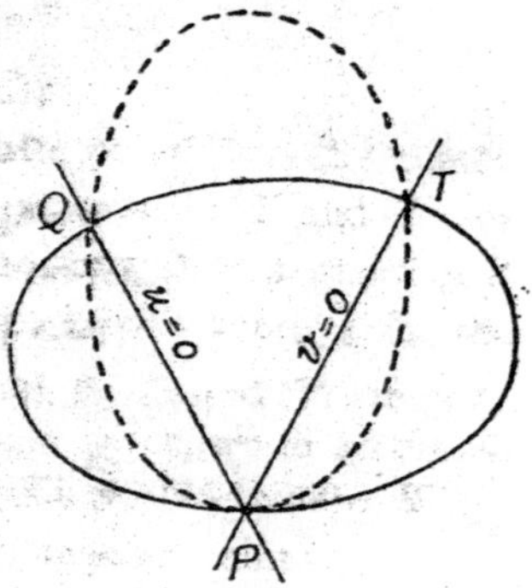

The conic $S=\lambda uv$ therefore passes through two coincident points at P and therefore touches the original conic at P and therefore has a contact of first order with $S=0$ at this point.

(ii) Let the straight lines $u=0$ and $v=0$ coincide. In this case point T also coincides at Q, so that the conic $S+\lambda uv=0$ touches the conic $S\equiv0$ at both points P and Q. The two conics are said to have double contact at P and Q and the equation to the conic through P and Q becomes

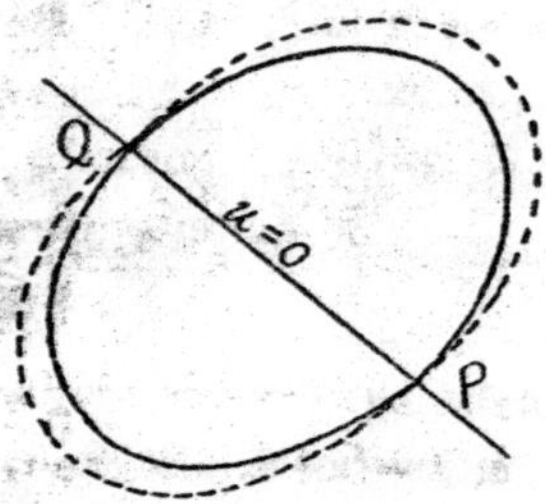

$$S=\lambda u^2.$$

If $u=0$, $v=0$ and $w=0$ be the equations to three straight lines, the conic $vw=\lambda u^2$ touches the conic $vw=0$ where $u=0$ intersects it, that is, it is a conic to which $v=0$ and $w=0$ are tangents and $u=0$ is the chord of contact.

(iii) Let P and Q coincide so that $u=0$ becomes a

tangent to the conic at P. Hence $S=\lambda uv$ touches the original conic $S=0$, where $u=0$ touches it and passes through R and T.

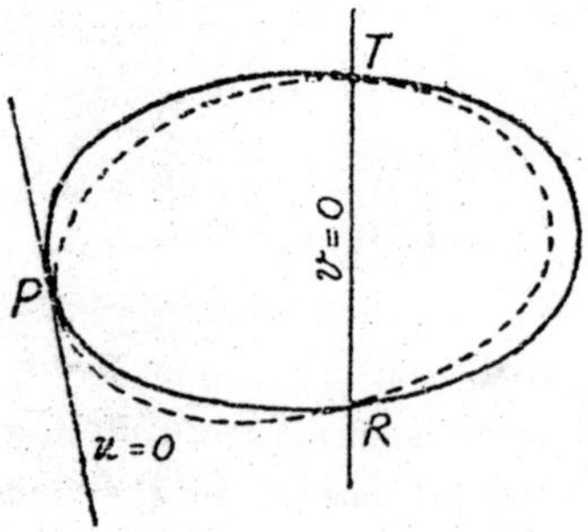

If in addition P and R also coincide then the conic passes through three coincident points at P and through T the point where the line $v=0$ through P cuts the conic $S=0$.

(iv) If $u=0$ and $v=0$ coincide and be tangent to the conic at P, the equation $S=\lambda u^2$ is a conic passing through four coincident points at P, and is said to have a contact of third order with $S=0$.

11·9. Pair of tangents to a conic.

Let the equation to the conic be

$$S\equiv ax^2+2hxy+by^2+2gx+2fy+c=0$$

and let the point from which the pair of tangents is drawn to the conic be (x_1, y_1).

The equation to the chord of contact of (x_1, y_1) with respect to the conic is

$$T\equiv axx_1+h(xy_1+x_1y)+byy_1+g(x+x_1)+f(y+y_1)+c=0. \quad \ldots(1)$$

The pair of tangents can be regarded as a conic which has a double contact with the conic $S=0$ at the points where the chord of contact intersects it. Hence its equation is

$$S+\lambda T^2=0. \quad \ldots(2)$$

But this conic passes through (x_1, y_1).

$$\therefore \quad S_1+\lambda S_1^2=0,$$

where $\quad S_1\equiv ax_1^2+2hx_1y_1+by_1^2+2gx_1+2fy_1+c=0$

or
$$\lambda=-\frac{1}{S_1}.$$

Hence the equation to the pair of tangents from (x_1, y_1) to the conic is

$$SS_1=T^2.$$

11·10. Director circle of a conic.

As in the previous section, the equation to the pair of tangents from (x_1, y_1) to the conic $S=0$ is given by

$$SS_1=T^2. \qquad \ldots(1)$$

If these tangents are at right angles to each other, the sum of coefficients of x^2 and y^2 in (1) should be equal to zero, which gives

$$(a+b)\ S_1-(ax_1+hy_1+g)^2-(hx_1+by_1+f)^2=0$$

or $$(x_1^2+y_1^2)\ (ab-h^2)+2x_1\ (bg-fh)+2y_1\ (af-gh)$$
$$+c\ (a+b)-g^2-f^2=0.$$

Hence the locus of (x_1, y_1) is the circle

$$(x^2+y^2)\ (ab-h^2)+2x\ (bg-fh)+2y\ (af-gh)+c\ (a+b)-g^2-f^2=0.$$

Cor. If the conic is a parabola, $ab=h^2$ and the locus becomes a straight line which is the directrix of the parabola.

11·11. Tangents from focus to a conic.

Let the focus be the origin and let a line drawn perpendicular to the directrix from the focus be the axis of x, so that the equation to the directrix is suppose $x+k=0$.

If e be the eccentricity of the conic, its equation becomes

$$x^2+y^2=e^2\ (x+k)^2$$

or $$x^2\ (1-e^2)+y^2-2e^2xk-e^2k^2=0. \qquad \ldots(1)$$

The equation of the pair of tangents from the origin to the conic (1) is

$$\{x^2\ (1-e^2)+y^2-2e^2xk-e^2k^2\}\ \{-e^2k^2\}=\{-e^2kx-e^2k^2\}^2$$

or $$x^2\ (1-e^2)+y^2-2e^2kx-e^2k^2=-e^2\ (x+k)^2$$

or $$x^2+y^2=0. \qquad \ldots(2)$$

In equation (2) the coefficients of x^2 and y^2 are equal and there is no term containing xy, so that it satisfies the analytical conditions for being a circle.

It can be shown that however the origin and axes of coordinates be changed, the equation of the pair of tangents from focus to the conic will always satisfy the conditions for being a circle.

11·12. Foci of a conic.

Let the conic be

$$S \equiv ax^2+2hxy+by^2+2gx+2fy+c=0. \quad \ldots(1)$$

Let (x_1, y_1) be the focus. Then the equation to the pair of tangents from the focus to the conic is

$$SS_1=T^2, \quad \ldots(2)$$

where $S_1 \equiv ax_1^2+2hx_1y_1+by_1^2+2gx_1+2fy_1+c$

and $T \equiv [x(ax_1+hy_1+g)+y(hx_1+by_1+f)+(gx_1+fy_1+c)]$.

In equation (2) the coefficients of x^2 and y^2 should be equal and the coefficient of xy should be zero.

These conditions give

$$aS_1-(ax_1+hy_1+g)^2=bS_1-(hx_1+by_1+f)^2 \quad \ldots(3)$$

and

$$hS_1=(ax_1+hy_1+g)(hx_1+by_1+f). \quad \ldots(4)$$

Equation (3) gives

$$\frac{(ax_1+hy_1+g)^2-(hx_1+by_1+f)^2}{a-b}=S_1$$

and equation (4) gives

$$\frac{(ax_1+hy_1+g)(hx_1+by_1+f)}{h}=S_1,$$

so that these equations can be rewritten as

$$\frac{(ax_1+hy_1+g)^2-(hx_1+by_1+f)^2}{a-b}=\frac{(ax_1+hy_1+g)(hx_1+by_1+f)}{h}$$

$$=S_1 \quad \ldots(5)$$

or

$$\frac{\left(\frac{\partial\phi}{\partial x}\right)^2-\left(\frac{\partial\phi}{\partial y}\right)^2}{a-b}=\frac{\frac{\partial\phi}{\partial x}\cdot\frac{\partial\phi}{\partial y}}{h}=4\phi,$$

where the equation to the conic is

$$\phi(x, y)=0.$$

Cor. The directrices are polars of the foci. Hence their equations can be easily obtained.

11·13. Axes of a conic.

Let the conic be

$$S \equiv ax^2+2hxy+by^2+2gx+2fy+c=0.$$

By the last section the equation

$$\frac{(ax+hy+g)^2-(hx+by+f)^2}{a-b}=\frac{(ax+hy+g)(hx+by+f)}{h}$$

represents a conic passing through the foci.

But it is a quadratic equation in

$$ax+hy+g \text{ and } hx+by+f.$$

Hence it can be solved for $\dfrac{ax+hy+g}{hx+by+f}$, so that it represents two straight lines, which are the axes of the conic.

11·14. Conic with a tangent and normal as axes of coordinates.

Let the conic be

$$S \equiv ax^2+2hxy+by^2+2gx+2fy+c=0.$$

The origin lies on the curve, so that $c=0$.

The point of intersection of x-axis with the conic is given by

$$ax^2+2gx=0.$$

But it is a tangent to the conic at the origin; hence both the roots should be zero, so that $g=0$.

The equation of the conic therefore becomes

$$ax^2+2hxy+by^2+2fy=0.$$

11·15. Condition that the line $lx+my+n=0$ may touch the conic $S=0$.

Let the point of contact be (x_1, y_1).

The equation of the tangent at (x_1, y_1) is

$$x(ax_1+hy_1+g)+y(hx_1+by_1+f)+gx_1+fy_1+c=0 \quad \ldots .(1)$$

This must coincide with the line

$$lx+my+n=0 \qquad \ldots(2)$$

Comparing coefficients, we get

$$\frac{ax_1+hy_1+g}{l}=\frac{hx_1+by_1+f}{m}=\frac{gx_1+fy_1+c}{n}=\lambda \text{ (say)} \qquad \ldots(3)$$

The above equations give

$$ax_1+hy_1+g-\lambda l=0, \qquad \ldots(4)$$

$$hx_1+by_1+f-\lambda m=0, \qquad \ldots(5)$$

$$gx_1+fy_1+c-\lambda n=0. \qquad \ldots(6)$$

Since (x_1, y_1) lies on the line (2), we get

$$lx_1+my_1+n=0. \qquad \ldots(7)$$

Eliminating x_1, y_1 and λ from (4), (5), (6) and (7), we get

$$\begin{vmatrix} a & h & g & l \\ h & b & f & m \\ g & f & c & n \\ l & m & n & 0 \end{vmatrix}=0$$

which when expanded gives

$$(bc-f^2)\, l^2+(ca-g^2)\, m^2+(ab-h^2)\, n^2+2\,(gh-af)\, mn$$
$$+2\,(hf-lg)\, nl+2\,(fg-ch)\, lm=0$$

or $\qquad Al^2+Bm^2+Cn^2+2Fmn+2Gnl+2Hlm=0$

where A, B, C, F, G and H are the cofactors of a, b, c, f, g and h in the determinant

$$\begin{vmatrix} a & h & g \\ h & b & f \\ g & f & c \end{vmatrix}.$$

11·16. Conic touching four given straight lines.

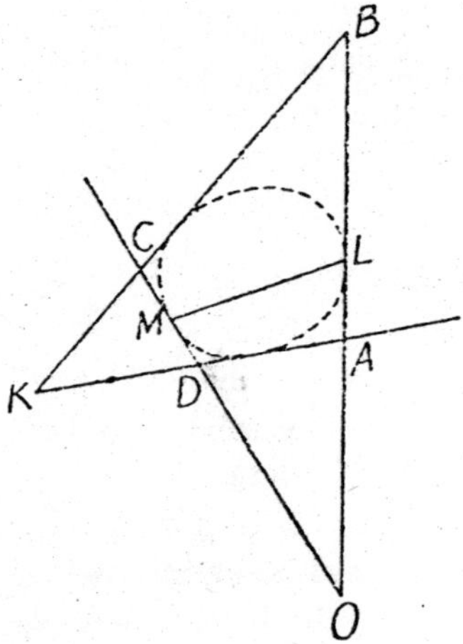

Let the four straight lines form a quadrilateral $ABCD$. Let the point of intersection of BA and CD be the origin O and let OAB and ODC be the axes of x and y respectively. Let the equations to the other sides BC and DA be $lx+my-1=0$

and $l_1x-m_1y-1=0$

respectively. Let L and M be the points of contact of the conic with the axes and let the equation to the line LM be

$$ax+by-1=0.$$

As the conic has double contact with the line

$$ax+by-1=0$$

and the axes are the original conic, its equation is

$$2\lambda xy=(ax+by-1)^2. \qquad \ldots(1)$$

The condition that the line BC touches it is given by the previous section as

$$\lambda=2\,(a-l)\,(b-m). \qquad \ldots(2)$$

Similarly DA will touch it if

$$\lambda=2\,(a-l_1)\,(b-m_1). \qquad \ldots(3)$$

a and b can be found in terms of λ from (2) and (3) as

$$\frac{2b-(m+m_1)}{m-m_1}=-\frac{2a-(l+l_1)}{l-l_1}=\pm\sqrt{\left\{1-\frac{2\lambda}{(l-l)_1\,(m-m_1)}\right\}}.$$

The values of a and b substituted from the above equations in (1) give the required equation of the conic

11·17. Circle of curvature.

The circle of curvature of a conic at a given point has three-point contact with the conic at that point. It is also called the osculating circle of the conic.

Let the conic be $S=0$.

The equation of the tangent to the conic at the point (x_1, y_1) is

$$T\equiv axx_1+h\,(xy_1+x_1y)+byy_1+g\,(x+x_1)+f\,(y+y_1)+c=0.$$

The equation of the conic passing through the point (x_1, y_1) and the points where $lx+my+n=0$ cuts the conic $S=0$ is given by

$$S+T\,(lx+my+n)=0. \qquad \ldots(2)$$

The conics obviously touch each other at (x_1, y_1) which is a pair of coincident points.

If line $lx+my+n=0$ also passes through (x_1, y_1), its equation becomes

$$l\,(x-x_1)+m\,(y-y_1)=0$$

and equation (2) then becomes

$$S+T\,\{l\,(x-x_1)+m\,(y-y_1)\}=0 \qquad \ldots(3)$$

which is a conic having three-point contact at (x_1, y_1).

If the coefficients in (3) are so chosen that it represents a circle, it is the circle of curvature of the conic.

11·18. Line at infinity.

We know that $ax+by+c=0$ is the straight line which cuts off intercepts $-c/a$ and $-c/b$ from the axes of x and y respectively. If a and b are zero, these intercepts are infinite in length and therefore the line $c=0$ is called the line at infinity.

We shall now find the geometrical meaning of a few equations.

(i) $S=\lambda u$ where λ is a constant and $u=0$ is the equation to a straight line.

The above equation can be written as

$$S=\lambda u\,(0.x+0\,y+1).$$

This represents a conic passing through the intersection of the conic $S=0$ with the straight line $u=0$ and the line at infinity. Since $S=0$ and $S=\lambda u$ have the same intersections with the line at infinity, they have their asymptotes parallel.

Let $S\equiv x^2+y^2-a^2=0$ be a circle.

Let $x^2+y^2+2gx+2fy+c=0$ be another circle.

$x^2+y^2+2gx+2fy+c=0$ can also be written as

$$x^2+y^2-a^2=-2gx-2fy-c-a^2$$

so that its equation is of the form

$$S=\lambda\mu$$

where $$u\equiv 2gx+2fy+c+a^2=0.$$

The two circles therefore meet the line at infinity in the same two imaginary points. These are called the circular points at infinity.

(ii) $S=\lambda$.

The above equation can be written as

$$S=\lambda\,(0.x+0\,y+1)^2.$$

This conic has a double contact with $S=0$ where the line at infinity meets it. Hence the tangents to both the conics at points where they intersect the line at infinity are the same. Both the conics have therefore the same asymptotes.

If $S=0$ is a circle $S=\lambda$, a concentric circle, the two have double contact at the circular points at infinity.

11·19. Solved Examples.

Ex. 1. *Find the condition that the pair of straight lines whose equation is $Ax^2+2Hxy+By^2=0$ may be parallel to conjugate diameters of the general conic.*

Let the lines given by

$$Ax^2+2Hxy+By^2=0 \qquad \ldots(1)$$

be the same as $y=mx$ and $y=m_1x$ so that (1) is equivalent to

$$B(y-mx)(y-m_1x)=0. \qquad \ldots(2)$$

Comparing coefficients in (1) and (2), we get

$$m+m_1=-\frac{2H}{B} \qquad \ldots(3)$$

and

$$mm_1=\frac{A}{B} \qquad \ldots(4)$$

$y=mx$ and $y=m_1x$ are parallel to conjugate diameters if

$$a+h(m+m_1)+bmm_1=0.$$

Substituting for $m+m_1$ and mm_1 from (3) and (4), the condition becomes

$$a-2h\left(\frac{H}{B}\right)+b\left(\frac{A}{B}\right)=0$$

or

$$aB+bA=2hH.$$

Ex. 2. *Prove that if a conic circumscribe a quadrilateral, the ratio of the product of the perpendiculars from any point P of the conic upon two opposite sides of the quadrilateral to the product of the perpendiculars from P upon the other two sides is the same for all positions of P.* *(Benaras 1958)*

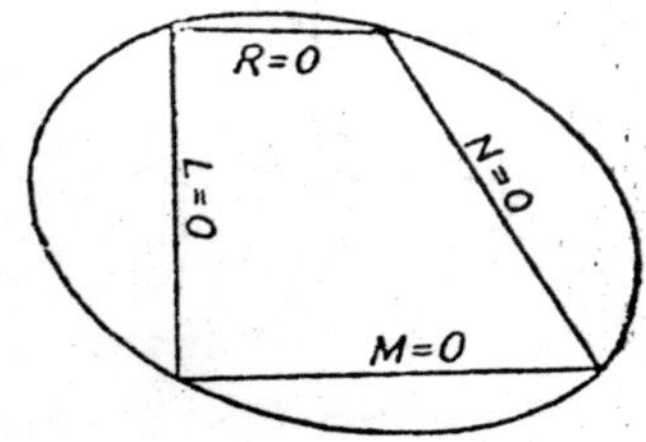

Let the four sides of the quadrilateral be $L=0$, $M=0$, $N=0$ and $R=0$.

$L=0$ passes through one pair of angular points and $N=0$ through the other pair.

Hence $LN=0$ is the conic passing through all the four angular points.

Similarly $MR=0$ is the equation to another conic passing through the four angular points.

Hence $LN=\lambda.MR$ is the equation to any conic through the four angular points.

We know that L is proportional to the perpendicular from any point (x, y) on the conic to the straight line $L=0$: therefore the relation $LN=\lambda.MR$ proves the above proposition.

Ex. 3. *If $S=0$ represent a conic, $T=0$ a tangent to it, and $u=0$ any other straight line, interpret the equations*

(i) $S+\lambda T=0$; (ii) $S+\lambda u^2+ku=0$; (iii) $S+\lambda Tu=0$.

Suppose $u=0$ intersects $S=0$ in A and B and $T=0$ touches it at P.

(i) $S+\lambda T=0$ represents a conic touching $S=0$ at P and having asymptotes parallel to those of $S=0$.

(ii) $S+\lambda u^2+ku=0$ is a conic which has a pair of parallel chords common with $S=0$, one of which is $u=0$, because $\lambda u^2+ku+c=0$ is a quadratic in u and will give two straight lines parallel to $u=0$.

(iii) $S+\lambda Tu=0$ is a conic which touches $S=0$ at P and passes through A and B, the points of intersection of $u=0$ with $S=0$.

Ex. 4. *Prove that the equation to the circle, having double contact with the ellipse*

$$\frac{x^2}{a^2}+\frac{y^2}{b^2}=1$$

at the ends of a latus rectum is

$$x^2+y^2-2a^3ex=a^2(1-e^2-e^4)$$

where e is its eccentricity. *(Benaras 1951)*

The equation to a conic having double contact with the ellipse $\frac{x^2}{a^2}+\frac{y^2}{b^2}=1$ at the ends of the latus rectum $x=ae$ is given by

$$\left(\frac{x^2}{a^2}+\frac{y^2}{b^2}-1\right)+\lambda(x-ae)^2=0.$$

If this is a circle, the coefficients of x^2 and y^2 are equal,

so that
$$\frac{1}{a^2}+\lambda=\frac{1}{b^2}$$

or
$$\lambda=\frac{1}{b^2}-\frac{1}{a^2}=\frac{e^2}{a^2(1-e^2)}.$$

The equation to the circle becomes

$$\left(\frac{x^2}{a^2}+\frac{y^2}{b^2}-1\right)+\frac{e^2}{a^2(1-e^2)}(x-ae)^2=0$$

or
$$\left(\frac{x^2}{a^2}+\frac{y^2}{a^2(1-e^2)}-1\right)+\frac{e^2}{a^2(1-e^2)}(x^2-2xae+a^2e^2)=0$$

or
$$x^2(1-e^2)+y^2-a^2(1-e^2)+e^2(x^2-2xae+a^2e^2)=0$$

or
$$x^2+y^2-2xa^3e-a^2(1-e^2-e^4)=0$$

or
$$x^2+y^2-2a^3ex=a^2(1-e^2-e^4).$$

Ex. 5. *If a system of conics be drawn having four-point contact with the conic $ax^2+2hxy+by^2+2fy=0$ at the origin, prove that the director circles of these conics form a coaxial system, whose limiting points are the origin and the point*

$$\left(-\frac{fh}{a^2+h^2}, \frac{af}{a^2+h^2}\right).$$ *(Bombay 1959)*

The tangent at the origin to the conic

$$ax^2+2hxy+by^2+2fy=0$$

is
$$y=0.$$

A conic having four-point contact with the given conic at the origin is therefore

$$ax^2+2hxy+by^2+2fy+\lambda y^2=0. \qquad \ldots(1)$$

The director circle of this conic is by article 12·10,

$$(x^2+y^2)\{a(b+\lambda)-h^2\}-2xfh+2yaf-f^2=0.$$

Putting $a(b+\lambda)-h^2=k$, this becomes

$$x^2+y^2-\frac{2xfh}{k}+\frac{2yaf}{k}-\frac{f^2}{k}=0$$

or
$$\left(x-\frac{fh}{k}\right)^2+\left(y+\frac{af}{k}\right)^2=\frac{f^2k+f^2h^2+a^2f^2}{k^2}$$

which is a coaxial system of circles.

This will have zero radius if

$$f^2(k+h^2+a^2)=0.$$

Either $f=0$ or $a(b+\lambda)-h^2+h^2+a^2=0$

or $$\lambda=-(a+b).$$

If $f=0$, the coordinates of the centre are $(0, 0)$.

If $\lambda=-(a+b)$, coordinates of the centre are

$$-\frac{fh}{a^2+h^2} \text{ and } \frac{af}{a^2+h^2}.$$

Ex. 6. *Prove that all conics through four given points have a pair of conjugate diameters parallel to the axes of the two parabolas through those points and the locus of their centres is a conic whose asymptotes are parallel to the axes of the two parabolas.*

We shall first find the equation to a conic through four points. Take the line joining two points as the axis of X and the line joining the other two points as the axis of Y. Let the straight lines whose equations are

$$lx+my-1=0 \text{ and } l_1x+m_1y-1=0$$

cut the axes in four given points.

Then $xy=0$ and $(lx+my-1)(l_1x+m_1y-1)=0$ are two conics passing through the four points and therefore the system of conics passing through the four points is given by

$$\lambda xy+(lx+my-1)(l_1x+m_1y-1)=0 \quad \ldots(1)$$

or $$(ll_1x^2+mm_1y^2+(ml_1+lm_1+\lambda)xy-(l+l_1)x$$
$$-(m+m_1)y+1=0 \ldots(2)$$

This equation will represent a parabola if

$$4ll_1mm_1=(ml_1+lm_1+\lambda)^2.$$

This is a quadratic in λ; hence two parabolas will pass through the four points. The parabolas are real if ll_1mm_1 is positive.

Terms of the second degree are squares of

$$\{\sqrt{(ll_1)}\,x\pm\sqrt{(mm_1)}\,y\}.$$

Hence the axes of the parabolas are parallel to the lines

$$\sqrt{(ll_1)}\, x \pm \sqrt{(mm_1)}\, y = 0.$$

The combined equation of the lines is

$$ll_1x^2 - mm_1 y^2 = 0.$$

The conic and the above lines satisfy the conditions of being conjugate diameters as

$$-ll_1mm_1 + ll_1mm_1 = 0.$$

Hence the first part of the proposition is proved.

The coordinates of the centre of the conic are given by

$$\lambda y + l\,(l_1x + m_1 y - 1) + l_1\,(lx + my - 1) = 0 \qquad \ldots(3)$$

and $$\lambda x + m\,(l_1x + m_1 y - 1) + m_1\,(lx + my - 1) = 0. \qquad \ldots(4)$$

Multiplying (3) by x and (4) by y and subtracting, we get

$$(lx - my)\,(l_1x + m_1 y - 1) + (l_1x - m_1 y)\,(lx + my - 1) = 0$$

or $$2ll_1x^2 - 2mm_1 y^2 - (l + l_1)\,x + (m + m_1)\,y = 0.$$

This is a conic whose asymptotes are parallel to the lines

$$ll_1x^2 - mm_1 y^2 = 0$$

which are parallel to the axes of the parabolas through the four points.

Ex. 7. *Find the equation of the circle of curvature of the parabola $y^2 = 4ax$ at the point $(at^2, 2at)$. Shew also that the circle of curvature meets the parabola again in the point $(at'^2, 2at')$ such that $t'^2 + 2tt' - 3t^2 = 0$.*

The equation of the tangent at $(at^2, 2at)$ is

$$ty = x + at^2 \qquad \ldots(1)$$

As the chords of intersection of a conic and a circle are equally inclined to the axes of a conic, one chord here becomes the tangent at $(at^2, 2at)$ and the other chord will be inclined to the axis by the same angle as the tangent is inclined to it.

The tangent is inclined to the axis by an angle whose tangent is $1/t$. Hence the inclination of the common chord to the axis is such that its tangent is $-1/t$.

The common chord passes through $(at^2, 2at)$. Hence its equation is

$$y-2at=-1/t\,(x-at^2)$$

or
$$ty+x=3at^2. \quad \ldots(2)$$

The conic having contact of the third order with the parabola is therefore

$$y^2-4ax+\lambda\,(ty-x-at^2)\,(ty+x-3at^2)=0.$$

As this is a circle, the coefficients of x^2 and y^2 should be equal, so that

$$1+\lambda t^2=-\lambda; \quad \therefore \quad \lambda=-\frac{1}{1+t^2}.$$

The equation of the circle of curvature becomes

$$y^2-4ax-\frac{1}{1+t^2}\,(ty-x-at^2)\,(ty+x-3at^2)=0$$

or
$$(1+t^2)\,(y^2-4ax)-(t^2y^2-x^2+3a^2t^4-4at^3y+2at^2x)=0$$

or
$$y^2+x^2-2ax\,(2+3t^2)+4at^3y-3a^2t^4=0.$$

As $(at'^2, 2at')$ is the other point of intersection, it lies on the common chord (2), so that

$$2att'+at'^2=3at^2$$

or
$$t'^2+2tt'-3t^2=0.$$

Examples XIX

1. A circle cuts the parabola $y^2=4ax$ in the points t_1, t_2, t_3, t_4. Prove that $t_1+t_2+t_3+t_4=0$.

2. Obtain the parabola which passes through $(0, 0)$, $(0, 4)$ and the points of intersection of the line $x=1$ with the circle $x^2+y^2=4$. *(Dehli 1947)*

3. Prove that the conic passing through the points $(l, 0)$, $(m, 0)$, $(0, n)$, $(0, p)$, (m, p) where l, m, n, p are positive and $m > l$, $p > m$, is an ellipse. If $m=2l$, $p=2m$, prove

that the centre is the point $\left(\frac{6l}{5}, \frac{6n}{5}\right)$.

4. Prove that the locus of centres of conics which touch the coordinate axes at distances a and b from the origin is the straight line $ay=bx$.

5. Pairs of tangents are drawn to the conic $\alpha x^2+\beta y^2=1$ so as to be always parallel to conjugate diameters of the conic $ax^2+2hxy+by^2=1$; show that the locus of their points of intersection is the conic

$$ax^2+2hxy+by^2=a/\alpha+b/\beta.$$

6. Two conics have double contact. Prove that the polar of the centre of one conic with respect to the other is parallel to the common chord. *(Lucknow 1946)*

7. Two circles have double contact with a conic, their chords of contact being parallel. Shew that the radical axis of the two circles is mid-way between the two chords of contact. *(Lucknow 1945)*

8. Prove that the locus of the poles of a given straight line with respect to the conics, which pass through four given points is a coinc. *(Utkal 1947, Bombay 1959)*

9. Prove that in general two parabolas can be drawn through four given points. *(Lucknow 1943)*

10. A conic has double contact with the parabola $y^2=4ax$. If the chord of contact passes through the vertex, and the conic passes through the focus of the parabola, prove that the locus of the centre of the conic is the parabola $y^2=a\,(2x-a)$. *(Lucknow 1946)*

11. Find the locus of the centres of rectangular hyperbolas which have contact of the third order with the parabola $y^2-4ax=0$. *(Patna Hon's 1946)*

12. Obtain the equation of the circle of curvature of the parabola $y^2=4ax$ at the vertex.

If any tangent is drawn to this circle and meets the parabola in point whose ordinates are y_1 and y_2, show that $\frac{1}{y_1}-\frac{1}{y_2}=\pm\frac{1}{2a}$.

(Lucknow Hon's 1945)

13. Find the equation of the parabola which has contact of the third order with the conic

$$ax^2+2hxy+by^2+2gx+2fy=0$$

at the origin.

(Utkal 1947)

14. Shew that the locus of points such that the chords of contact of tangents drawn from them to the conic $ax^2+2hxy+by^2=1$ subtends a right angle at the centre is the conic $x^2(a^2+h^2)+2h(a+b)xy+y^2(h^2+b^2)=a+b$.

15. Two conics $S_1=0$, $S_2=0$ have a pair of common chords $u=0$, $v=0$ such that $S_1-S_2\equiv uv$. Prove that the conic $\lambda^2u^2-2\lambda(S_1+S_2)+v^2=0$ has double contact with both $S_1=0$ and $S_2=0$.

16. Prove that in general two parabolas can be drawn through the points of intersection of the conics

$$ax^2+2hxy+by^2+2gx+2fy+c=0$$

and $$a'x^2+2h'xy+b'y^2+2g'x+2f'y+c'=0,$$

and their axes will be at right angles if

$$\frac{h}{a-b}=\frac{h'}{a'-b'}.$$

17. Prove that the locus of the pole of the axis of x with respect to the circle of curvature at any point of the parabola $y^2=4ax$ is

$$(x-2a)^3y^2=12a(x^2-ax+a^2)^2.$$

18. The circle of curvature of the rectangular hyperbola $x^2-y^2=a^2$ at the point $(a\operatorname{cosec}\theta, a\cot\theta)$ meets the curve again in the point $(a\operatorname{cosec}\phi, a\cot\phi)$. Show that

$$\tan\frac{\phi}{2}\tan^3\frac{\theta}{2}=1.$$

19. Prove that the foci of the conic whose equation is $ax^2+2hxy+by^2=1$ lie on the curve

$$\frac{x^2-y^2}{a-b}=\frac{xy}{h}=\frac{1}{h^2-ab}.$$

20. Find the equation of equiconjugate diameters of the conic $ax^2+2hxy+by^2=1$.

21. Shew that any two concentric conics have in general one and only one pair of common conjugate diameters.

22. Conics are drawn through four given points. Prove that the polar of a fixed point with respect to the conics passes through a fixed point. *(Benares 1960)*

23. Prove that the centres of rectangular hyperbolas which have contact of the third order with a given parabola lie on an equal parabola. *(Karnatak 1958)*

24. Show that in general there are two points on a given line conjugate with respect to each conic passing through the four fixed points. *(Bombay 1959)*

25. Find the equation to the parabola which has four-point contact with the conic $ax^2+2hxy+by^2+2gx+2fy=0$ at the origin. Show that the axis of the parabola makes with the tangent at the origin an angle

$$\tan^{-1}\frac{fF+gG}{fG\sim gF},$$

where F, G are the cofactors of f and g respectively in

$$\begin{vmatrix} a & h & g \\ h & b & f \\ g & f & c \end{vmatrix}.$$ *(Poona 1959)*

26. Show that the foci of the conic $ax^2+2hxy+by^2+2gx+2fy+c=0$ are given by the equations $Fx+Gy-H=Cxy$

and $2Gx-2Fy-A+B=C(x^2-y^2)$, where

$$D\equiv\begin{vmatrix} a & h & g \\ h & b & f \\ g & f & c \end{vmatrix}, \quad A\equiv\frac{\partial D}{\partial a}, \ldots F\equiv\frac{1}{2}\frac{\partial D}{\partial f}, \text{ etc.}$$

Find the real foci and the corresponding directrices of the conic $x^2+xy+y^2+4x-7y+7=0$. *(Poona 1960)*

27. A family of conics have double contact with a given conic at the extremities of a given chord. Prove that the locus of the centres of conics of the family is the diameter of the given conic conjugate to the given chord. *(Poona 1960)*

11·20. Confocal conics.

Conics having the same foci are said to be confocal.

The foci of a central conic lie on its principal axes. Therefore all confocal central conics will have a common centre and the same principal axes.

Let $\frac{x^2}{l^2}+\frac{y^2}{m^2}=1$ be an ellipse confocal with the ellipse

$$\frac{x^2}{a^2}+\frac{y^2}{b^2}=1.$$

As the distance from the centre to the focus is the same in both these ellipses,

$$l^2-m^2=a^2-b^2=\text{constant}$$

or
$$l^2-a^2=m^2-b^2=\lambda \text{ (say)}.$$

$$\therefore \quad l^2=a^2+\lambda \quad \text{and} \quad m^2=b^2+\lambda.$$

The equation to confocal ellipses therefore becomes

$$\frac{x^2}{a^2+\lambda}+\frac{y^2}{b^2+\lambda}=1$$

which for different values of λ represents a system of conics confocal with the ellipse $\frac{x^2}{a^2}+\frac{y^2}{b^2}=1$.

11·21. Tracing of the confocal system.

The equation of a system of confocal conics is

$$\frac{x^2}{a^2+\lambda}+\frac{y^2}{b^2+\lambda}=1.$$

If λ is positive the curve is an ellipse.

If λ increases the principal axes of the curve increase and their ratio tends to 1, when λ becomes very large, so that a circle of infinite radius is the limiting form of one of the confocals.

If λ is negative, the principal axes will decrease as λ increases, and the ratio $\dfrac{b^2+\lambda}{a^2+\lambda}$ also decreases as λ increases, so that the ellipse becomes flatter and flatter until λ is equal to $-b^2$, when the minor axis vanishes, and the major axis becomes $2(a^2-b^2)$, that is, the distance between the foci. Hence the ellipse reduces to the line joining the foci in the limiting form.

If $b^2+\lambda$ is negative that is $\lambda < -b^2$, the curve is a hyperbola.

When $b^2+\lambda$ is small, the conjugate axis is small, so that the hyperbola is very flat. When λ tends to $-b^2$ the curve tends to the hyperbola whose transverse axis is $2\sqrt{(a^2-b^2)}$ and conjugate axis is 0. Hence the two straight lines obtained by producing the line joining the foci both ways is the limiting form of the hyperbola.

When $\lambda \to -a^2$, the transverse axis of the hyperbola tends to zero. In the limit, the two branches of the hyperbola tend to coincide with the y-axis.

If λ is negative and numerically greater than a^2, the curve is imaginary.

The figure is as shown below :—

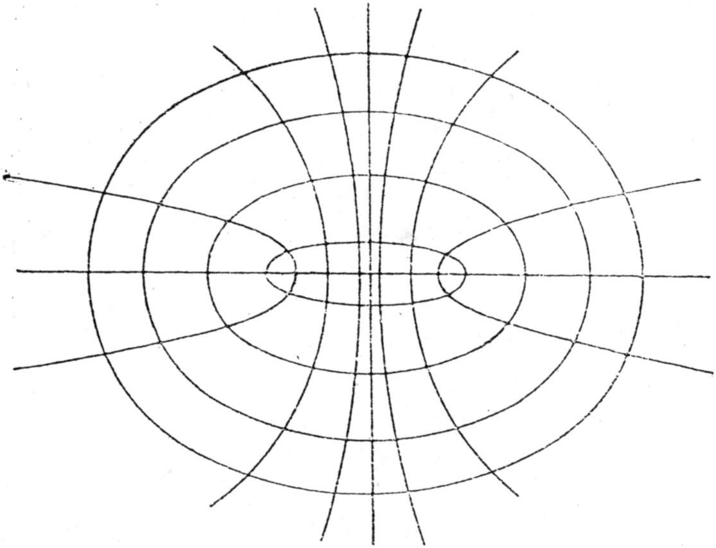

11·21. Confocals through a given point.

Let the equation of the conic be

$$\frac{x^2}{a^2}+\frac{y^2}{b^2}=1.$$

Then the equation to the confocal conic is

$$\frac{x^2}{a^2+\lambda}+\frac{y^2}{b^2+\lambda}=1,$$

where λ is a parameter and let $a>b$. A member of the confocal system passes through (x_1, y_1) if

$$\frac{x_1^2}{a^2+\lambda}+\frac{y_1^2}{b^2+\lambda}=1$$

or $$(a^2+\lambda)(b^2+\lambda)-x_1^2(b^2+\lambda)-y_1^2(a^2+\lambda)=0. \qquad \ldots(1)$$

This is a quadratic equation in λ and therefore gives two values of λ.

Let $f(\lambda) \equiv (a^2+\lambda)(b^2+\lambda) - x_1^2(b^2+\lambda) - y_1^2(a^2+\lambda) = 0.$

Now $f(-a^2) = x_1^2(a^2-b^2) = x_1^2a^2e^2$ which is positive and $f(-b^2) = y_1^2(b^2-a^2) = -y_1^2a^2e^2$ which is negative.

Hence one root of $f(x)=0$ lies between $-a^2$ and $-b^2$. If λ_1 is the root, $a^2+\lambda_1$ is positive and $b^2+\lambda_1$ is negative. Therefore the confocal for this value λ_1 is a hyperbola.

Now $f(\lambda)$ tends to ∞ as $\lambda \to \infty$ and $f(-b^2)$ is negative. The other root lies between ∞ and $-b^2$. Hence

$$-a^2 < -b^2 < \lambda_2,$$

if λ_2 is the other root, so that $a^2+\lambda_2$ and $b^2+\lambda_2$ are both positive. The confocal corresponding to λ_2 is therefore an ellipse.

Hence two members of a confocal system pass through a given point, of which one is an ellipse and the other a hyperbola.

11·23. Confocal conics through a point cut at right angles.

Let the two confocals be

$$\frac{x^2}{a^2+\lambda} + \frac{y^2}{b^2+\lambda} = 1 \qquad \ldots(1)$$

and

$$\frac{x^2}{a^2+\lambda_1} + \frac{y^2}{b^2+\lambda_1} = 1. \qquad \ldots(2)$$

Since (x_1, y_1) lies on the conics,

$$\frac{x_1^2}{a^2+\lambda} + \frac{y_1^2}{b^2+\lambda} = 1 \qquad \ldots(3)$$

and

$$\frac{x_1^2}{a^2+\lambda_1} + \frac{y_1^2}{b^2+\lambda_1} = 1. \qquad \ldots(4)$$

Subtracting (4) from (3), we get

$$\left(\frac{x_1^2}{a^2+\lambda} - \frac{x_1^2}{a^2+\lambda_1}\right) + \left(\frac{y_1^2}{b^2+\lambda} - \frac{y_1^2}{b^2+\lambda_1}\right) = 0$$

or

$$(\lambda_1-\lambda)\frac{x_1^2}{(a^2+\lambda)(a^2+\lambda_1)} + (\lambda_1-\lambda)\frac{y_1^2}{(b^2+\lambda)(b^2+\lambda_1)} = 0$$

or $$\frac{x_1^2}{(a^2+\lambda)(a^2+\lambda_1)}+\frac{y_1^2}{(b^2+\lambda)(b^2+\lambda_1)}=0. \qquad \ldots(5)$$

The tangents to the conics (1) and (2) at (x_1, y_1) are

$$\frac{xx_1}{a^2+\lambda}+\frac{yy_1}{a^2+\lambda_1}=1 \qquad \ldots(6)$$

and $$\frac{xx_1}{b^2+\lambda}+\frac{yy_1}{b^2+\lambda_1}=1. \qquad \ldots(7)$$

These two tangents are at right angles to each other if

$$\frac{x_1^2}{(a^2+\lambda)(a^2+\lambda_1)}+\frac{y_1^2}{(b^2+\lambda)(b^2+\lambda_1)}=0$$

which is the same as (5).

Hence the proposition.

11·24. Propositions on confocal conics.

1. *One and only one conic of a confocal system will touch a given straight line.*

Let the equation of the confocal system be

$$\frac{x^2}{a^2+\lambda}+\frac{y^2}{b^2+\lambda}=1. \qquad \ldots(1)$$

Let the given straight line be

$$lx+my+n=0. \qquad \ldots(2)$$

The straight line (2) will touch the conic if

$$(a^2+\lambda)\,l^2+(b^2+\lambda)\,m^2=n^2.$$

The above equation gives only one value of λ; hence only one conic touches the given straight line.

2. *The difference of the squares of perpendiculars drawn from the centre on any two parallel tangents to two given confocals is constant.*

Let the equation of the conics be

$$\frac{x^2}{a^2}+\frac{y^2}{b^2}=1 \quad \text{and} \quad \frac{x^2}{a^2+\lambda}+\frac{y^2}{b^2+\lambda}=1.$$

Let the two straight lines be

$$x\cos\alpha+y\sin\alpha=p, \qquad \ldots(1)$$

and $$x \cos \alpha + y \sin \alpha = p'. \qquad \ldots(2)$$

These touch the conics respectively if

$$p^2 = a^2 \cos^2 \alpha + b^2 \sin^2 \alpha$$

and $$p'^2 = (a^2 + \lambda) \cos^2 \alpha + (b^2 + \lambda) \sin^2 \alpha,$$

so that $p'^2 - p^2 = \lambda$ which is a constant.

3. *The point of intersection of two perpendicular tangents, one to each of two given confocals lie on a circle.*

Let the confocals be

$$\frac{x^2}{a^2+\lambda_1} + \frac{y^2}{b^2+\lambda_1} = 1 \text{ and } \frac{x^2}{a^2+\lambda_2} + \frac{y^2}{b^2+\lambda_2} = 1.$$

The straight lines perpendicular to each other and touching the given confocals are

$$x \cos \alpha + y \sin \alpha = \sqrt{[(a^2+\lambda_1) \cos^2 \alpha + (b^2+\lambda_1) \sin^2 \alpha]}$$

and $$x \sin \alpha - y \cos \alpha = \sqrt{[(a^2+\lambda_2) \sin^2 \alpha + (b^2+\lambda_2) \cos^2 \alpha]}.$$

Squaring and adding the above equations the locus of their point of intersection is

$$x^2 + y^2 = \lambda_1 + \lambda_2 + a^2 + b^2$$

which is a circle.

4. *The locus of the pole of a given straight line with respect to a system of confocal conics is a straight line.*

Let the given line be

$$lx + my + n = 0 \qquad \ldots \quad \ldots(1)$$

and let the equation of the confocal conic be

$$\frac{x^2}{a^2+\lambda} + \frac{y^2}{b^2+\lambda} = 1. \qquad \ldots \quad \ldots(2)$$

Let (x_1, y_1) be the pole of (1) with respect to (2).

Then the polar of (x_1, y_1) with respect to the conic is

$$\frac{xx_1}{a^2+\lambda} + \frac{yy_1}{b^2+\lambda} = 1 \qquad \ldots \quad \ldots(3)$$

which is the same as (1).

Comparing coefficients in (1) and (3),

$$\frac{x_1}{a^2+\lambda} = -\frac{l}{n} \text{ and } \frac{y_1}{b^2+\lambda} = -\frac{m}{n}.$$

Eliminating λ between these equations, we get

$$-\frac{nx_1}{l}-a^2=-\frac{ny_1}{m}-b^2.$$

Hence the locus of (x_1, y_1) is

$$\frac{nx}{l}-\frac{ny}{m}+a^2-b^2=0$$

which is a straight line perpendicular to the given line.

12·25. Solved Examples.

Ex. 1. *Show how to express the coordinates (x_1, y_1) of a point P in terms of the parameters λ_1, λ_2 of the two confocals passing through P and belonging to the family*

$$\frac{x^2}{a^2+\lambda}+\frac{y^2}{b^2+\lambda}=1$$

If 2ϕ be the angle between the tangents from P to the ellipse,

$$\frac{x^2}{a^2}+\frac{y^2}{b^2}=1,$$

show that $$\tan 2\phi=\frac{2\sqrt{(-\lambda_1\lambda_2)}}{\lambda_1+\lambda_2}.$$

Let the two confocals be

$$\frac{x^2}{a^2+\lambda_1}+\frac{y^2}{b^2+\lambda_1}=1 \qquad \ldots \quad \ldots(1)$$

and $$\frac{x^2}{a^2+\lambda_2}+\frac{y^2}{b^2+\lambda_2}=1. \qquad \ldots \quad \ldots(2)$$

As they pass through (x_1, y_1),

$$\frac{x_1^2}{a^2+\lambda_1}+\frac{y_1^2}{b^2+\lambda_1}=1 \qquad \ldots \quad \ldots(3)$$

and $$\frac{x_1^2}{a^2+\lambda_2}+\frac{y_1^2}{b^2+\lambda_2}=1. \qquad \ldots \quad \ldots(4)$$

Solving the above equations for x_1 and y_1, we get

$$\frac{x_1^2}{(a^2+\lambda_1)(b^2+\lambda_2)}-\frac{x_1^2}{(a^2+\lambda_2)(b^2+\lambda_1)}=\frac{1}{b^2+\lambda_2}-\frac{1}{b^2+\lambda_1}$$

or $$x_1^2\left\{\frac{(\lambda_2-\lambda_1)(b^2-a^2)}{(a^2+\lambda_1)(a^2+\lambda_2)(b^2+\lambda_1)(b^2+\lambda_2)}\right\}=\frac{\lambda_1-\lambda_2}{(b^2+\lambda_2)(b^2+\lambda_1)}$$

or $$x_1^2=-\frac{(a^2+\lambda_2)(a^2+\lambda_1)}{b^2-a^2}.$$

$$\therefore\quad x_1^2=\frac{(a^2+\lambda_1)(a^2+\lambda_2)}{a^2-b^2}$$

Similarly $y_1^2=\dfrac{(b^2+\lambda_1)(b^2+\lambda_2)}{b^2-a^2}$.

Equation to the tangents from (x_1, y_1) to the ellipse

$$\frac{x^2}{a^2}+\frac{y^2}{b^2}=1$$

is $$\left(\frac{x^2}{a^2}+\frac{y^2}{b^2}-1\right)\left(\frac{x_1^2}{a^2}+\frac{y_1^2}{b^2}-1\right)=\left(\frac{xx_1}{a^2}+\frac{yy_1}{b^2}-1\right)^2.$$

The angle 2ϕ between the tangents is given by

$$\tan 2\phi=\frac{2\sqrt{\left[\dfrac{x_1^2y_1^2}{a^4b^4}-\left(\dfrac{y_1^2}{a^2b^2}-\dfrac{1}{a^2}\right)\left(\dfrac{x_1^2}{a^2b^2}-\dfrac{1}{b^2}\right)\right]}}{\dfrac{y_1^2}{a^2b^2}-\dfrac{1}{a^2}+\dfrac{x_1^2}{a^2b^2}-\dfrac{1}{b^2}}$$

$$=\frac{2\sqrt{\left(\dfrac{x_1^2}{a^4b^2}+\dfrac{y_1^2}{a^2b^4}-\dfrac{1}{a^2b^2}\right)}}{\dfrac{x_1^2}{a^2b^2}+\dfrac{y_1^2}{a^2b^2}-\dfrac{1}{a^2}-\dfrac{1}{b^2}}$$

$$=\frac{2\sqrt{\left[\dfrac{1}{a^2b^2}\left(\dfrac{x_1^2}{a^2}+\dfrac{y_1^2}{b^2}\right)-\dfrac{1}{a^2b^2}\right]}}{\dfrac{1}{a^2b^2}(x_1^2+y_1^2-a^2-b^2)}$$

$$=\frac{2ab\sqrt{\left(\dfrac{x_1^2}{a^2}+\dfrac{y_1^2}{b^2}-1\right)}}{x_1^2+y_1^2-a^2-b^2}$$

$$=\frac{2\sqrt{(b^2x_1^2+a^2y_1^2-a^2b^2)}}{x_1+y_1^2-a^2-b^2}.$$

Now $b^2x_1^2+a^2y_1^2-a^2b^2$

$$=\frac{b^2(a^2+\lambda_1)(a^2+\lambda_2)}{(a^2-b^2)}+\frac{a^2(b^2+\lambda_1)(b^2+\lambda_2)}{b^2-a^2}-a^2b^2$$

$$=\frac{b^2a^4+a^2b^2(\lambda_1+\lambda_2)+b^2\lambda_1\lambda_2-a^2b^4-a^2b^2(\lambda_1+\lambda_2)-a^2\lambda_1\lambda_2}{(a^2-b^2)}-a^2b^2$$

$$=-\frac{(a^2-b^2)\lambda_1\lambda_2}{(a^2-b^2)}$$

$$=-\lambda_1\lambda_2.$$

Also $x_1^2+y_1^2-a^2-b^2$

$$=\frac{(a^2+\lambda_1)(a^2+\lambda_2)}{a^2-b^2}+\frac{(b^2+\lambda_1)(b^2+\lambda_2)}{b^2-a^2}-a^2-b^2$$

$$=\frac{a^4+a^2(\lambda_1+\lambda_2)+\lambda_1\lambda_2-b^4-b^2(\lambda_1+\lambda_2)-\lambda_1\lambda_2}{a^2-b^2}-a^2-b^2$$

$$=\frac{(a^2-b^2)(\lambda_1+\lambda_2)}{(a^2-b^2)}$$

$$=\lambda_1+\lambda_2.$$

$$\therefore \quad \tan 2\phi=\frac{2\sqrt{(-\lambda_1\lambda_2)}}{\lambda_1+\lambda_2}.$$

Ex. 2. *If two conics confocal with a given conic pass through a given point P, and PQ, PR the normals at P to the confocals meet the polar of P with respect to the given conic in Q and R, then*

$$PQ=\frac{-\lambda_1}{l_1},\ PR=\frac{-\lambda_2}{l_2},$$

where l_1, l_2 are the perpendiculars from the centre to the tangents at P to the confocals and λ_1, λ_2 the parameters of the confocals.

Let (x_1, y_1) be the coordinates of P and let the given conic be

$$\frac{x^2}{a^2}+\frac{y^2}{b^2}=1. \qquad \ldots(1)$$

The confocals are

$$\frac{x^2}{a^2+\lambda_1}+\frac{y^2}{b^2+\lambda_1}=1 \qquad \ldots(2)$$

and
$$\frac{x^2}{a^2+\lambda_2}+\frac{y^2}{b^2+\lambda_2}=1. \qquad \ldots(3)$$

The equation of the tangent at (x_1, y_1) to the confocal (1) is

$$\frac{xx_1}{a^2+\lambda_1}+\frac{yy_1}{b^2+\lambda_1}=1 \qquad \ldots(4)$$

The equation of the normal to the confocal at P is

$$\frac{y_1(a-x_1)}{b^2+\lambda_1}-\frac{x_1(y-y_1)}{a^2+\lambda_1}=0$$

or
$$\frac{x-x_1}{\dfrac{x_1}{a^2+\lambda_1}}=\frac{y-y_1}{\dfrac{y_1}{b^2+\lambda_1}}$$

or
$$\frac{x-x_1}{\dfrac{l_1x_1}{a^2+\lambda_1}}=\frac{y-y_1}{\dfrac{l_1y_1}{b^2+\lambda_1}}=r$$

where l_1 is the length of the perpendicular from the centre upon the tangent to (2), and r is the distance of any point (x, y) on the normal to (x_1, y_1), so that

$$l_1=\frac{1}{\sqrt{\left\{\dfrac{x_1^2}{(a^2+\lambda_1)^2}+\dfrac{y_1^2}{(b^2+\lambda_1)^2}\right\}}}.$$

The coordinates of any point Q on the normal are

$$x_1\left(1+\frac{l_1r}{a^2+\lambda_1}\right), y_1\left(1+\frac{l_1r}{b^2+\lambda_1}\right).$$

If this point lies on the polar of P with respect to the conic, its equation being

$$\frac{xx_1}{a^2}+\frac{yy_1}{b^2}=1,$$

we get

$$\frac{x_1^2}{a^2}\left(1+\frac{l_1r}{a^2+\lambda_1}\right)+\frac{y_1^2}{b^2}\left(1+\frac{l_1r}{b^2+\lambda_1}\right)=1.$$

As (x_1, y_1) lies on the confocal (2),

$$\frac{x_1^2}{a^2+\lambda_1}+\frac{y_1^2}{b^2+\lambda_1}=1.$$

In view of this the above equation becomes

$$\frac{x_1^2}{a^2}\left(1+\frac{l_1 r}{a^2+\lambda_1}\right)+\frac{y_1^2}{b^2}\left(1+\frac{l_1 r}{b^2+\lambda_1}\right)=\frac{x_1^2}{a^2+\lambda_1}+\frac{y_1^2}{b^2+\lambda_1}$$

or
$$\frac{\lambda_1 x_1^2}{a^2(a^2+\lambda_1)}+\frac{\lambda_1 y_1^2}{b^2(b^2+\lambda_1)}+\frac{x_1^2 l_1 r}{a^2(a^2+\lambda_1)}+\frac{y_1^2 l_1 r}{b^2(b^2+\lambda_1)}=0$$

or
$$(\lambda_1+l_1 r)\left\{\frac{x_1^2}{a^2(a^2+\lambda_1)}+\frac{y_1^2}{b^2(b^2+\lambda_1)}\right\}=0$$

which means that
$$\lambda_1+l_1 r=0$$

or
$$r=-\frac{\lambda_1}{l_1}.$$

Similarly we can prove that

$$PR=-\frac{\lambda_1}{l_2}.$$

Ex. 3. *TP, TQ are tangents one to each of two confocal conics, whose centre is C; show that if the tangents are at right angles to one another, CT will bisect PQ.* (*Aligarh 1944*)

Let the confocals be

$$\frac{x^2}{a^2}+\frac{y^2}{b^2}=1$$

and
$$\frac{x^2}{a_1^2}+\frac{y^2}{b_1^2}=1.$$

Let the tangents be

$$\frac{xx_1}{a^2}+\frac{yy_1}{b^2}=1 \text{ and } \frac{xx_2}{a_1^2}+\frac{yy_2}{b_1^2}=1.$$

The straight line CT passes through the intersection of the above tangents and the origin. Hence its equation is given by eliminating the constant between them, so that ts equation is

$$x\left(\frac{x_1}{a^2}-\frac{x_2}{a_1^2}\right)+y\left(\frac{y_1}{b^2}-\frac{y_2}{b_1^2}\right)=0.$$

This will pass through the middle point of PQ if

$$\left(\frac{x_1+x_2}{2}\right)\left(\frac{x_1}{a^2}-\frac{x_2}{a_1^2}\right)+\left(\frac{y_1+y_2}{2}\right)\left(\frac{y_1}{b^2}-\frac{y_2}{b_1^2}\right)=0,$$

or $$\frac{x_1^2}{a^2}+\frac{x_1x_2}{a^2}-\frac{x_1x_2}{a_1^2}-\frac{x_2^2}{a_1^2}+\frac{y_1^2}{b^2}+\frac{y_1y_2}{b^2}-\frac{y_2y_1}{b_1^2}-\frac{y_2^3}{b_1^2}=0. \quad ..(1)$$

But as (x_1, y_1) lies on $\frac{x^2}{a^2}+\frac{y^2}{b^2}=1$

and (x_2, y_2) lies on $\frac{x^2}{a_1^2}+\frac{y^2}{b_1^2}=1$, we have

$$\frac{x_1^2}{a^2}+\frac{y_1^2}{b^2}=1 \text{ and } \frac{x_2^2}{a_1^2}+\frac{y_2^2}{b_1^2}=1.$$

Hence equation (1) becomes

$$x_1x_2\left(\frac{1}{a^2}-\frac{1}{a_1^2}\right)+y_1y_2\left(\frac{1}{b^2}-\frac{1}{b_1^2}\right)=0.$$

But the conics are confocal.

$$\therefore \quad a_1^2-a^2=b_1^2-b^2$$

and the above equation reduces to

$$\frac{x_1x_2}{a^2a_1^2}+\frac{y_1y_2}{b^2b_1^2}=0,$$

which is the condition for the tangents TP and TQ to be at right angles.

Ex. 4. *Prove that two conics $ax^2+2hxy+by^2=1$ and $a'x^2+2h'xy+b'y^2=1$ can be placed so as to be confocal if*

$$\frac{(a-b)^2+4h^2}{(ab-h^2)^2}=\frac{(a'-b')^2+4h'^2}{(a'b'-h'^2)^2}.$$

(Karnatak 1958)

Let the equation to the two conics referred to principal axes be $$\alpha x^2+\beta y^2=1$$
and $$\alpha' x^2+\beta' y^2=1.$$

If these conics are confocal,

$$\frac{1}{\alpha}-\frac{1}{\beta}=\frac{1}{\alpha'}-\frac{1}{\beta'}$$

or
$$\frac{\beta-\alpha}{\alpha\beta}=\frac{\beta'-\alpha'}{\alpha'\beta'}. \quad \ldots(1)$$

But by invariants,

$$\alpha+\beta=a+b. \quad \ldots(2)$$

$$\alpha'+\beta'=a'+b'. \quad \ldots(3)$$

$$\alpha\beta=ab-h^2. \quad \ldots(4)$$

$$\alpha'\beta'=a'b'-h^2. \quad \ldots(5)$$

Now
$$\beta-\alpha=\sqrt{\{(\alpha+\beta)^2-4\alpha\beta\}}$$
$$=\sqrt{\{(a+b)^2-4\ (ab-h^2)\}}.$$

Similarly $\beta'-\alpha'=\sqrt{\{(a'+b')^2-4\ (a'b'-h'^2)\}}.$

Substituting in (1), we get

$$\frac{\sqrt{\{(a+b)^2-4\ (ab-h^2)\}}}{(ab-h^2)}=\frac{\sqrt{\{(a'+b')^2-4\ (a'b'-h'^2)\}}}{(a'b'-h'^2)}$$

or
$$\frac{(a+b)^2-4\ (ab-h^2)}{(ab-h^2)^2}=\frac{(a'+b')^2-4\ (a'b'-h'^2)}{(a'b'-h'^2)^2}$$

or
$$\frac{(a-b)^2+4h^2}{(ab-h^2)^2}=\frac{(a'-b')^2+4h'^2}{(a'b'-h'^2)^2}.$$

Examples XIII

1. Find the conics confocal with $x^2+2y^2=2$ which pass through the point (1, 1). *(Lucknow 1946)*
2. Prove that the locus of points lying on a system of confocal ellipses, which have the same eccentric angle ϕ, is a confocal hyperbola whose asymptotes are inclined at an angle 2ϕ.
3. Prove that the locus of the point of contact of tangents drawn from a given point to a system of confocals is a cubic curve, which passes through the given point and through the foci.

If the given point lies on the major axis, prove that the cubic reduces to a circle. *(Utkal Hon's 1946)*

4. TP, TQ are tangents one to each of two confocal conics; show that if the tangents are at right angles to one another, the line PQ always touches a third confocal conic.

5. If the confocals through (x_1, y_1) to the ellipse

$$\frac{x^2}{a^2}+\frac{y^2}{b^2}=1$$

are $\dfrac{x^2}{a^2+\lambda_1}+\dfrac{y^2}{b^2+\lambda_1}=1$ and $\dfrac{x^2}{a^2+\lambda_2}+\dfrac{y^2}{b^2+\lambda_2}=1$,

show that (i) $\dfrac{x_1^2}{a^2}+\dfrac{y_1^2}{b^2}-1=-\dfrac{\lambda_1\lambda_2}{a^2b^2}$,

(ii) $x_1^2+y_1^2-a^2-b^2=\lambda_1+\lambda_2$. *(Aligarh 1946)*

6. Show that the equation of the pair of tangents from P to $\dfrac{x^2}{a^2}+\dfrac{y^2}{b^2}=1$ referred to the noamals to the confocals through P as axes is

$$\frac{x^2}{\lambda_1}+\frac{y^2}{\lambda_2}=0$$

where λ_1 and λ_2 are the parameters of the confocals through *P*.

7. Tangents are drawn to the parabola $y^2=4x\sqrt{(a^2-b^2)}$ and on each is taken the point at which it touches one of the confocals

$$\frac{x^2}{a^2+\lambda}+\frac{y^2}{b^2+\lambda}=1.$$

Prove that the locus of such points is a straight line.

8. Tangents are drawn to the ellipse

$$\frac{x^2}{a^2}+\frac{y^2}{b^2}=1$$

from any point *T* on a given hyperbola confocal with the ellipse. If 2θ be the angle between the tangents, prove that $\sin\theta$ varies inversely as *CD*, where *CD* is the semi-diameter conjugate to CT of the ellipse through T confocal with the given one.

9. Prove that the locus of the points of contact of parallel tangents to a system of confocals is a rectangular hyperbola.

10. Prove that the product of the four normals to an ellipse from a point P is $\frac{\lambda_1\lambda_2(\lambda_1-\lambda_2)}{a^2-b^3}$, where λ_1, λ_2 are the parameters of the confocals to the given ellipse which pass through P, and a, b the semi axes of the given ellipse.

11. Two tangents to the ellipse $\frac{x^2}{a^2}+\frac{y^2}{b^2}=1$ at points whose eccentric angles are θ and ϕ, intersect in a point on a confocal ellipse whose parameter is λ. If the tangents are at right angles,

$$\cos(\theta-\phi)=\frac{a^2b^2-\lambda^2}{a^2b^2+\lambda^2}.$$

12. If 2θ is the angle between the tangents from a point P to the parabola $y^2=4a(x+a)$ and b and $-c$ are the parameters of the two confocals through P, show that

$$\tan 2\theta=\frac{c+a}{b-a}.$$

13. If λ, μ be the parameters of the confocals which pass through two points P, Q on a given ellipse, show (i) that if P, Q be extremities of conjugate diameters, then $\lambda+\mu$ is constant, and (ii) that if the tangents at P and Q be at right angles, $\frac{1}{\lambda}+\frac{1}{\mu}$ is constant.

14. The rectangle contained by the perpendiculars let fall on a straight line from its pole with respect to the ellipse $\frac{x^2}{a^2}+\frac{y_1^2}{b^2}=1$ and from the centre of the ellipse

is constant $(=\lambda)$; prove that the straight line touches the confocal

$$\frac{x^2}{a^2+\lambda}+\frac{y^2}{b^2+\lambda}=1.$$

15. A parabola is drawn having its axes parallel to a given straight line and having double contact with a given ellipse; prove that the locus of its focus is a hyperbola confocal with the ellipse and having one asymptote in the given direction.

16. Show that if a conic confocal with $\frac{x^2}{a^2}+\frac{y^2}{b^2}=1$ touches $x\cos\alpha+y\sin\alpha=p$, its equation is

$$\frac{x^2}{a^2+\lambda}+\frac{y^2}{b^2+\lambda}=1$$

where $\lambda+a^2\cos^2\alpha+b^2\sin^2\alpha=p.$

17. The confocal hyperbola through the point on the ellipse

$$\frac{x^2}{a^2}+\frac{y^2}{b^2}=1$$

whose eccentric angle is α has for its equation

$$\frac{x^2}{\cos^2\alpha}-\frac{y^2}{\sin^2\alpha}=a^2-b^2.$$

18. The straight lines OPP', OQQ' cut an ellipse in P, P' and Q, Q' respectively and touch a confocal ellipse; prove that $OP.OP'.QQ'=OQ.OQ'.PP'$.

19. If from a fixed point tangents be drawn to one of a given system of confocal conics, and the normals at the points of contact meet in Q, shew that the locus of Q is a straight line.

20. A triangle circumscribes an ellipse and two of its angular points lie on a confocal ellipse; prove that the third angular point lies on another confocal ellipse.

ANSWERS

Exercise I

3. $\left(\frac{l^2+m^2}{l+m}, \frac{l^2+2lm-m^2}{l+m}\right); \left(\frac{l^2-2lm-m^2}{l-m}, \frac{l^2+m^2}{l-m}\right)$

9. 3   11. $\sqrt{7}l$ 12. $\frac{1}{4}c^2\sqrt{3}$

13. (i) $r^2=a^2$ (ii) $\theta=\tan^{-1}p$ (iii) $r=2m\cos\theta$ (iv) $r^2=a^2\cos 2\theta$

14. (i) $x^2+y^2=a^2$ (ii) $y=px$ (iii) $(x^2+y^2)^3=4a^2x^2y^2$
(iv) $(2x^2+2y^2-ax)^2=a^2(x^2+y^2)$ (v) $x^2-y^2=a^2$

15. $x^2+y^2=a^2-c^2$

16. (i) $4x^2(k^2-4a^2)+4k^2y^2=k^2(k^2-4a^2)$ (ii) $2(x^2+y^2+a^2)=k$
(iii) $(k^2-1)(x^2+y^2+a^2)+2ax(k^2+1)=0$

17. $r^2-6r\cos(\theta-\pi/3)+9-k^2=0$

18. $x^2+2ay=a^2$ when a is the distance of the fixed point from the fixed straight line and the axes of coordinates are taken to be the lines through the fixed point parallel and perpendicular to the given line.

Exercise II

1. (a) $y=x\tan\theta-c$ (b) $x=a, y=b$ (c) $y-x=1$.
2. (a) 5 ; 30° (b) $\frac{1}{2}$; 120°
3. (a) $x/5-y/6=1$ (b) $x+y=1$; $x-y=3$
4. (i) 30° (ii) $6\sqrt{3}$ sq. units (iii) 3
5. (i) $bx\cos\left(\frac{\theta+\phi}{2}\right)+ay\sin\left(\frac{\theta+\phi}{2}\right)=ab\cos\left(\frac{\theta-\phi}{2}\right)$
(ii) $2x-(t_1+t_2)y+2at_1t_2=0$ 6. 30°   7. $3\sqrt{2}$

Exercise III

1. (a) $x^2+y^2-x-8y+13=0$ (b) (5, 6)
2. (a) $(\sqrt{13}+3\sqrt{2})x+(\sqrt{13}-2\sqrt{2})y-5\sqrt{13}+7\sqrt{2}-10\sqrt{26}=0$
(b) $\left(-\frac{80}{37}, -\frac{7}{37}\right)$
3. (a) $y-b=\frac{b'-b}{a'-a}(x-a)$; $y-b=\frac{b'-b}{a-a'}(x-a')$

(b) $\left[\pm\frac{a}{a+b}\{\sqrt{(a^2+b^2)}+b\},\ 0\right]$

5. $y-a=0$; $4x-3y+3a=0$ 7. $y=x$; $x+3y=4$; $2x+y=3$

10. $2x+16y+13=0$ 11. $(1, 8)$

12. $x+y=2$; $x=y$

18. $(l-l')\,x+(m-m')\,y=0$; $(l+l')\,x+(m+m')\,y+n+n'=0$

Exercise IV

1. $r\cos(\theta+\alpha)=l$ 2. $\theta=\alpha$ 4. $\sqrt{2}$

5. $\theta=\tan^{-1}\left(\frac{A'-A}{B-B'}\right)$

6. $r=\frac{r_1r_2\sin(\theta_1-\theta_2)}{\sqrt{\{r_1^2+r_2^2-2r_1r_2\cos(\theta_1-\theta_2)\}}}$

$\theta=\tan^{-1}\left(\frac{r_2\cos\theta_2-r_1\cos\theta_1}{r_1\sin\theta_1-r_2\sin\theta_2}\right)$

Exercise V

2. $y^2=4x\operatorname{cosec}^2\alpha$ 3. $\frac{1}{2}\tan^{-1}\frac{2h}{a-b}$ 5. $30°$

6. $2xy+a^2=0$

Exercise VI

6. $4\,(a'h+bh')\,(ah'+b'h)+(aa'-bb')^2=0.$

Exercise VII

1. $\lambda=\frac{9}{2}$ or 3 ; $x-4y+2=0$; $2x-y+2=0$

11. $x^2\,(c^2-a^2m^2)+2a^2mxy+y^2\,(c^2-a^2)=0$

14. $\frac{2a\sqrt{(x'^2+y'^2-a^2)}}{2a^2-x'^2-y'^2}$.

Exercise VIII

1. $x^2+y^2+2y=1$

2. $x^2+y^2-(x+y)\,\frac{(h^2+k^2)}{h+k}=0$

3. $x^2+y^2-ax-by=0$ 4. $x^2+y^2-4x\pm5y+4=0$

5. $(-58,\ -24)$; 65 7. $2x^2+2y^2-2x+3y-2=0$

11. $x^2+y^2-17x-19y+50=0$ 15. $8x^2+8y^2-25x-3y+18=0$

Exercise IX

1. $(c-f^2)\, c^2+2fglm+(c-g^2)\, m^2-2n\,(gl+fm)+n^2=0$
2. $y=(x-a)\tan\alpha$
3. $x^2+y^2-2hx-2ky=0$
7. $a\cos^2\alpha+b\sin^2\alpha\pm\sqrt{(a^2+b^2\sin^2\alpha)}$
12. $x\,(my-n)=ly^2$
13. $(x^2+y^2-2hx)\,(y-k)+2xk\,(x-h)=0.$

Exercise X

1. $3x-2y+1=0$; $2\sqrt{(4-\frac{16}{13})}$ 5. $(-2,\ -1)$
8. $\alpha x-\mu y=0,\ \ x^2+y^2+\mu\,(x/a+y/b+2)=0,$ μ being a parameter.
9. $x^2+y^2+x+3y+\frac{1}{2}=0$
10. $x^2+y^2-4+\lambda\,(2x+4y-2)=0$
11. $(1,\ 2)$; $(3,\ 1)$
19. $6x+4y-8=\pm\sqrt{6}\,(x-12)$
20. $x=-1,\ 3y=4x-5,\ y=1,\ 3x+4y=5$
21. $3x\pm y\sqrt{7}=8$; $x\pm y\sqrt{15}=8$; $(\frac{8}{3},\ 0)$; $(8,\ 0)$
22. $\pi/6$ or $\pi/6\pm\cos^{-1}(-\frac{1}{3})$
26. $r^4-r^2\,(4a^2\sin^2\beta+4ap\cos\beta)+4a^2p^2=0$
 $p=2a\cos^2\beta/2$ or $-2a\sin^2\beta/2.$
29. $(0,\ \alpha)$; $(2a,\ \alpha).$

Examples XI

1. (i) Vertex $(-2,\ 2)$, focus $(-1\frac{1}{2},\ 2)$, axis $y=2$, directrix $x=-\frac{5}{2}$
 (ii) Vertex $(4,\ \frac{11}{2})$, focus $(4,\ 5)$, axis $x=4$, directrix $y=6$
 (iii) Vertex $\left(a,\ \dfrac{a}{4}\right)$, focus $\left(a,\ -\dfrac{3a}{4}\right)$, axis $x=a$, directrix $y=\dfrac{5a}{4}$
2. $x^2+4y^2-4xy+4x+2y-1=0$
3. Axis $y+b=0$ vertex $\left(\dfrac{c-b^2}{2a},\ -b\right)$

4. Latus rectum $\frac{7}{16}$, vertex (3, 1), focus ($\frac{41}{16}$, 1)

5. $9y^2=4ax$ 7. $8a\sqrt{3}$

9. $(3a, -2a\sqrt{3})$, $(3a, +2a\sqrt{3})$ 13. $(\frac{1}{2}, 2)$

17. $x(x-h)^2+y(x-h)(y-k)+a(y-k)^2=0$

21. $y^2-ax+3a^2=0$ 22. $(3a, 0)$.

24. $al^3+2alm^2+m^2n=0$

Examples XII

2. $4(y^2-4ax)=(x+4a)^2 \tan^2 \alpha$

12. $(2a+b)y^2=4a^2x$ 13. $(y^2-4ax)^2=c^2(4a^2+y^2)$

15. $(x+a)^2-y^2=4a^2$.

Examples XIII

1. $4(\alpha-2a)^3=27a\beta^2$; $4(x-2a)^3=27ay^2$

6. $\left(h-6a, -\frac{k}{2}\right)$ 8. $y^2=a(x-a)$

10. $16(xy)^{2/3}(x^{2/3}+y^{2/3})=b^2$

14. $\left(\frac{a}{a^2+2ab\cos\omega+b^2}, \frac{b}{a^2+2ab\cos\omega+b^2}\right)$

$x(b+a\cos\omega)+y(a+b\cos\omega)=\cos\omega$

Exercise XIV

1. $17x^2+17y^2-2xy-2x-110y+161=0$

2. $\frac{4x^2}{81}+\frac{4}{45}y^2=1$

3. $\sqrt{\frac{3}{5}}$, $(\pm\sqrt{6}, 0)$; $8\sqrt{(\frac{1}{10})}$.

4. Centre (1, 5), eccentricity $\sqrt{7}/4$, foci $(1\pm\sqrt{7}, 0)$, latus rectum, $\frac{9}{2}$

5. Centre (2, −1), eccentricity $\sqrt{\frac{2}{3}}$, foci $(2\pm\sqrt{\frac{2}{3}}, 0)$, major axis=1, minor axis=$\frac{1}{\sqrt{3}}$

7. $\frac{1}{\sqrt{2}}$ 8. $\frac{x^2}{a^2}+\frac{y^2}{b^2}=\frac{y}{b}$

10. $\sqrt{\frac{1}{2}}$, $\frac{1}{2}x^2+y^2=1$ referred to the major axis as x-axis and the minor axis as y-axis.

23. $a^2x^2+2abxy \operatorname{cosec} 2\theta+b^2y^2=(a^2-b^2)^2 \cos^2 2\theta$ where sum of eccentric angles at points P and Q is 2θ.

Examples XV

1. (a) $y=x/3,\ y=-2x$ (b) $2x^2+3y^2=48$
3. $\phi=\tan^{-1}(-\frac{4}{3})$
10. $\left(\frac{x^2}{a^2}+\frac{y^2}{b^2}\right)^2=\frac{(x^2+y^2)}{(a^2+b^2)^2}$
19. $2(a^2x^2+b^2y^2)^3=(a^2-b^2)(a^2x^2-b^2y^2)^3$.

Examples XVI

1. $(2\sqrt{10}-3\sqrt{13})\,x-(3\sqrt{10}+\sqrt{13})\,y=0$ and $(2\sqrt{10}+3\sqrt{13})\,x-(3\sqrt{10}-\sqrt{13})\,y=0$
2. $7x^2+12xy-2y^2-2x+14y-22=0$ 3. (2, 1)
6. $6x^2-xy-y^2-23x+4y+15=0$
7. $6x^2+13xy+6y^2-38x-37y+34=0$
8. $x^2+3xy+2y^2+2x+3y=0$
13. $\frac{xx_1}{a^2}-\frac{yy_1}{b^2}=\frac{x_1^2}{a^2}-\frac{y_1^2}{b^2}$.

Examples XVII

2. $(-1, 2)$ and $(-9, 10)$; $x-y-1=0$ and $x-y+23=0$.
3. (1, 1) ; $2\sqrt{2}$, 2 ; $x+2y-3=0$, $2x-y-1=0$.
4. $\left(12-\frac{121}{3-\sqrt{130}}\right)x-7y=0$ and $\left(12+\frac{121}{3-\sqrt{130}}\right)x-7y=0$.
5. (2, 1)
6. Parabola ; axis $2x+2y-1=0$; tangent at the vertex $4x-4y+5=0$; latus rectum $\sqrt{\frac{1}{2}}$; the parabola is on the origin side of the tangent at the vertex.
7. Parabola ; axis $4x+3y-1=0$; tangent at the vertex $3x-4y=0$; latus rectum $\frac{1}{5}$; the parabola is on the negative side of the tangent at the vertex.
8. Ellipse ; centre $(-2, 3)$; major axis $x+y-1=0$, length $4\sqrt{3}$; minor axis $x-y+5=0$, length 4.

9. Ellipse ; centre (2, 3) ; major axis $2x-y-1=0$, length $2\sqrt{6}$: minor axis $x+2y-8=0$, length 4 ; foci $\left(2\pm\frac{2}{\sqrt{5}}\right.$; $\left.3\pm 2.\frac{2}{\sqrt{5}}\right)$.

10. Rectangular hyperbola ; centre (4, 2) ; transverse axis $x-2y=0$, length $2\sqrt{10}$; asymptotes $3x-y-10=0$ and $x+3y-10=0$; eccentricity $\sqrt{2}$.

11. $(2x+3y-5)(5x+3y-8)=36$.

12. Parabola ; axis $5x-12y+1=0$; tangent at the vertex $12x+5y+2=0$; latus rectum $\frac{1}{13}$; parabola lies on the positive side of the tangent at the vertex.

13. Transverse axis $x+y+4=0$, length $2\sqrt{\frac{2}{7}}$; conjugate axis $x-y+4=0$, length $2\sqrt{\frac{2}{3}}$.

14. Parabola, axis $x+2y=0$, tangent at the vertex $2x=y$; latus rectum $2x-y+\frac{3}{2}=0$, length $\frac{6}{5}\sqrt{5}$, focus $(-\frac{3}{5}, \frac{3}{10})$.

15. Hyperbola ; centre $(-2, 2)$; transverse axis $x-y+4=0$, length $2\sqrt{2}$; conjugate axis $x+y=0$, length $2\sqrt{\frac{2}{5}}$.

16. Hyperbola; centre $(-7, -4)$; transverse axis $x-y+3=0$, length 12 ; conjugate axis $x+y+11=0$, length $4\sqrt{3}$.

17. Ellipse ; centre $(-\frac{18}{25}, \frac{1}{25})$, major axis $3x+4y+2=0$, length $2\sqrt{5}$; minor axis $4x-3y+3=0$, length $\sqrt{10}$.

18. Hyperbola ; centre $(-1, 2)$; transverse axis $x+2y-3=0$, length $\sqrt{2}$; conjugate axis $y-2x-4=0$, length $\frac{2}{3}\sqrt{3}$.

19. $4x-3y+2=0$; $(-\frac{1}{5}, \frac{2}{5})$; 8, $3x+4y=11$; (1, 2).

20. Foci $(1, -1)$, $(-1, 2)$; directrices $2x-3y-1=0$, $2x-3y+4=0$.

21. Axis $2x-4y-1=0$; tangent at the vertex $16x+8y-5=0$; vertex $\left(\frac{7}{20}, -\frac{3}{40}\right)$; $\frac{2\sqrt{5}}{5}$.

22. $(3x-4y+7)(4x+3y+1)=7$; centre $(-1, 1)$; transverse axis $x+7y-6=0$, length $\frac{2}{5}\sqrt{14}$; conjugate axis $7x-y+8=0$, length $\frac{2}{5}\sqrt{14}$; asymptotes $3x-4y+7=0$ and $4x+3y+1=0$.

23. $2x^2+3xy-2y^3-\frac{25}{18}(17x+16y)=0$; $x^2-y^2=\frac{245}{9}$.

25. Ellipse; major axis $2x=y$ length 2; minor axis $x+2y=0$, length 1 ; centre $(-1, 1)$.

27. Ellipse ; centre $(2, -3)$; major axis $4x+3y+1=0$, length 6 ; minor axis $3x-4y-18=0$, length 4.

31. Focus $(-\frac{1}{25}, -\frac{11}{50})$, directrix $8x-6y+1=0$.

32. Hyperbola; centre $(-1, 1)$; transverse axis $x-y+2=0$, length $\dfrac{4\sqrt{3}}{3}$; conjugate axis $x+y=0$, length 4.

33. Ellipse ; centre $(\frac{3}{2}, 2)$; major axis $2x-y-1=0$, length $\sqrt{11}$; minor axis $2x+4y-11=0$, length $\dfrac{2\sqrt{11}}{3}$; foci $\left(\dfrac{3}{2}\pm\dfrac{\sqrt{11}}{6},\ 2\pm\dfrac{\sqrt{11}}{3}\right)$.

37. Hyperbola ; centre $(-\frac{1}{5}, \frac{7}{5})$; semi axes 2 and $4\sqrt{3}$; asymptotes $8x-y+3=0$ and $4x+7y-9=0$; eccentricity $\sqrt{\frac{13}{3}}$.

Examples XVIII

9. $\dfrac{l \sec \alpha}{r}=1+(e \sec \alpha)\cos \theta.$

25. $r=\dfrac{le \cos \theta}{e^2-1}.$

Examples XIX

2. $4x^2+4xy+y^2-7x-4y=0.$

11. $y^2+4ax+8a^2=0.$

12. $x^2+y^2-4ax=0.$

13. $\{(af-gh)\,x-(\log-fh)\,y\}^2+2\,(af^2+bg^2-2fgh)\,(gx+fy)=0.$

20. $ax^2+2hxy+by^2-\dfrac{2\,(ab-h^2)}{a+b-2h\cos\omega}\,(x^2+y^2+2xy\cos\omega)=0.$

25. $ax^2+2hxy+by^2+2gx+2fy+\dfrac{h^2-ab}{af^2+bg^2-2fgh}\,(gx+fy)^2=0.$

26. $\left(\dfrac{-31\pm\sqrt{649}}{6},\ \dfrac{37\mp\sqrt{649}}{6}\right).$

Directrix is

$$x\,(2x_1+y_1+4)+y\,(x_1+2y_1-7)+4x_1-7y_1+14=0,$$

where $(x_1,\ y_1)$ are coordinates of the focus.

Examples XX

1. $3x^2-y^2\pm\sqrt{5}\,(x^2-y^2)=2.$